AF522655

ENCYCLOPAEDIA OF NATURE OF GEOGRAPHY

ENCYCLOPAEDIA
OF
NATURE OF GEOGRAPHY

(Volume 2)

A.Z. Bukhari

ANMOL PUBLICATIONS PVT. LTD.
NEW DELHI - 110 002 (INDIA)

ANMOL PUBLICATIONS PVT. LTD.
4374/4B, Ansari Road, Daryaganj
New Delhi - 110 002
Ph.: 23261597, 23278000
Visit us at: www.anmolpublications.com

Encyclopaedia of Nature of Geography

First Published, 2005

ISBN 81-261-2443-1 (Set)

[Responsibility for the facts stated, opinions expressed, conclusions reached and plagiarism, if any, in these volumes is entirely that of the Editor. The Publishers bear no responsibility for them whatsoever.]

PRINTED IN INDIA

Published by J.L. Kumar for Anmol Publications Pvt. Ltd., New Delhi and Printed at Mehra Offset Press, Delhi.

Content

Preface

The universe, which we live in is full of geographical dimensions. Our world has incorporated in itself, varied social, cultural, political, religious and economical aspects of specific significance. Following such wonderful features, natural wealth and the economic progress made, some countries are termed as developed, while some as developing and the rest as underdeveloped. However, all nations of the world, despite all the odds have been interacting with each other for millennia. They have cooperated, and are still co-operating and would continue to cooperate with each other in future also.

Undoubtedly, the world has diverse and dissimilar physical features and people have different complexions, different cultures and different ways of living. It is only due to different geographical situations, but we share all these with each other in many ways. We should thank God Almighty for giving us such beautiful natural gifts, differing widely from each other, to be able to fulfil our diverse kinds of basic requirements. Due to geographical variations, we find at some places, abundance of coal and at others, abundance of gold and so on. Our geographical conditions have compelled us to go on interacting with each other and also being dependent on each other.

Different natures of soils, plateaus, mountains , islands, icelands, flora, fauna, water resources, river systems, landscapes, climate, monsoon etc., are our geographical assets and we are fortunate enough that we have such kind of possessions.

This beautiful world of ours has diverse cultures, religions, languages, and people of different castes and communities and different physical features as well as complexions. Still we ensure unity and integrity in diversity, solemnly. Almost all the countries in the world have got their own political systems and are developing in their respective ways.

Geography is a vast subject and a perfect discipline in its own right. It has various streams and branches. Studying Geography is not an easy task in the modern context. Each stream demands separate books on selected topics. Hence, there is still a dearth for good books. This endeavour is in the same direction. Here, we offer a series of books on all branches and streams of the discipline — Geography. This encyclopaedic study is a milestone in the area of the subject.

The Editor is confident that he has accomplished a good job, for which there has been a room for a long time. Hopefully, this courageous effort of his would be accorded a warm welcome by all concerned — scholars, researchers, students and professionals, alike. And that achievement is not small.

Editor

1

Geography, a Science

Geography's Character

The question that is more commonly placed at the beginning of methodological treatments we have intentionally reserved for the end. We wished to avoid any attempt to determine the character of geographic science by logical deductions from assumptions. For that form of reasoning can have no greater validity than the basic assumptions, and in this case we could have but little reliance in the assumptions. We therefore proceeded by examining directly the subject whose character we wish to determine-namely the field of geography. We traced the development of the nature of the field, as understood by its significant students, since its beginning as a modern field of study in the late eighteenth century. We examined the logical foundation that such students have given to geography, as the study of the areal differentiation of the earth. In the light of both these considerations, the historical development and the logic of its concept, we studied how geography selects the phenomena that it regards as significant in areal differentiation, and how it may conceive of the areal divisions that it is to study. This survey should now enable us to determine inductively what kind of a science this subject is. It should enable us to answer the questions that geographers have often discussed and which Douglas Johnson, and later Colby, have raised more specifically, as to whether geography is, in this or that particular sense, a science, entitled "to take its place in learned circles and in public esteem with other basic sciences".

The reader will observe from the heading of this section, however, that we still persist in begging the question whether geography is a science of any kind. Finch has considered this question directly and, to my mind, effectively, If an issue is made of this point, it commonly resolves itself into a debate over the meaning of a particular word, a word that is not essential to our further thought. In particular we feel justified in ignoring the various undefined concepts of the term "science" upon which many recent methodological discussions have been based. It is a common, but nonetheless naive and erroneous, assumption of many students of the natural sciences that the nature of "science" is sufficiently well known as to require no statement. When they do attempt to state it, their definition commonly excludes the social studies without indicating in what other kind of knowledge these are to be included. But many physicists and chemists define "science" in terms that could not include zoology or geology; I have even heard such a definition stated by an eminent geologist who did not appear to realize what a small part of his field could be included under it.

Whatever may be gained from the discussion of such a question, there is no need for it here. One would gladly avoid the problem by using some other term possessing less emotional connotation than "science" seems to have acquired for some of our colleagues in other fields, if another suitable word were at hand. In a recent public address, a renowned natural scientist, whose position as university president requires him to consider the problems of the social fields, declared that the "scientific approach" was not applicable in the social studies and regretted that we had no word like the German word *Wissenschaft* to apply to such studies; but he failed to note that German students seem to get along fairly satisfactorily without any word corresponding to our word "science"—that is, other than the word *"Wissenschaft."*

If we were to use the word in a definitive sense, it would be necessary to attempt to determine its meaning specifically. But our need here is simply for a convenient handle' to apply to that genera form of knowledge that is distinct from either common sense knowledge or from artistic and intuitive knowledge. For that form of knowledge our language provides only the word "science" and we will therefore use it in that sense without, for the moment, wishing to claim that geography is entitled to any of the distinctions granted to "science" as used in some other, undefined sense. At most, one may add, any such claims influence only those who make them; no matter how logically we might demonstrate in theory that geography should be granted a title, the esteem that presumably goes with that title will be granted only in recognition of more solid contributions.

Though we consider, then, any question of titles as of little importance, it 'is of very great importance for us as geographers to know what sort of a study geography is. We have noted many disagreements among geographers that have been produced by the desire among some to make geography into a certain kind of science-the only kind, perhaps, that they would call by that name. But any efforts that require geography to change its essential character must be in vain; we cannot make over geography in any fundamental way, we can only fulfil that which it has been and is. Ignoring, therefore, any question as to what geography should be, let us consider what kind of a study it is.

If the classification of the sciences were in fact, as is frequently, supposed, analogous to that of the species of organic life, we could expect to derive the character of geography in major part from a consideration of the generic character of the order and genus of sciences to which it belonged, and simply add to that the specific differences between geography and other sciences of the same genus. But Hettner reminds us that no branch of science is in reality a separate and distinct science. There is only one science, which human limitations require us to divide more or less arbitrarily. The classification of these parts of science involves, therefore, difficulties similar to those which we found in classifying the area of the world that are simiply part of a single whole.

To be sure, there are geographers who assert that they are interested primarily in "the physical aspects of geography," but one will look hard and long to find any of them who do not contribute published studies involving human aspects of the subject. Fortunately, when such students become concerned with a particular area, they quite forget that they have labelled themselves "physical geographers" and proceed to study all features interconnected in the area.

Indeed, it is somewhat misleading to overemphasize the position of geography as "a bridge, between the natural and the social sciences." Though penck has used this analogy a number of times he would be one of the first to insist that, insofar as there is a gulf between the two groups, the gulf is of man's making, it is not present in the reality that science is to study. We cannot, however, accept his further inference that the concept of scientific laws has been developed only on the one side of this artificial gulf and the bridge of geography is needed to carry it across to the social sciences on the other. A concept of this kind requires no bridges. On the other hand, Penck may mean that scientific laws in the social sciences can be developed on a sound basis only if they are connected, through geography, to the natural sciences. Even in this sense we would be claiming too much, for the social sciences have other connections with the facts and relationships of the non-human world, notably through human physiology and psychology.

Whatever conclusions may be drawn with respect to that question, geography is not to be thought of as a connecting link between two groups of sciences, but rather as a continuous field intersecting all the systematic sciences concerned with the world. It therefore has not two but many facets, as Schluter observed; the difference in methods between studies of climate and of landforms is in many respects greater than the difference between the study of natural vegetation and of cultivated crops.

Learning from Geography

The most that we can learn about the nature of geography from the conventional classification is that geography necessarily shares in whatever difficulties or limitations the social sciences are heir to, and that, on the other hand, it shares in part in the greater ease with which facts -and relationships can be determined if the human element is not involved. Since the developments of the last generation have destroyed the faith in absolutes of the nineteenth century physicists, we know that there is here no difference An kind, but only in degree, between the two groups and among the different sciences in each group., Furthermore, this is a difference which applies only in general, not necessarily in the particular instance. Failure to recognize this fact has led many geographers to presume that geographic work had- a major degree of soundness if its feet were established in the natural sciences, regardless of how wildly it might leap from there to conclusions in the uncertain atmosphere of the social sciences. In reality, few facts of the natural environment can be established with such a degree of certainty as the rate of population growth in the United States, or the areas included within the dominions of the political states of the world.

We can secure much more insight into the character of geography if we consider it in terms of the classification which we discussed in the fourth section of this paper. According to Kant, Humboldt, and Hettner, it is necessary to look at science as a whole from different points of view. From one point of view, all reality may be regarded as a collection of many different kinds of phenomena which can be sorted into groups according to the kinds of objects with which they are concerned. The student who approaches science from this point of view endeavours to learn everything he can about the phenomena of one particular group of objects regardless of where and when they may be found. Since it is possible to classify all objects, roughly, as animate and inanimate, of non-human (natural) or human origin, this "systematic" point of view permits a fairly clear subdivision into different "systematic sciences."

In the reality which science is to study, however, the phenomena are not arranged according to the classification which the systematic point of view constructs. Consequently this point of view gives an incomplete view of reality. If phenomena were simply piled and mixed together in reality without meaning, it would perhaps be sufficient simply to state that fact. We know, however, that there are significant relations between the different kinds of phenomena that are found together in any particular section of reality, and also between phenomena in different sections of reality. That is, there is some degree of system or order in the actual arrangement of phenomena in reality. To comprehend reality more fully, therefore, we must not only study phenomena, but must also study the different sections of reality in order to understand the character of each section in comparison with the character of other sections. To understand the character of any section of reality we must attempt to comprehend the integration of phenomena of different kinds that are actually integrated in it.

Although this integration can be stated theoretically in the singular, the nature of reality forces us to take two separate points of view. The whole of reality may be divided into sections in terms of either space or time. Though a single section combines these—here and now is one point in reality it becomes practically, if not theoretically, impossible to consider simultaneously differences in time and differences in space. Only if the phenomena are relatively simple, as in astronomy, or the data relatively meager, as in paleogeography, have efforts to combine the two met with success. The consideration of sections of reality in terms of time is the historical point of view, represented by historical geology, prehistory, and history in the narrower sense. The consideration of sections of reality in terms of space is the chorological point of view, represented by astronomy and geography.

Everyone of these historical and chorological sciences must study all !be kinds of phenomena that are found in its particular sections of reality. Theoretically these could include phenomena of all the systematic fields, whether physical, biological, or social. Only special circumstances limit the range within certain of these fields. W. M. Davis recognized this common characteristic of geography, history, and astronomy. "Dealing with things or events of many kinds in definite relations to time or place, they cannot have the singleness of content which subjects like mathematics and physics and chemistry possess." Astronomy, he continues, is essentially the mathematics, physics, and chemistry of the universe, and only the fact that evidence of organic life has not been found in the heavens has prevented the astronomer from overlapping into biology, or even more might add, the social sciences. Likewise, it is only the circumstance that natural conditions on the earth have changed but little in historical (not *human)* times that largely limits history-as distinct from "prehistory"—to human phenomena. Nevertheless the eruption of Vesuvius is a phenomenon of concern not only to the geologist but perhaps even more to the historian-as is indicated by the fact that the reader knows at once to which eruption we refer. Likewise anyone studying the history of Holland in the Middle Ages must consider the changes consequent upon the formation of the Zuider lee.

It maybe particularly instructive to glance at that special division of the historical view of science known as historical geology. The innocent layman might suppose that one could study the inanimate rocks of the earth's crust without overlapping into the fields that study the phenomena of life. But since the historical geologist is the only scientist who is presented with material for studying the history of the world in remote times, he finds that he must include historical botany, zoology, and human anatomy, and even to some extent historical social anthropology.

This consideration of the nature of different kinds of science should enable us to meet "the ,oft-discussed assertion" of which Colby speaks in his presidential address, namely, "that geography has no distinctive phenomena" at the centre of its interest, as have, for example, soil science, botany, and chemistry". The geographer need not hesitate to acknowledge the truth of that assertion/ even though it establishes an essential difference in character between his field of study and the systematic sciences like chemistry, botany, or political science. The group of sciences among which geography is thereby classified should not, one would suppose, prove humiliating to the geographer.

No Claimer

Geography does not claim any particular phenomena as distinctly its own, but rather studies all phenomena that are significantly integrated in the areas which it studies, regardless of the fact that those phenomena may be of concern to other students from a different point of view. The astronomer has no monopoly of the study of the stars, he is not disturbed if physicists and chemists study the elements of the stars. Similarly, geography need not look for any concrete objects as its own. The rocks which the historical geologist uses for his data are equally the concern of the dynamic geologist and his pet fossils are proper objects of study for the botanist, zoologist, or anthropologist. Likewise the historian is not disturbed if told that his field is an aggregate of economics, political science, and sociology.

Finally, geography does not distinguish any particular kind of facts as "geographic facts." As Barrows has often insisted, any particular fact—meaning a primary fact, not a relationship loosely considered as fact, nor a deduction from relationships-is not a "chemical fact," a "geological fact," or an "economic fact" ; it is simply a fact, and any branch of science may use it. It is only because various kinds of facts are more commonly studied in certain sciences than in others that these conventional, but misleading expressions are in common use. Thus the facts concerning the price of wheat in different places and different times may be considered most frequently in economics, and therefore are called "economic facts," but they could equally well be called "historical facts" or "geographic facts." Geography in particular cannot accept either the popular misconception which classifies under "geographic facts" only the facts of location, or, the misconception common in scientific circles which considers this term as including, in addition to the facts of location, only the facts of natural phenomena. In the broadest sense, just as all facts of past time are historical facts, so all facts of the earth surface are geographical facts. And just as history does not use all facts, but only those—of whatever kind-that are "historically significant," so geography will determine which facts it will, utilize, not according to their substance, but according to their geographical significance, *i.e.*, their relation to the areal differentiation of the world.

To state, for example, that Vesuvius is (and was) a volcano located at 40° 49' N., 14° 46' E., is to state a fact which is no more geographic than geologic or historical-it is of course simply a fact. In the systematic geography of volcanoes, we are concerned with this fact in its relation on the one hand to the zone of diastrophic action that runs through the Mediterranean region, and, on the other hand, to the fertile ash soils of the neighbouring Canlpagna, to the ruins of Pompeil and buried Herculaneum, the hazards of life of the population of the area, and the landscape effect of the volcanic mountain in the level plain.

In sum, then, geography, like history, is to be distinguished from other branches of science not in terms of objects or phenomena studied, but rather in terms of fundamental functions. If the fundamental functions of the systematic sciences can be described as the analysis and synthesis of particular kinds of phenomena, that of the chorological and historical sciences might be described as the analysis and synthesis of the actual integration of phenomena in sections of space and time.

Both history and geography might be described as naive sciences, examining reality from a naive point of view, looking at things as they are actually arranged and related, in contrast to the more sophisticated but artificial procedure of the systematic sciences which take phenomena of particular kinds out of their real settings.

It is not surprising, therefore, to find both history and geography developed as fields of study in the earliest period of scientific thought. Furthermore, it was natural enough that each of these should have become a "mother of sciences." The attempt to integrate all kinds of phenomena in space or time leads to the discovery of many kinds of phenomena, any of which may then appear to be worthy of study in themselves; indeed the attempt to understand their significance in a total integration requires that they be studied in themselves. Consequently we may expect this evolutionary process to continue indefinitely, so long as the new kinds of phenomena discovered are deemed worthy of study in their own right. Thus, if geographers have discovered the phenomena of house types and can demonstrate that they are sufficiently significant, we may expect some branch of systematic science to make these objects a subject of special study.

On the other hand it should not be supposed, as has often been done, that the recognition of the independence of daughter fields thereby reduces the extent of the field which geography or history is to study. On the contrary the mother field remains exactly what it was before. Furthermore, as Richthofen observed and as Hettner has repeatedly emphasized, the progress in these related fields enriches the materials to be studied in geography. Just as the development of economics and political science has greatly increased the ability of historians to interpret history, so modern geography has benefited enormously from the development of systematic physiography, climatology, soil science, etc., and should benefit from the findings of economics and other social sciences. What contribution geography can make in return will be considered later.

The failure to understand that geography is to be defined essentially as a point of view, a method of study-just as all science is a method of study—has caused many to suppose that the growth of the daughter sciences had left nothing for the parent science to do. Attempts have been made to save the day by claiming for geography a particular type of phenomena, such as relationships between man and nature, or by searching for new objects of study which no one else has previously considered worth studying, or by attempting to metamorphose abstract concepts of area into concrete objects. Each of these efforts, to a greater or less extent, has caused geography to depart temporarily from its path of development in directions which have proved, or will prove, to lead either into fields that other sciences will not cede to geography, or into the bog of mystical thinking.

One answer to the question at the head of this section, therefore, is that geography is a study which looks at all of reality found within the earth sure face from a particular point of view, namely that of areal differentiation. This might be called the *position* of geography as a field of knowledge. More significant to the general question is the *character* of geography as a field in which knowledge is acquired.

General Nature of Science

To understand the character of geography as a field in which knowledge is acquired, it is necessary to understand the essential character of the whole field of knowledge of which it is a part. We are concerned here, not with all knowledge, but with that sort of knowledge—by whatever name one chooses to call it-that is distinguished from either commonsense knowledge or from artistic perception "by the rigour with which it subordinates all other considerations to the pursuit of the ideals of certainty, exactness, universality and system".

Geography attempts to acquire knowledge of the world in which we live, both facts and relationships, which shall be as objective and accurate as possible. It seeks to present that knowledge in the form of concepts, relationships, and principles that shall, as far as possible, apply to, all parts of the world Finally, it seeks to organize the dependable knowledge so obtains in logical systems, reduced by mutual connections into as small a number of independent systems as possible. It is in terms of the manner in which geography pursues these ideals that we will attempt to describe its character as a field of study.

It should be noted, in general, that our definition of the kind of knowledge of which geography is a part is based not on what is known, that is, on what has been learned, but rather on the pursuit of knowledge—i.e., the fundamental principles governing the manner in which the unknown is to be learned. Different branches of this "kind of knowledge" differ in the degree to which they have been able to approach the ideals stated; no branch of science can claim to have attained perfect certainty or exactness, actual universality, or complete organization of all its knowledge in a single system.

If one compares these ideals, as the fundamental requirements for that form of knowledge, which hereafter for convenience we will call "'science," with the ideals of artistic perception (in whatever form one finds them stated by students of art), it is clear that there can be no logical combination of the two and no transition from one to the other. The artist certainly does not subordinate all other considerations to an ideal of exactness, nor to one of certainty. He may require that his work express a fundamental universality but that concept does not control the details of his work. Likewise he may require that an individual work of art should be organized, but does not seek a common organization of all similar works of art. In other words, we are dealing here with two essentially different approaches ;both artists and geographers may attempt to acquire and present knowledge of an area of the earth, but neither can adopt the ideals of the other without sacrificing his own.

That geography seeks to make its knowledge of the world as accurate and certain as possible is an assumption which presumably would not be questioned by professional geographers, It is not, however, a correct corollary of this assumption that geography should arbitrarily limit itself to particular kinds of facts because they appear to be subject to more accurate and certain means of measurement than others. Even if it were possible to demonstrate -as would not be the case-that the facts and relationships of immaterial phenomena could never be determined as accurately or certainly as those of either visible or material phenomena, no principle of science would require us to limit ourselves to those phenomena that could be studied somewhat more accurately and certainly, excluding those that could be studied somewhat less accurately and certainly. On the contrary, in order that our knowledge of, say, the cultural character of an area may be made as certain as possible, we are required to consider all the facts that bear upon that knowledge, whether the material products of cultural ideas or the immaterial manifestations of those ideas. Both sets of facts must, of course, be observed as accurately and certainly as possible.

Likewise these ideals admit of no limitation in the specific methods of observation in geography, but rather require that we utilize all methods that will lead to more accurate and certain knowledge. Thus, because statistical data are neither sufficiently accurate nor sufficiently detailed, we must employ the technique of field work-both direct observation and personal interview—to check and supplement the knowledge gained from statistical data. On the other hand, even if it were physically possible to make a complete field observation of every part of a region, the observations obtained reflect temporal conditions, commonly of a single season, which, as R. E. Dodge noted, may lead to erroneous generalizations concerning the continuous use of the land. Consequently the findings of field observations must be checked with statistical data, even though these were very indirectly based on observation.

The scientific ideal of certainty commands that the terms and concepts of description and relationships be made both as specific and as certain as possible-we cannot develop a sound structure on a marsh foundation in which ambiguous concepts shift their meaning whenever pressure is applied to them, and concepts apparently specific prove to be but dubious analogies.

Accuracy and Certainty

The ideals of accuracy and certainty apply not only to the manner in which primary facts are established and to the formulation of fundamental concepts and technical terms, but also to the processes of mathematical and logical reasoning by means of which we induce relationships of observed facts, and thereby deduce further conclusions as to facts. When a single student studies a particular scientific problem, no matter how hard he strives to live up to these ideals, no matter how critically he examines his own work, there remains the possibility of error, whether through carelessness or through subjective influences affecting his observations and reasoning. Since this is a generally accepted axiom in those sciences whose facts can be measured with the highest degree of accuracy and whose relatively simple phenomena make logical reasoning most certain, how much more uncertain are the findings of one student in geography! In order to make possible a higher degree of accuracy and certainty, therefore, it is a recognized principle of all science that studies should be carried on, organized, and presented in such a manner as to provide an accumulation of evidence of different students on the same problem. Every scientific study should, that is, be scholarly, by which we mean both that it should make use of all previous scholarly studies bearing on the problem and that it should be presented in a form usable by subsequent students.

It is one of the handicaps of the social sciences, and in part also of geography, that in much of their work it is not possible for later students to have access to the primary data used in any study; much must be taken on faith in the professional ability and reliability of the individual student. Consequently it is all the more essential in these fields that similar studies of other students should be utilized and referred to, not as a matter of professional courtesy, but for the sake of accumulating evidence in order that results may be more certain.

Science will accept on faith from any scientist no more than is absolutely necessary.

In order that subsequent students may reexamine the findings of a particular study whether in order to utilize them further or simply to check or correct them it is essential that there be a clear understanding of the generic concepts employed, of the methods of observation and reasoning by which the results were obtained, and, finally, of the relation of the study to the field as a whole. Obviously such a clear understanding will be easiest if the field has developed standard techniques and organization; if it has not, they must be specifically indicated in each study.

It is not the function of this paper to pass Judgement on the scientific quality of work actually produced by geographers-it is their ideas about the nature of geography, and the consequences of those ideas that concern us. On the latter basis, however, it is a fair question to raise, whether the research publications in geography indicate that either geographers in general, or the editors who to some extent, control their publications, have accepted these standards as essential to geography standards dictated by the scientific ideals of accuracy and certainty.

With respect to these ideals as essential principles in the pursuit of knowledge, there can be no differences among the different branches of science, but only differences in degree of attainment. Undoubtedly the attainments of geography in this respect are not of such a high degree as to tempt us to compare it with other sciences in such terms. But neither need we feel humiliated if those who choose to make such comparisons should assign our field to the "lower classes." In any competition the understanding observer will judge attainments not only in terms of degree of success but also in terms of the relative difficulties of the tasks undertaken.

The consideration of the manner in which geography may pursue tile other two scientific ideals, universality and system, is of the greatest importance in understanding the character of geography as a field of study. Since these are not so readily understood as the two ideals which we have just considered, each will require detailed consideration.

Generic Concept

If Galileo's most famous experiment had merely demonstrated that when he, Galileo, dropped two specific objects of different weight from the Leaning Tower of Pisa they fell together at the same rate of speed, that fact would have found but a small place in scientific knowledge. Its great importance, of course, was that subsequent experiments showed that he had illustrated a universal, a relation that was true regardless of where, when, or by whom the weights were dropped. Few will question that it is an essential function of science to seek for such universals. On the other hand, if later experiments had shown that the same results were not obtained elsewhere or at other times, the fact that they did take place on that one occasion, if substantiated as a fact, even though never explained, would represent a bit of scientific knowledge. "The business of science," as Barry puts it, "is to learn as much as it can". Though it seeks for universals, it cannot ignore certain and accurate knowledge which it is unable to express which perhaps never can be expressed in terms of universals. One might say that it is an axiomatic ideal of science to attain complete knowledge of reality expressed as completely as possible in terms of universals, but in any case expressed in some way.

While everyone recognizes the importance of universals in science, it is a common error to overlook that part of our scientific knowledge which cannot, as yet at least, be expressed in universals. Many have assumed that science was concerned exclusively in the development of laws and principles. This concept, according to Hettner, represents an outgrowth of the great development of laws and principles in astronomy, physics, and chemistry in the last century. Indeed, it came to be assumed that physics and chemistry were exclusively abstract sciences, concerned only with laws and principles. More recently, however, any scientists and philosophers of science have recognized that no branch of science concerned with reality—as distinct from theoretical mathematics-can limit itself to laws and principles. Though science strives for universals, these do not exhaust the study of reality, there is always an individual remainder that is not described or explained. If this be ignored our knowledge is less complete than it might be, a *quod est absurdum* which no science can accept. "Research," as Lehmann says, "is a seeking for the most complete presentation possible".

The contrast between that aspect of science which can be expressed in universals and that which is concerned with the individual object or phenomenon as worthy of study in itself, received the particular attention, at the turn of the century, of two German philosophers, Windelband and Rickert. They appear to have classified the different branches of science on this basis, distinguishing between "nomothetic" or law-making sciences, and "idiographic" sciences, those concerned with the *einmalige,* the unique. In a rough way this seemed to conform to the conventional division between the natural sciences and the social sciences.

It seems clear, however, that these two aspects of scientific knowledge are present in all branches of science. The common idea that generalizations and laws themselves are the purpose of science is characterized by Hettner as an extraordinary adherence to medieval scholastic realism. On the contrary, they are "merely the means to the ultimate purpose, which is the knowledge of actual reality, the individual facts, either conditions or events". The astronomer develops laws of celestial mechanics not in order to prove that the universe is governed by law which is a philosophical rather than a scientific thesis—but in order that he can rightly understand the motions of the heavenly bodies. He does not forget his interest in the latter as individuals. The students 'who have mapped the surface of the visible side of the moon or who study the rings of Saturn are no less astronomers if they have not thereby been able to establish laws.

Nevertheless it is clear that in some sciences the nature of the phenomena studied permits of a much greater development of nomothetic knowledge, whereas in others the greater differences between the phenomena studied, and also their greater importance individually to man, requires the students in those fields-whether they will or no-to concern themselves in large degree with the unique. This is, then, a significant respect in which there are important and permanent differences of degree among the different branches of science.

A comparison of various branches of science from this point of view may aid us in understanding the character of geography, may help us to perceive what kind of science, in this particular respect, geography is.

Although we have noted that all branches of science are concerned to some extent with the study of the unique as well as of universals, there are some sciences in which the actual work of many students may be largely, if not completely, absorbed in the search for universals, the study of the unique being left to other students or to other divisions of science. I presume that this may be the case in physics and chemistry, particularly in their theoretical aspects. To a lesser extent the same is true of the biological sciences and psychology, and also of certain branches of economics and, perhaps, sociology. (It is both impolitic and dangerous to discuss the nature of a field in which one has not been trained. I might say, however, that the statements in this and the following paragraphs are based on discussions with colleague in the appropriate fields.)

On the other hand, there are certain branches of most of the sciences in which the students are very largely concerned with the study of the individual object, the unique. We have already noted the importance of this form of study in astronomy. Idiographic studies are certainly of great, if not major, importance in geology. This is obviously the case in paleontology. It is true of mineralogy in so far as the student concentrates on certain mineral deposits for the purpose of securing complete knowledge of those deposits rather than of developing new principles of mineral deposition. Even more does the statement apply to the great and certainly valuable work of the geological survey in mapping the areal geology of particular area (though this might logically be considered a branch of geography carried on by geologists).

Economic Factors

Among the social sciences, economics has developed its nomothetic knowledge possibly to a greater extent than some of the natural sciences listed above. In political science... probably only those branches commonly referred to as "political theory" are primarily nomothetic. Most of the students of comparative government, for example, are concerned not merely with general principles of government, but perhaps even more-are concerned to know and understand the differences between individual governments. In their persons, these students combine the namothetic and idiographic interests, whereas in physics or biology or perhaps economics, these may be separated between different students. Since the conditions which have resulted in this situation in political science are very similar to conditions in geography it may be instructive to analyze them.

The first condition is that for most of the phenomena of political science a generic description provides only a bare outline of common features, the individual specimens differing notably with respect to other features. The phenomena are much more complex in structure than those of physics, or even of economics, and, though less complex than the specimens of organic life, do not have the high degree of similarity that results from the common origin of specimens, species, genera, and orders in the biological sciences. Even though it be true that, to the research physician-in contrast perhaps with the student of human physiology—"every human body is different and unique, every spine to some extent different from every other one," these differences are very much smaller than those found among, say, the individual governments of the same type In other words, a generic description of human spines gives a much *more* nearly complete picture of anyone spine than is possible in a similar approach to governments. If the political scientist is to learn as much about governments as the physiologist learns about human spines he will have to give a much greater share of his attention to the unique characteristics of each specimen.

Individual Governance

In addition, more attention will be paid to an individual government, because of its greater importance, in comparison with an individual human spine. This second condition holds generally the phenomena of political science are, as single cases, so important (to man) that it is necessary that each be understood as completely as possible. Whatever the personal preferences of a student may be, mankind will not be satisfied with a science that studies dictatorships, for example, only to the extent that generic concepts and social laws can be developed, and declines to consider the differences between the present government of Italy and that of Germany on the grounds that these are unique cases.

A third consideration is of practical import. With millions of human spines in the world, a single scientist, even a whole group of scientists, may be kept fully occupied in considering only those aspects of the spine that are relatively universal, and may be able to derive sufficient universal knowledge thereby to justify his self-limitation, leaving to others the consideration of individual characteristics. But the political scientist cannot find, even in the great storehouse of history, more than a relatively small number of dictatorships of all kinds to study, hardly more than a handful of cases of a particular type of dictatorship. Under these conditions, to limit himself to the study of generic features and the development of principles, and to decline to study individual specimens as unique objects would be absurd. The reverse consideration is perhaps even more significant; human physiology cannot find time to study the individual characteristics of all human spines, whereas individual studies of all the governments in the world would not be impossible for the field of political science.

It is obvious that the same considerations apply in large part, and with even greater force, in history. To be sure, an individual historian, or particular group of historians, might limit themselves to considering the history of constitutions, of wars, or even of battles and from these studies derive laws or principles of value to the general historian interested in the study of periods of history and the sequence of events. Whether such historians would not thereby become students of military science or of constitutional government is a question which is not for us to consider; certainly such work represents but a small part of history.

If, now, we consider our, own field there can be little question that geography is very much concerned with the study of individual phenomena. Indeed some critics might feel that geographers tend to lose themselves entirely in the consideration of the unique, regardless of its importance. If we were to study each and every feature of the earth surface that could be shown to have geographic significance, without considering their relative importance, it is clear that geography would become hopelessly lost in an insurmountable mass of detail. *Not* only every small district might be studied, but every single village, every mountain, every little rill on the face of the earth.

Individual Features

On the other hand, a geographic description that is confined to generic terms is inadequate. Even a brief and superficial description of a major area must name and describe a great number of individual features. The Alps and the Himalayas, Mount Rainier or Vesuvius, the Amazon, the Mississippi, or the Niagara, states like Germany or the United States, cities like London, Paris or New York-if included within the area described—these will be considered not merely as examples of types or as illustrations of principles, but for their own individual significance.

One reason for this, as we have already seen, is that any generic description, however detailed, is completely inadequate to depict these individual features. When a biologist has described the nucleus of a certain type of cell he has provided a fairly complete picture of the nucleus of any individual cell of that type, whereas a full knowledge of the 'character of the commercial core of Chicago would give one but little understanding of the corresponding district in Minneapolis. It is only a very general similarity which justifies the citizens of the latter *city* in calling their commercial core, in imitation of the larger city, "the loop" ; actually the street car lines do not loop in Minneapolis, as do the street and elevated car lines in Chicago.

If the previous example appears to depend on details of secondary importance, consider such geographic features as the Great Lakes System or New York City. There are, to be sure, a number of great lakes in Africa, but, as a system of connected fresh-water seas, the Great Lakes of North America have no counterpart on earth—or, so far as we know, in the universe. When one considers New York in its fundamental aspects as the focal point of trade of the greater part of a continent, and then looks for its counterpart simply in this one respect ignoring entirely its local character, one finds no other city of the type; Shanghai, Buenos Aires, and Hamburg are all centres of trade for extensive areas but no one of them functions for its particular continent as New York functions for North America.

The informed reader will recognize that while these are conspicuous cases they represent a rule that applies to most geographic features, even to those of slight importance. Because these features are not adequately described in terms of types, geographers, as well as laymen, fall back on analogies. Possibly these have their value; but if taken too seriously they are deceptive. To call Hankow, "the Chicago of China," or the Pittsburgh-Cleveland area, "the Ruhr of America," is to give a false as well as true impression.

Our examples have likewise illustrated the other consideration which requires geography to examine many phenomena as individuals-namely, their great importance as individual phenomena. By this is meant not merely their utilitarian importance in the world, but also their importance to the earth surface on a purely academic basis. Even if Niagara Falls had no material value, whether actually or potentially, a geographic science would not be satisfied 'with a description which simply classified it as of a certain type of falls along with hundreds of others. In itself, in all its individual characteristics, it is a phenomenon of scientific concern to man.

The concern for the unique in geography is not limited to phenomena but applies also to the relationships between phenomena. Thus, if the relationship of Winnepeg to the lakes and barren land on the north and the international boundary on the south is essentially unique, we do not disregard it on that account.

It might appear as though geography were as largely limited to the study of the unique as is history, and, until the last century, that was no doubt the case. As Hettner notes, "geography was confined largely to such idiographic description. . . general concepts were to be found only in the crude form expressed by such common type-words as mountain, valley, *city*, etc." Insofar as the situation was unavoidable, such a statement would represent no criticism of geography as a science, any more than it does of history. Barry's statement may also be expressed in negative form: it is not the business of science to learn what cannot be learned.

On the other hand, a geography which was content with studying only the individual characteristics of its phenomena and their relationships and did not utilize every opportunity to develop generic concepts and universal principles would be failing in one of the main standards of science. We can therefore agree with Hettner that "the greatest scientific advance of geography has been the fact that, by taking over and developing the results of the systematic sciences—earlier in one branch of the field, later in, others—geography has gone over to general observations depending on generic concepts".

Undoubtedly the development of generic concepts has progressed much further in respect to the natural phenomena of concern in geography than in respect to the cultural phenomena, partly because of the close relations which geography has had in most countries with geology and other natural sciences, and also because of the greater development during the past century of the systematic sciences concerned with natural phenomena as compared with those concerned with cultural phenomena. Nevertheless, examples of generic concepts could be cited from every section of systematic geography.

Thanks to such generic concepts it is possible to express many characteristics of a particular feature in one word or phrase—or in a symbol, such as Cfb—so that one can give a relatively brief description which can easily be kept in mind. Furthermore, the use of such generic concepts has made it possible to develop principles of relationships between different factors which are found repeated in different parts of the world. These principles can then be used in studying the integration of phenomena in specific areas.

While general principles have been most fully developed in connection with the, relationships between various natural features, progress has also been made toward general principles concerning the relations between any particular cultural item, such as a crop, a particular kind of factory, or a city, and all the other features that are significantly related to it. Such principles have been developed in every phase of economic geography, and, to a lesser extent, in other cultural aspects of geography.

To some degree, therefore, geography may be called a generalizing or nomothetic science. Both Hettner and Penck find that in this respect geography has a great advantage over history, though it would be an error to overlook the use in history of generic concepts and principles contributed by the systematic social sciences.

Few geographers are satisfied, however, with the present state of development of principles in geography-either in terms of their quantity or of their reliability. Insofar as more principles or more reliable principles are possible, no one, of course, should be satisfied. Professional geographers, however, may feel less discouraged if they perceive clearly the difficulties and positive limitations that are placed in their way in the effort to develop generalizations and principles. More than that, the clear recognition of those facts may prevent misdirected efforts that can only yield fallacious principles.

Jame's Approach : James seeks the explanation for our difficulty in the fact that the phenomena which we observe are so much larger than the observer that we cannot see the woods for the trees, or, to use his excellent simile like a microbe crawling over the face of a newspaper photograph, we see only the details of the printed dots—the larger design of the photograph lies beyond our range of vision. Granted the force of this difficulty, have not geographers long since discovered the solution which he recommends, namely the use of the map to bring the larger relationships within the range of our vision? And yet we 'have not advanced very far in the establishment of principles.

Major Difficulties

One major difficulty lies in the fact that the integration of phenomena which we must study in areas is an integration of a large number of independent, or semi-independent factors. Consequently we seldom have to do with simple relationships—e.g., rainfall to soil, temperature to crops, etc. Theoretically we might follow the logic of the systematic sciences; by assuming that all other conditions remain the same but we have only the laboratory of reality in which to study these features, and in that laboratory the other elements do not remain the same, except perhaps in a very small number of cases, and we have no way of making them remain the same. Indeed, even if we knew the theoretical principles governing the relation of each individual factor to the- total resultant, in the case of such complex resultants as cultural features, a principle which attempted to state the sum total of all the relationships, each in its proper proportion, would be far too complicated for us to be able to use. This is a general difficulty that applies not only to all 'the more complicated aspects of the social sciences, but also to many phenomena in the natural sciences. Even if one knew all the principles and had all the data, the solution would be involved in a mathematical equation so complicated that no finite mind could solve it. Even when geography has attained a maturity of techniques and methods, it will, as Colby suggests, hesitate to predict resultants of such a complexity of factors.

The second major difficulty, definitely limiting the applicability of such rough principles as may be developed in geography, results from the insecure foundation on which the principles rest, namely the generic concepts. This limitation is present, to be sure, in every branch of science, in greater or less degree. In those sciences that are able to perform controlled experiments, specimens may be selected that conform almost exactly to the generic type involved in the principle; to the extent to which actual specimens in reality differ from the generic type, the principle will riot apply. In geography we seldom have a sufficient number of specimens of any generic type to permit us to discriminate in our selections. Further, as we have seen, our specimens do not conform closely to a generic type, but only within very wide limits. This wide divergence of the individual within the type is illustrated by the marked lack of agreement among geographers, not on the classification of a minority of doubtful specimens, but in regard to the very types, and systems of types, themselves. Since sound principles can only be developed if we have a sound system of generic concepts, we must consider for a moment how such concepts are developed in geography.

We indicated earlier that many, if not most, of the generic concepts in use in geography had been brought in from other branches of knowledge. One group, perhaps the larger, are those whose expression in words of common speech reflects the fact that man studied geography before he ever heard of such a thing as science. While the concepts of common sense are likely to be crude, many of these have been found to be adaptable to scientific thought. Ocean, lake, river, mountain, plain, city, port, farm, crops—for all of these the ordinary meaning, as used in common speech, can be sharply defined for scientific use. To invent, in their place, technical terms with which to impress the outsider would be pretentious. On the other hand, many similar terms in the language are used in such different ways that it may not be possible to pin them down to definite technical meanings.

A second large group of generic concepts has been brought into geography from the systematic sciences, notably from geology. Though many of these have also proved useful, it is erroneous to presume that their eminent scientific authority assures their logical utility in geography. The common rule of scientific work that requires the students in one field to take cognizance of the findings in neighbouring fields and, unless they can show errors, to assume that the facts and relationships established in the other branch of science are correct, does not apply to a classification system, which cannot be called either correct nor false, but simply more or less useful. If, for example, geologists have classified landforms into generic types based on their past history rather than on their present forms, their concept of a "plateau" need not necessarily be accepted by the geographer simply because it is a "scientific concept" in contrast, say, to the time-honoured concept of the same word in common speech, and as used by geographers long before the geologists gave it a new and different meanings.

Finally, geographers themselves have developed various systems of types into which they may classify the features which they study. They have been particularly concerned to supplement generic concepts based on individual elements with generic concepts based on element-complexes. These new systems of classification often appear in opposition to those taken over from common thought or from the neighbouring systematic sciences, so that there has been a great deal of argument as to which generic concepts are proper for the science of geography to use. This is not, as many have thought', simply a matter of preference either for the "new" or for the "old." . The question is more fundamental: from the particular point of view of geography, which concepts are of greatest value? Since this has caused much argument, and is clearly of great practical importance in the development of the field—in contrast perhaps with some of the questions that have required our consideration-it merits a brief analysis.

The fact that it is impossible to include all the characteristics of any object in a single generic concept means that any particular object may be classified simultaneously in many different types according to the bases of classification. But these types may be sufficiently alike so that, in view of the limitations of our vocabulary, we use the same terms for different concepts. Take the very simple case of coal. The paleobotanist, I presume, will classify coal as a vegetable product, to the stratigrapher it is simply a kind of rock, and the student of mining will say it is a mineral. Since coal is, in fact, simultaneously all of these things, it is included in each of these concepts though no one of them of course completely expresses what coal is.

Consequently the classification of objects into generic types cannot be determined simply on the naive basis of "the inherent characteristics of the objects themselves." In terms of all their characteristics, no classification is possible. Which characteristics will be considered in developing generic concepts will therefore depend on the purpose for which the generic concepts are developed. Generic concepts are not an end in themselves but merely a scientific tool whose particular purpose varies according to the point of view of the different branches of science.

In general, the purpose of developing generic concepts is to provide a single statement of a collection of common characteristics shared by objects which otherwise differ. This is useful for the primary purpose of describing reality in that it provides a shorthand method of description which likewise enables us more easily to grasp and retain descriptions. Furthermore such precise shorthand collectives are extremely useful, if not practically indispensable, in developing principles of relationships between objects. In both cases it is clear that the generic concepts should express those characteristics of objects which are of greatest importance according to the particular point of view under which they are being studied. For example, a student in the science of heat engineering may ignore, in his classification of coals, various characteristics that have no effect on the burning qualities of coal, however essential these characteristics may be to the paleobotanist.

We have, therefore, the fundamental basis for judging different systems of generic concepts in geography which classification most completely and clearly expresses those characteristics that are of most importance from the point of view of geography, whether for geographic description or as a basis of establishing principles of geographic relationships.

To the extent that the point of view in geography is different from that of other branches of science, whether systematic or historical, there is no presumption in favour of generic concepts introduced into geography from other sciences; on the contrary it might be safer to presume the opposite. On the other hand, the generic concepts which we have found in common speech are usually developed from the same naive point of view toward reality that geography also expresses. To the geographer, as well as to the sailor, the whale is a fish or—if this be regarded as a misuse of a word-the whale is to be included under the genera concept of ocean animals rather than land animals.

Many of the concepts which geography has inherited from the geologic physiography of the past century were developed by students whose primary concern was in the processes of formation of landforms, presumably as a part of earth history. It was, therefore, appropriate that they should attempt, however difficult it may be, to classify landforms in terms of those characteristics which resulted from, and therefore were the key to, their genesis.

If the geographer were likewise directly and primarily concerned with the genesis of landforms, he would, without hesitation, accept the same generic concepts; but on this basis geography would be difficult to distinguish from dynamic physiography. Conceivably, however, the geographer might require a genetic classification as the best means for expressing to the fullest the character of the landform. This, if I understand it correctly, is the line of reasoning followed by Hettner. The geographer seeks to establish types of landforms or climate "according to the entirety of their characteristics". This, we found, cannot be done directly. Conceivably it might be produced by a genetic classification (though that would appear as a first step rather than a final step, as Hettner suggests). But is this possible? Let us assume different landforms, each of which represents a resultant form developed by separate processes of folding, elevation, vulcanism, and erosion. A system of classification could be constructed in which these various processes determine the major and minor sub-divisions, but again we would have the unanswerable question of which processes are major, which are minor.

Penck's View : Penck discussed this difficulty, from the point of view of geomorphology itself, as early as 1906. He observed that there had appeared to be the finest harmony between structure and surface form, so that geomorphology seemed to have won a sure base for its concepts in tectonics. Thus "in place of the original geographic concept of a mountain area as a sum of unevennesses, many had adopted a tectonic concept according to which a mountain area was a strip of strongly folded land." But more careful morphological investigations had shown that the relations between tectonics and geomorphology were not so intimate as one had at first supposed, and one must conclude that the form of the land which the geographer is to study and explain is something different from the structure of the land which the geologist studies. More recently, Penck has illustrated the same conclusion by comparing the Bohemian massif and the Central massif of France. Both, he says, are products of bursting *(zerborstcn);* they have great similarity in composition and full relationship in their later geological development, and yet geographically ate as different as convex and concave forms.

A complete genetic description, to be sure, would arrive ultimately at a complete description of form, but a genetic classification allowing for all the processes that may in any area be significant-in all possible arrangements in time-would not be a classification of types. It would only be an indirect way of describing an end result that could more briefly and accurately be described by direct consideration of the present characteristics of the landform.

If, then, any classification limits us to some characteristics, forces us to ignore others, which are the characteristics that are of most concern to the geographer seeking to study an area in terms of the integration of its phenomena? Surely those characteristics that are significantly related to ether phenomena of the present earth surface. In other words, he is more concerned with the form or physiognomy of the landform as a functioning factor in the total complex of areal phenomena than as an end product of its own genetic causes.

No doubt there are cases where the characteristics of primary concern are the same from both points of view, volcanoes or canyons, for example. In such cases, both groups of students will use the same generic concepts. Where this coincidence of interest is lacking, geographers are under no compulsion to follow the direction of the genetic physiographer, or the geologist. In German geography the struggle for independence in scientific approach to the study of landforms took shape in opposition to the methods of W. M. Davis. According to both Ule and Burger, the distinctly geographic point of view was most effectively presented by Passarge, as early as 1912, and," throughout his work since, he has emphasized the description of present landforms as functioning factors in the region (or landscape). In passing, one may add a note of defence for Davis-if defence be needed. Lecturing at the Sorbonne, according to the testimony of Lehmann, Davis stated that "if it proves very difficult to work out the development of the landform, so that, in the problems of the past one loses sight of the landscape, the geographer will do better to describe the land simply with the help of the older orographic presentation".

No doubt most American geographers today would accept Sauer's conclusion that "there is no necessary relation between the mode of origin of a relief form and its functional significance". Frequently, however, they fail to draw the logical corollary that type concepts based on genesis may have little significance for function. They forget, for example, that to tell us that an area is covered with "Wisconsin moraine" gives us no precise description of either its surface form or its parent rock content. Geographers have freed themselves from the genetic classification, Sauer notes, most fully in the study of climates. Possibly the delay in establishing a corresponding freedom in the study of landforms may be due to the fact that most geographers, in this country as well as in Germany, have been trained particularly in geology and geomorphology.

It should not need to be stated (but since the writer has been misunderstood in oral discussion of this point, he may be permitted to clarify his standpoint) that the suggestion that geographers need to free themselves from a geological point of view is in no sense an attack on geologists. One may feel that the latter have abused a good geographic term in calling a "plateau" what they mean to 'describe as a "former plateau," but there are too many difficulties in terminology to justify making an issue on that. The concepts which the geologist has developed are not subject to attack by the geographer; presumably they are entirely suitable for the purposes of studying reality from the geological point of view. All that is said here is merely that there is no reason to presume that the concepts of the geologist are suitable—and many reasons for believing the opposite-for the purposes of studying reality from the geographical point of view.

To establish sound principles of the relation between, say, cultivation or soil wash on the one hand, and landforms on, the other, it is necessary to have generic concepts which express measured characteristics of landforms that are significant to cultivation and soil wash, rather than characteristics expressed by types of origin. In this case, the problem is simple at the first step, but very complicated beyond that. The essential characteristic of the landform, from the point of view just stated, is its slope, but its slope is a surface in solid geometry which varies constantly and irregularly at every point and in every direction at every point. Recent developments in the measurements of slopes by intervals, however, indicate that this problem does permit of partial solution.

In brief, in the naive point of view toward reality which geography shares with the common man, phenomena are significant in terms of their relations to other present phenomena of geographic significance rather than in terms of their origins. Generic concepts and systems of classification will therefore be more useful in geography if based on the functional rather than the genetic aspects of phenomena.

Principles of Relations

In spite of the difficulties involved, geography has been able to develop principles of the relations between the variable elements of area. To a lesser extent, as yet, have the relations between the variations in element-complexes been put into general principles, This should also be possible, even if more difficult, In general the study of element-complexes is still iargely concerned with the first step, the development of types; but even in this stage—as represented by the two systems of classification of rural areas discussed in the previous section-generic relationships are at least suggested. When we consider, however, not the elements, nor the element-complexes of areas, but the areas themselves, can we hope to develop general principles? This question lies at the very heart of geography.

It is axiomatic that a necessary condition for the development of general principles is the construction of generic concepts or types. It is presumably on this account that Passarge's suggestion of a systematic classification of areas into types has made such an impression on many students: it appears to offer the possibility of developing universals in the very core of geography, thus raising the "scientific" quality of our work.

Our approach to this question will be more ingenuous if we remember that, while it is essential to science to *seek* for universals, it is in no way essential that they be found; that, on the contrary, important parts of scientific knowledge-in every branch of science—cannot be expressed in universals. By placing a special value on the end product, scientific principles, we are in danger of passing lightly over the fundamental bases for them. We will not be able to establish principles by first calling what are not objects, objects, and then classifying them as generic types without considering carefully whether they are generic types. We may deceive ourselves with schematic outlines of classification and with schematic explanations based on vaguely expressed principles, but the inapplicability of these principles to the reality of the world will show our substructure to be counterfeit. As Kroeber observed-in discussing a different branch of science—"all schematic explanations seem essentially a symptom of a discipline's immaturity".

We need not repeat the reasons why any efforts to establish a world classification of areas in terms of the totality of their characteristics is doomed to failure. All the areas of the world can be classified generically in accordance with any particular element-complex, but the sum total of element complexes varying in different areas cannot be classified in a single system of generic types. For our present purposes, however, it is not necessary to classify all areas; if we could find some areas that could, as wholes, be classified in generic types, we might hope to go on to generic principles.

No matter how small the number of areas we consider, however, the essential difficulty remains the same. The integration of phenomena which we must study in any area is not a complete unit integration, but rather consists of a set of somewhat related but somewhat separate element-complexes, some of which are but parts of integrations extending into other areas-in the full sense, extending over the whole world.

If we may borrow James's simile of the microbe studying the newspaper photograph, we may say—with some exaggeration-that in the geography of any area, it is a though several separate photographs, each with its own design and each of different size and shape, had been superimposed in printing, each cut arbitrarily to fit. The geographer, in the position of the microbe, may be able to reconstruct the pattern of each separate design and classify it generically, but how can he classify the sum total of the separate designs in the particular superposition in which they are found? If the same combination of element-complexes, overlapping in the same way, occurred in more than one area, one could establish a generic type of this very complex form. While it is conceivable that such repetition may occur-just as it is conceivable that a hand 'of bridge may be repeated-it would be necessary to find numerous cases of repetition to provide us with generic concepts of regions from which we could develop principles.

We are in danger of confusing ourselves by using the word "region" as a more convenient term than "area of a certain type." When we say that the Po Plain and the Middle Danube Plain are ,"agricultural regions" of the same type as the American Com Belt, we only mean that these are areas within which' we find approximately the same agricultural element-complex. Even if we found exactly the same agricultural element-complex, the three areas could not be called specimen areas of the same species ; there is, and can be, only one "Com Belt" in the world. Any element-complex of an area appears in other areas, but the combination of all of them in the actual mixture as actually found, occurs but once on the earth.

The reader may feel that, while this conclusion holds for regions of such large size as those just considered-since the world can include but a relatively small number of such large areas-if we consider smaller regions, localities, perhaps, we should be able to establish types. A colleague who has been good enough to read this paper in manuscript suggests that he could select three localities, one in Mongolia, one in Patagonia, and one in the Great Plains, which are so similar that they could be compared, "if not to peas in a pod, at least to peas in pods in different gardens." Undoubtedly we have in this case very similar combinations of many of the same elements and element-complexes, but can we say that these common features are, in each case, the most important features of the area? Almost all systems of land use types, including that developed by the colleague who made this suggestion, classify these areas in different major world types. Even if we omit the nomadic area in Mongolia and consider only the commercial grazing localities in Patagonia arid the Great Plains, on what basis can we say that the marked differences in development of roads, railroads, and urban communities are of minor importance in comparison with the similarity of other features?

The previous example appears to lead to type localities because it involves areas dominated by one particular element-complex-including both cultural and natural elements. The situation becomes clearer if one takes localities of more average complexity. Thus a locality in the Po Plain may appear, in certain respects, like one in the American Com Belt, and so one is tempted to classify both as "peas rather than as tomatoes," but in other, no less important respects, the locality of the Po Plain is like one in the. Neapolitan plain, and on that basis, we must, so to speak, classify it also as a tomato. Further, in respect to all those features in the Po Plain locality that result from its location at the foot of the Alps, on main routes from northern Europe to peninsular Italy, in proximity to the Adriatic and the Tyrrhenian Seas, this locality is essentially different from all localities in other parts of the world.

In sum, the uniqueness of the region is of a very different order from the uniqueness of each human spine, or of each pea in a pod. Each of these is unique in characteristics that can unquestionably be called minor while major characteristics are identical, whereas a region is unique in respect to its total combination of major characteristics. In this sense we may agree with those who speak of the "individuality" of areas, though we may find another term used' by French geographers, "personality," too suggestive of an organic whole.

In a very interesting discussion of this problem, Gradmann has suggested that when one has studied a locality in terms of all its interrelated phenomena one has the impression of a complete and single picture which one therefore feels should be expressible in just the right word or brief expression. But "what we possess as a mental picture cannot be passed on to others in its finished state. Each one must work it out himself." The author must lead the reader through the entire analysis and synthesis so that he may himself come to the ultimate perception. His concluding comment should not be spoiled by translation: *"Damit wird reichlich Wasser in den Wein unserer jungen Begeisterung gegossen, und es ist niemandem zu verargen, wenn ihm der Trank furs erste nicht recht munden will"*.

Limited Application

There remains, however, a method of limited application by which it might appear possible to recognize localities as specimens of the same type. Within any single large area, for example the Corn Belt, or the Austrian Alps, do we not find a definite type of locality in repeated examples which are as like as peas in a pod? Cannot these be considered, as specimens of the same type of area, defined even in terms of total characteristics? One sees at least a resemblance to the specimens of a biological species, since the features of the different localities within a major region have more or less common origins.

If we assume that the particular combination of natural elements is essentially the same in different localities of a single large area and that the cultural development was controlled by essentially the same human factors, there still remain two significant differences. Even within this limited range, we cannot ignore the significance of relative location. The localities near the centre of the region will differ from those nearer the periphery in a number of ways that may be of great importance. Further, if these localities are specimens, their characteristics must include size and shape—characteristics which are not merely of academic interest, but may well affect the development of features, notably urban features, within them. But how can we determine the size and shape of the localities? Only on a very approximate and somewhat arbitrary basis—which would in any case give us differences as great as though the peas in a pod included objects formed like peas, pumpkins, and goose-neck squashes.

This last consideration reveals the essential error in our analogy. The localities of a larger region do not represent independent specimens of a species but simply similar parts of a whole, whose similarities are based in major part on the fact that they are but parts of a whole. Indeed that "whole," the larger region, is not really a Whole, but only a part of the single complete Whole which we have, the whole world.

In other words, the attempt to develop generic concepts of areas, as distinct from generic concepts *about* areas—and, on that basis, to compare areas in themselves and develop principles of their relations-rests on the fallacious assumption of the area as an actual object or phenomenon. We are misled by our terminology. When we say that certain areas belong to the same type, we can only mean that they contain one or more elements, or element-complexes, each of which is of the same type in all the areas. That is, having classified the phenomena found in areas in various systems of generic types, we then label an area that includes one or more of those types in terms of the types it contains. We are not actually classifying the area, but only one or-more of its characteristics. The area itself is not a phenomenon, any more than a period of history is a phenomenon; it is only an intellectual framework of phenomena, an abstract concept which does not exist in reality. It cannot, therefore, be compared as a phenomenon with other phenomena and classified in a system of generic concepts, on the basis of which we could state principles of its relations with other phenomena. Indeed we cannot properly speak of relationships (other than purely geometric) between areas, but only between certain phenomena within different areas. Likewise, the area, in itself, is related to the phenomena within it, only in that it contains them in such and such locations.

It may be objected that we have considered the word "area" too literally, that what is meant by area is simply the sum total of all interrelated phenomena found within an abstractly limited space. This sum total is, of course, an actuality, but it is not a phenomenon: the combination of more or less related phenomena, some of which are incomplete, since parts lie outside the area, does not form a phenomenon-is not a something that has relations as a unit to similar units, other than the purely geometric relation of location. (To clarify any confusion resulting from earlier use of terms, we should note that, if we are justified by certain authorities in calling a sum of interrelated elements that form a *relatively* closed total, a *unit-but* not a Whole-we must not then consider this loose unit as having the attributes of a precise unit—*e.g.*, of forming a single phenomenon that has relations as a unit with other similar units. To return to geographic ground, we know that the relations that we may loosely speak of as relations between areal units are, in fact, nothing but relations between some of the elements in one area and some in another.)

The conclusion that areas, as such, cannot be studied in terms of generic concepts, but can only be regarded as unique in their *einmalige* combinations of interrelated phenomena, leads some writers to the conclusion that the study of regions is no proper subject for a science. Since we refuse to define a science-though we have referred the reader to the views of such students of cognition as Cohen, Barry, and Kraft-we cannot debate this question. We may repeat, however, that every branch of science is, to a greater or less extent, concerned with the unique. From the standpoint of general culture, as Grano suggests, there is scarcely anything unique that deserves so much attention as the totality of interrelated phenomena in area that forms the milieu of man.

On the other hand the conclusion that we cannot consider areas themselves in generic concepts and principles does not mean that we cannot utilize generic concepts that express marked similarities in the characteristics of different areas. On the contrary, it is of great value to discover element-complexes and combinations of several element-complexes repeated in different areas. Both for the purpose of simplifying the enormous task of learning the complicated character of the different areas of the world, and in order to develop principles of relationships to aid our understanding of that world, we shall want not only relatively limited element-complexes that express some character of many areas, but also the most complete element-complexes possible to express many characteristics of perhaps few areas. These are of great value, even though no single generic concept can express all the characteristics of an area, even though anyone system of element-complexes may not apply to all areas of the world, and even though a single, very complex combination of element-complexes may be found in but very few areas.

For our present purposes, therefore, we can utilize many element-complexes which we found inapplicable to a system of world division. Thus, though it is not possible logically to divide the world in terms of the combination of relief, soil, and drainage, we do find many localities in which a particular integrated combination of these factors is present—e.g., "black bottoms," as a particular type of floodplain (whether we have as yet a classification of floodplains suitable" for geographic purposes is another question). Likewise we can recognize and utilize element-complexes that include both natural and cultural elements. When one has described the terraced vineyards of steep valley walls, whether of the Italian Alps or the Rhine Gorge, one has provided a major part of the description of similar localities found anywhere between those places. Similarly, one may construct a generic description of a port at the mouth of a navigable river in the humid tropics-the warehouses and factories constructed of imported materials by imported techniques in striking contrast to the background of primitive forest, the residence district of the foreign controlling group in contrast to the native quarters. In its general outline such an element-complex will be found repeated at perhaps hundreds of points on tropical coasts.

Generic descriptions of this kind are of unquestioned value, both for description and for interpretation—i.e., the development of principles of relations. It is only necessary to remember that they cannot include all, even of the major, characteristics of the localities described, nor can they be arranged to form a single system that will describe the world, even in outline.

Knowledge Organisation

In any field of science, knowledge must be organized into systems in order that the student of any particular problem may have ready at hand the knowledge of facts, generic concepts and principles that bear upon his problem. If all the knowledge in a field cannot be organized into a single system, it is necessary that the several systems be interrelated as far as is possible.

In geography, knowledge is organized into systems in two quite different ways. We may divide the phenomena of areal differentiation into major groups each consisting of closely related phenomena, and thus develop *specialized* branches of the whole field of geography. These include, then, physical geography, economic geography-which may be further subdivided into agricultural geography, the geography of mining, of manufacturing, etc. political geography, and what might be termed sociological geography. It is, of course, an error to include historical geography in this group, for that is, in itself, a complete geography of any past period.

On the other hand all geographical knowledge, including that in each of its specialized branches, may be organized according to the two different points of view required in studying the areal differentiation of the world: the view of any particular variable phenomenon in the relations of its differentiation to that of other variables over the world, and the view of the total character of all the variables within the area.

The organization of geographic knowledge in terms of the individual phenomena studied is called "general geography" by European geographers, or "systematic geography" in this country. The organization according to areas is most commonly referred to as "regional geography." (The use of the term "special geography" for this form of organization is fortunately no longer common. In German geography the usual term is *"Landerkunde."*)

These two different methods of dividing the field of geography cannot be combined on a single plane. Each of the specialized branches of geography is represented in both systematic and regional geography-in systematic geography by separate studies of single elements or element-complexes, in regional geography by a *part* of a regional study, limited to a particular group of related aspects. Thus the study of the element-complexes of concern to agriculture—complexes of land use and of natural conditions, etc.— a study in regional agricultural geography—e.g., Colby's study of the raisin production of south central California.

It is clear that other groupings of geographic features into special fields are possible, as in the geography of settlements *(Siedlungsgeographie)* or in urban geography. The geography of any particular city is clearly a special form of regional geography, whereas the studies of individual urban features as found repeated in many cities would be included in systematic geography. Further, because a city, like a farm, represents an actual element-complex that may be considered as a distinct unit, systematic urban geography may study the cities of the world, or any part of it, in terms of their differential character in relation to other geographic differences.

That no place is provided for "mathematical geography" and cartography may require a word of explanation. In climatology we are concerned with areal differences that result from the relation of the planet earth to the sun, but for the facts of that relationship we depend upon astronomy. Likewise, the study of the exact shape and measurement of the earth presents problems to astronomy and geodesy. The problem of projecting a sphere, or the geoid, on a plane surface, the science of projections, is essentially a problem in applied mathematics, which is of no less concern to astronomy than to geography. Consequently, the chapter on "mathematical geography" that often forms the introductory chapter of a geography text, commonly consists, as Hettner has noted, almost entirely of astronomical and other non-geographical material. Though such an arrangement may seem logical, it may be questioned whether the most effective method of introducing students to the field of geography is to begin with detailed studies from other fields-as though a text in biology should begin with a detailed study in chemistry, leading up to biochemistry. Finally, cartography, which is a technique rather than a science, is a form of knowledge of service in many Sciences. Because it is mere essential to geography than in any other science, and has been developed to highest extent in geography, it is both natural and reasonable that it should be most closely associated with our science, but it is no more a branch of geography, logically speaking, than is statistics a branch of economics.

Study of World

Geography, as the study of the world, necessarily includes a large number of different aspects that are, or may be, represented by specialized fields. Nevertheless it is a well-known fact, which can be tested by any random observation of the literature, that by far the greater part of the work in the field whether in systematic or regional geography—consists of either physical (natural) or economic geography. Urban geography theoretically overlaps beyond these special fields, in actual practice consists of but little more. The fairly sizeable literature in political geography is as yet of minor importance and of sociological geography we have almost none. This situation might be regarded as a natural result of the particular manner in which geography has developed in the past century, or lit might be suggested that geographers have commenced with the more obvious and simpler problems, leaving the more complex ones for later study. Overlooking the dubious psychological assumption involved in the latter explanation, there are reasons of much more permanent validity.

In the study of the areal differentiation of the world, the interest of geography in each of the many features which contribute to that differentiation is in proportion to its relation to the total. Each of the natural features varies notably in different parts, and its variations are significantly related to those of some other natural features and many cultural features. In general, the marked differences in the natural environment of different parts of the world, and the partial dependence of most cultural features on the natural environment, is adequate demonstration of the axiom that physical geography is of fundamental importance in geography as a whole.

The justification for the notable concentration of geographic work of recent decades on economic geography is not so obvious. To many students this may appear to represent an emphasis on studies of presumably practical value that appears foreign to the spirit of science. Undoubtedly, much of the work in this field is stimulated by such extraneous considerations; if it is definitely designed to serve practical purposes as in land use surveys, etc.—it finds its justification as an applied form of geography. Undoubtedly any science receives stimulating reactions from the work of those who endeavour to apply its knowledge and methods to particular problems, but there is an essential incompatibility between the two forms of science., theoretical and applied. The latter is defined and determined in character by the nature of the problem which it has to study in any particular case. A particular, complex problem will not fit into any theoretical branch of science, but calls for all forms of knowledge and techniques applicable to it. Even if the work be divided among a group of different scientists, the division can hardly follow, even in principle, the divisions of theoretical science.

Further, it may be claimed that much of the interest in economic geography represents primarily an interest in economic phenomena, which are studied more or less in their geographical aspects. Granted that this may be true for many studies-notably those concerned with the world distribution of certain products, or with the economic situation of particular countries, even though studied in terms of a "geographic basis"—there remain, nevertheless, fundamental reasons that require all the parts of geography concerned with human phenomena to recognize economic geography as of fundamental importance.

In our examination of a long list of cultural features, we found that the cultural features whose differences in different areas were of the greatest geographic significance—i.e., in terms of their relation to other areal differences, both natural and cultural-were for the most part economic features. They do not by any means include all economic features, since many of these are of very little geographical significance, though they may be of great economic significance. Consequently geographers are justified in regarding human, or cultural geography very largely in terms of economic geography.

Furthermore, since economic geography requires detailed consideration of the natural features to which economic phenomena are related, a regional study in economic geography constitutes the greater part of a full study in regional geography. To put it simply, most people live where they live—rather than move elsewhere or die—not because they like the climate, the politics or the customs, but because they are able there to make a living; the manner in which they make a living, and consequently the manner in which they live in general, are, in large part, determined by the interrelation of economic and natural features which it is the function of economic geography to study.

Emphasis on Agriculture

On the same basis we justified the major emphasis in geography on agricultural (in the sense of land use) geography. The features of urban areas are, to a much greater extent, undifferentiated in different areas. Needless to say, however, this does not mean that the regional geography of even a predominantly rural area is complete if the cities are omitted or barely mentioned ; the relations between the agricultural areas and the cities is of essential importance in understanding the character of the area.

On the other hand, in at least two major parts of the world-namely the two parts of greatest concern to European and American geographers—areal differentiation is represented in great part by the differential development, in intensity and in character, of urban development. "For these areas in particular, geography is greatly concerned with the geography of manufacturing, upon which the urban development is largely based, and with the study of the cities themselves as the most extraordinary features of areas that man has produced.

The conclusion that physical and economic geography make up the major portion of geography as a whole does not for a moment suggest that the other parts are to be ignored. In the first place there are many features of economic geography that cannot be correctly interpreted without an understanding of their relation to areal differences in culture, in the narrower sense of the word, and in political organization. These are therefore geographically significant features, and, in order that their relationships to others may be known with certainty, they need also to be studied systematically.

That major areal differences in culture may be of great importance in relation to other geographic features is familiar to anyone who has traveled in eastern Europe, or has crossed the Rio Grande, or been to mid-latitude South America, not to mention the vast differences to be observed in areas of Sino-Japanese and Indian culture. The differences of culture of a secondary order, however, appear to have but minor effect. A careful study of the maps of crop production in Europe reveals the importance of a boundary running from the North Sea to the Alps, east of which rye and potatoes are far more important than in areas of similar natural conditions to the west; the line follows approximately the Franco-Germanic cultural (not national) boundary. The difference mentioned, however, is clearly minor. Likewise, the more obvious differences in architecture, in customs, etc., can hardly be regarded as geographic differences of the first magnitude. Consequently, a detailed systematic study of the geography of peoples whatever its value for other purposes, would appear to offer a minor contribution to geography as a whole.

On the other hand, just as the importance of the geography of mining is found, not in itself, but in its significant relation to the far more important geography of manufacturing, so the geography of peoples, even in great detail, is of major concern to the geography of states. While the study of states in the sense of independent political units-is perhaps but a lesser part of political science, the geography of states constitutes the major part of political geography. This conclusion results from a particular characteristic of the state that is often lost sight of in the discussions of the place and function of political geography as a part of the field as a whole.

Social Organisation

As a social organization any political form, whether a state, a government, or other commonalty, is a feature that differs, to be sure, in different parts of the world, but the differences have but little relation to those of other features. The proposition that the mountainous islands of the Hellenic archipelago constituted the necessary background for the development of the democracies of the Greek city states is a proposition in the geographic aspects of political science; it would not be suggested by a systematic study in the geography of mountainous islands or of democracies. If the "mother of parliaments" is located on a fair-sized island, close to, but separated from the mainland, an island of certain conditions of climate, relief and soil, her daughters appear to thrive in' areas of radically different natural environment and of very different economic geography. In the study of political organizations considered from this point of view, geography, as Penck suggests, may have little place other than to offer supplemental suggestions to the political scientist.

On the other hand, the geographer has a direct interest in the state as a division of the earth surface. The fact that division, as Penck insists, is made by man and is not inherent in the nature of the earth, is immaterial to us, since we have found that any division of the earth can only be made by man. Neither can we accept his description of the earth surface of a state as "merely the stage of man's (political) action, to be sure a stage that influences it"; such a description represents only the political scientist's point of view of the state area. For geography, the state is an area in which certain conditions are universally true in contrast with those of other state-areas. It is, therefore, an area of homogeneity in certain very important respects, and so forms the simplest as well as the most definitely delimited of all geographic phenomena. Further, as we observed in our examination of the concept of regions as units, the state is the only area larger than a city which is organized as a functioning areal unit Whole. Unlike the abstract concept of the "region," the state area is, in many respects, a concrete unitary object; it is a piece of the earth's surface sharply defined and separated from other pieces, with which its relations are, in many respects, the relations between whole units. Like other concrete objects, the state-area has size, form, and structure. Indeed, it is in the consideration of the structure of a state that our concept of regions becomes of practical rather than merely academic importance.

To many geographers, particularly in this country, the concept of the state as an areal unit may appear remote from geography, not merely because it can be observed in the visible landscape only with difficulty, if at all, but also because of a wide spread impression that man's political structure is something extraneous, essential neither to the earth nor to man. On the contrary, as Schluter has noted, "the state (in the widest sense) is no younger than man, but rather older. Man could not become a human being without the protection of an association which contains the seed of the state. Man has been from the beginning the *zoon politikon'* of Aristotle". For man, as a political animal, it has been just as natural to create a state as a farm or a city, and the state which he creates, as East concludes, "whatever else it is . . .additionally and inevitably a geographical expression and as such forms part of the subject matter of geographical science".

Since there has been considerable controversy over the relation of this part of geography to the field as a whole, we may examine briefly the significance of the state-area as a subject of geographic study. The notable differences in size and form of the state areas of the world is one of the most obvious facts of areal differentiation of the earth surface. It is, therefore, one of the characteristics of the world which needs to be understood in a subject devoted to the study of the differences in different parts of the world. That it could be considered as a fact of minor importance is refuted by the reality of the power of states in controlling not only the political, but also the economic life within their areas. Finally, this important fact of areal differentiation is significant in geography if it is significantly related to other features of areal differentiation.

It is of course false to conceive of the division of any part of the earth into state-areas as "natural"—i.e., as determined by the natural conditions; since the phenomenon itself is cultural, neither the state nor any of its elements, such as boundaries, can be natural. But one cannot compare the political map of Europe or Asia with the corresponding relief map without realizing that there are close relations between the two forms of differentiation. Even where the political map may seem highly arbitrary, as in South America, a consideration of the map of population density, and then of the maps of natural elements that explain it, will show that it is only the outer boundaries that are arbitrary; the essential division into states is very significantly related to the natural geography of the continent.

The state-areas differ not only in the obvious features of size, form and locational relations to each other but also in their internal structures. The differences in structure in different state-areas is directly related to the regional structure of the area in terms of physical, economic, and ethnological geography.

In the reverse direction, the economic differentiation of the world with which we are concerned in economic geography is notably affected' by the efforts of all states-more marked in some than in others, but present in all to organize all the cultural features of its area into a more or less homogeneous and closed unit.

The most obvious product of these efforts is tariff walls, which, though visible only in the minute form of boundary stones and frontier stations, may have a greater effect on the geography of production and trade than the highest mountains or the widest oceans. Many of the less obvious effects are perhaps, in total, even more important. Thus, the regional geography of the Paris Basin can in itself provide but a minor explanation of one of the most important features within it, the city of Paris, in size and character out of all proportion even to the large fertile plain of Northern France. Only the successful effort of the state of France, in past centuries, to bring all its regions, from the North Sea to the Mediterranean, into an organized unit with Paris as the centre can account for that particular geographic phenomenon. Obversely, the agricultural and industrial development of each of the regions of France has been notably affected by its inclusion in this political-economic areal unit.

Political Geography

The position of political geography, as primarily the geography of states, is therefore not to be considered as a remote peripheral location, but rather as one in close relation to the major aspects of geography. On the other hand, the special field of *Geopolitik,* which has had such a notable development in post-war Germany, represents a very broadly defender quite undefined field in which geography, in terms of political geography; is utilized for particular purposes that lie beyond the pursuit of knowledge. It represents, therefore, the application of geography to politics and one's estimate of its value and importance will depend on the value that one assigns to the political purpose it is designed to serve. Since it is designed to serve national politics from the German point of view, its positive value from that point of view may be considered as offset by its negative value from the point of view of other countries. In its influence On the world's thinking, this branch of geography, which produced the concept of the *"Lebensraum"* of the state, would appear, at the moment, to be the single most influential branch of geography. (The history and methodological problems of *Geopolitik,* as well as of political geography, proper, have been treated previously by the writer. Since then, Ancel, in France, has attempted to rescue the term geopolitics *(Geopolitique)* for international science, and East, in England, has published a significant study; the most comprehensive study of political geography is that of Maull.

When one considers the theoretically possible field of sociological geography from the point of view which we have been following, it may appear doubtful that the development of studies of that character can make important contributions to geography. In areas of primitive development one might study the geography of clothes, or implements, or conceivably of manners and customs, and religions. For the important parts of the world, however, such studies would apparently have little geographic significance. Men wear hats in Chicago that allow their ears to freeze because the winters are not cold in London. The urban architecture of Midwestern United States is significant chiefly in revealing the lack of indigenous culture. Areal differences in religion are of some importance in political geography and perhaps in a few cases in economic geography, but in general this is a phenomenon that shows relatively little relation to other features of areal differentiation. Even it we take the case most commonly cited, that of Mohammedanism as the religion of the nomads of the steppes; we find that it flourishes also under tropical rains in the intensively cultivated paddy fields of Bengal or Java.

Prejudgements of future developments in science, however, are among the most dangerous of predictions. Many aspects of culture, other than those of economic and political facts, are significantly related to regional differentiation. Certainly it would be unwise to dismiss this field on the basis of examples selected to prove the lack of significant relation. The studies of settlement forms, house types, etc., that have been made in Europe and in some parts of this country-and with particular enthusiasm recently, I am told, in Japan-indicate the possibility of systematic contributions to geography as a whole.

In each of the specialized branches of geography which we have discussed, the pursuit of knowledge is dominated by the same ideals that we found applicable to geography as a whole. The degree to which these ideals can be attained varies in the different fields as it does in the corresponding fields of systematic science. In general, no doubt, the degree of accuracy and certainty may be highest in physical geography and decreases in the various branches of cultural geography, more or less in the order in which we discussed them. Such a comparison however is by no means universally true. Some economic facts can be established with far more accuracy and certainty than many facts of natural phenomena, and few facts in geography can be established with such a high degree of certainty and exactness as those concerning the extent of state-areas.

In the development of universals the greatest contrast in geography is not found among the special fields but rather between systematic geography in all its branches, and regional geography, whether partial or complete. It is particularly on this account that the separate development of systematic geography is so important, both in itself and in its relation to regional geography.

The division between these two different points of view is made necessary by the very nature of geography. The areal differentiation of the earth surface, which geography is to study, is a differentiation expressed in a great number of individual features whose variations are in part related to each other, in part independent. Geography, therefore, must study the areal differentiation of each of these features over the world, not as a part of the systematic study of that particular object, but as a study of one form of areal differentiation of the earth surface; this is systematic (or general) geography. it is clear, however, that a full understanding of the differences between areas cannot be obtained by simply adding together the appropriate sections of systematic geography. It is necessary to study the totality of interrelations of all geographic features found together at one place; this is regional geography. In the historical section we saw that this difference between the two points of view within geography, first stated by Varenius, was present in the work of Humboldt, and formed, as Wagner noted, an inevitable dualism in the field-not in respect to materials, but in respect to methods of approach. More recently the philosopher Kraft, in examining our field, has substantiated this statement. In general, in modern geography there appears to be agreement on the distinction between the two points of view, as outlined by Richthofen and following him, more clearly, by Hettner. The relative importance of the two aspects, however, remains as it has always been a subject of continuous controversy.

Two Methods

The relation between the two methods of organization has been illustrated by Hettner in the following manner. If we consider the variations of the geographic elements as though contained in surfaces parallel to the earth's surface, then all of them together would form a series of surfaces at different levels above the earth's surface. In any particular part of systematic geography, one studies all the variations in a single surface, in relation, it may be, to variations in the surfaces of the other elements. In regional geography, however, we strike a limited section vertically through all the surfaces, in order to comprehend the totality of their characteristics in a single area.

These two methods of organizing geographic knowledge are not only interconnected in every part but are also by no means so distinct in practice as is frequently supposed. A systematic 'study need 'not cover the whole world but may be limited to a continent or to any area within which there are variations in the feature studied. (Consequently the term "general geography" is unfortunate.) If, then, all the features of a small area are studied systematically in succession, the mere addition of the total series of these systematic studies does not form a study in regional geography, but simply the systematic geography of a limited area. The essential difference is in the point of view. In regional geography the study is focussed on the individual localities or districts, which, whether smaller or larger, are conceived (arbitrarily) as units. The purpose then is to comprehend the particular geographic character of each of these units that is, the particular manner of formation of all the geographic factors in their causal connections.

As we have implied a number of times, the terms inherited in geography for these two divisions of the field have been found unsatisfactory by many students. For work in which individual categories of phenomena are studied over extensive areas, or the whole world, Varenius' term *"geographia generalis"* is almost universally used in Germany *(allgemeine Geographie)* as well as in France *(geographie generale)*. As Hettner and others have found, however, this is misleading, not only because such studies may be limited to but a part of the earth, but particularly because they consider the, features of the earth surface by individual categories. Richthofen at one time suggested the term "analytic geography", but Hettner rightly objects that both analysis and synthesis are required in both forms of geography. Schluter nevertheless appears to have adopted the term.

In his earliest considerations of this question, Hettner, we noted, endeavoured to express more clearly the close relation of the two aspects of geography by using in place of "general geography" the term *"vergleichende Landerkunde"*—taken, he tells us, from the title of a course given by Richthofen. Though he did not use this term in his subsequent methodological treatments, Hettner has recently returned to it as the title for a comprehensive work covering the whole of general geography, except for the omission of the seas. Whatever conclusion German students may ultimately come to, American geographers will no doubt agree with Penck in finding this usage unfortunate. On the other hand, the term now widely used in the American literature *"systematic geography,"* finds ample precedent in the writings of many German geographers. Thus Richthofen described this form of geography as organized *"auf Grund systematischer Principien"*, Hettner called it *"die systematische Darstellung"*, and Penck has called it the method *"nach systematischem Gesichtspunkt"*. Unfortunately the fact that the term is not of Germanic origin, and its German equivalent is unsuitable, apparently excludes its use as the title for a major part of the field of geography in that country.

The other term introduced by Varenius, *"geographia specialis,"* has been very largely replaced, in Germany by *"Landerkunde,"* or more recently by *"Landerkunde"* (ignoring here any differences between the two), and in practically all non-German lands by the appropriate form of "regional geography." Any number of German students however have found *"Landerkullde"* unsuitable, both because it excludes the seas and because it suggests areal divisions much larger than those now commonly studied as regions. As early as 1831, Frobel suggested "region" in place of "Land" Supan wished to return to *"Spezialgeographie"* (or *spezielle Geographie"*); Waibel would prefer either that term or *"regionale Geographie"* and Lautensach favours *"regionale Geographie"*. Since neither of these terms, however, is of German origin, there is little likelihood that German geographers of the present period will change to them.

Division of Field

The fact that the field of geography may be divided in two different, and intersecting, directions frequently leads to confusion. We may clarify the situation by classifying a few examples; if we limit these to studies by the present writer there will be no danger that anyone be offended. The study of types of political boundaries is a systematic study in political geography in which the attempt is made to construct generic concepts and to suggest some general principles whereas the study of boundaries in Upper Silesia is a partial regional study-regional political geography. Similarly the study of the Upper Silesian Industrial District is a partial regional study limited largely to a part of the economic geography of the area. Finally, the study of Austria-Hungary is a study in regional political geography of a past period, that of the beginning of this century—*i.e.*, a study in historical geography, limited to political regional geography. In other words, we recognize no "boundary between economic and regional geography," such as Pfeifer apparently would have us observe. Each overlaps the other. But likewise it follows—as possibly he intended to say—that regional geography is not to be thought of as complete if it is limited to economic regional geography.

Relative Analysis

The conclusion that systematic and regional geography are two coordinate forms of organization of knowledge in our field may have raised a question in the reader's mind-as it did in the writer's—concerning the position among the sciences to which we have assigned geography in general. If geography is to be classified as a choreographic science along with astronomy, and these are to be included with the historical fields as integrating sciences, should we not logically expect a corresponding division in each of these fields between the systematic and the sectional (regional or periodic) approach? At first glance, at least, one might suppose that almost all of astronomy would be included under the systematic approach, almost all of history in the study of periods. Insofar as this is the case, does it cast doubt on our thesis of the logical similarity of geography to these fields?

Although astronomy and geography are both concerned with the study of spatial integrations-things related to each other in space the character of the spaces that they study and the things interrelated within them are so completely different that no amount of logical similarity should lead us to expect similar results in the developments of the two fields. For much the greater part of his work the astronomer may consider celestial space as extraordinarily simple, consisting on the one hand of homogeneous ether-which in much of his work, he may regard simply as empty space and on the other hand of masses of inanimate matter, most of which are sharply defined units widely separated and therefore related to each other essentially as whole units. Within any celestial region—e.g., that occupied by the solar system the problem of integration is little, if anything, more than the systematic problem of the relations between these unit masses, a problem primarily concerned with their effects on each other's position and motions, studied in terms of but two factors, gravity and free motion in space. Even within systematic astronomy, however, not all the findings can as yet be reduced to scientific laws. Though the motion of the sun in reference to other stars has been measured, the laws determining that motion have not been constructed, possibly never can be.

Astronomy does include studies corresponding to those of regional geography. These are represented most clearly by the detailed examination of those units in the solar system that are near enough so that differences between different parts may be observed. In the same category, though different in character, are studies of the groupings of stars in our universe, and the detailed examination of individual stellar nebulae.

If astronomy is largely concerned with systematic studies because of the relative simplicity of its subject matter, exactly the reverse is the case in the historical sciences. Of this group, historical geology shows most clearly the distinction between the systematic and the periodic approach. The study of climatic changes in past ages, of changes in mountain development, and, in general, the changes in the continental landforms, or the evolution of the horse—all represent systematic studies in the history of the earth. In contrast are studies which attempt to provide a generalized picture of associated phenomena of climate, landforms, and vegetable and animal life of the Upper Mississippian or any other past period in the history of the earth.

The comparison to which we have most repeatedly referred throughout our discussion of the nature of geography is, of course, that with history in the ordinary sense of the history of "historic times.' Various students, however, have suggested that the comparison can only be related to regional geography, that history lacks systematic studies.

We should not be misled by the fact that history is commonly taught only in terms of what we may call "periodic history." In their research problems historians frequently concentrate on the development and changes in some very restricted group of phenomena through a succession of years. Such studies may treat the development of a particular form of constitution, the growth of labour legislation, the changes in the price of wheat in England, or the development of roads in Minnesota.

Nevertheless, so far as an outsider may judge, the work of this character has by no means the importance to the field of history as a whole that systematic geography has to geography. In particular it has not yielded to history generic concepts and principles that are nearly as definite as those developed in systematic geography.

If one compares the particular problems studied in the two fields, as shown in their publications, it is obvious that historians are concerned with phenomena whose interrelations are far more complex than those commonly studied in geography. The logical basis for this difference is not so obvious; indeed our fundamental assumption of the relation of the two fields leads logically to the conclusion that the same phenomena may be studied in each field; history may consider areal phenomena and geography may consider historical events.

Neither history nor geography, however, need consider all the phenomena that are found in the sections of reality which they study, but only those phenomena which differ significantly in different sections of time or space, respectively. In each case the attention is chiefly focussed on those phenomena which differ most and whose differences are most significant to the total differentiation. In the total reality with which both history and geography are concerned-namely the phenomena of the world in historic times there is one major group of phenomena, the natural phenomena which are causally of fundamental importance to all the other phenomena, but which, while differing markedly in different areas of the world, differ but slightly in different periods of historic time. This, of course, constitutes a great difference between history, in the narrower sense, and pre-history, not to mention paleontology.

Areal Difference

In consequence, the areal differences that are of greatest importance in geography are either differences in the natural features themselves or in cultural features which are closely related to the natural features. We would have a similar situation in history only if such features as climate and landforms had varied as radically through historic times at the same place, as they vary over the world at the same time. In other words, if the natural environment of England since the time of Caesar had varied from humid to arid, from polar to tropical, from plains to mountains, the agricultural history of England would represent the most important part of its history, and history would long since have developed the systematic branches of climatic history, landforms history, etc. Indeed, if Ellsworth Huntington's thesis of the historical importance of even minor variations in climate should be substantiated, it would be not only logical but necessary for history to develop a systematic study of climatic history the study of the relations of climatic changes to other historical features.

In any case, the exceptional character of the example just cited tests the rule the relative fixity of natural conditions during historic times results in a notable degree of constancy in those cultural features which are most closely related to natural conditions. The manner of land use in any area may remain much the same for centuries, in China, for millenia. Cities do not pass through a generic *process* of youth, maturity, and old age to death; they may continue in approximately the same condition for indefinite periods of time.

Consequently the phenomena which show the most notable differences in relation to time are cultural phenomena less closely related to natural conditions-commonly, therefore, phenomena of much more complicated character-such as manners and customs, political organization, inventions, etc. Furthermore, not only are these phenomena in themselves more complex than those with which geography is most concerned, but their interrelations through different periods of time are more complex than the interrelations of the principal geographic phenomena in different areas. Indeed, in most cases the character of one period of history largely determines the character of the next, whereas the character of one area in geography has commonly but minor effect on the character of its neighbours. It is not surprising, therefore, that historians are more clearly aware of the fictional nature of their divisions of time than are geographers of their corresponding divisions of area.

On the other hand there are some sudden breaks in historical development that produce changes almost as great as the change from sea to land in geography, namely, when new discoveries or inventions, or the migration of peoples, introduce a new culture into an area. The frontier of settlement in America during the past several centuries was not only a line marking great geographic contrasts but, as it passed through any region, it represented an historical revolution in the adaptation of man to nature. The historian of this revolution, therefore, must understand the principles governing the relation between cultural and natural features in order to study history. Much the same is true of such historical problems as the industrial revolution, and the associated agricultural revolution in Europe.

The historian who is concerned with these problems, involving less complex features than those that form most of the material of history, presumably will not hesitate to make systematic studies wherever possible and to explain the relationships where he can. The fact that most historical events may be too complex to permit of definite explanation should not lead to a dogma that no historical events can be interpreted. Unfortunately, however, situations comparable to those mentioned above, in which the fundamentals of man's adaptation to nature are notably changed, are relatively few in history and most of them took place at such an early date that the historian has scanty reliable data from which to study them. Thus, a systematic study of the history of "frontiers"—in the sense of a border of progressive settlement—should consider not only the frontiers in the New World and in Siberia, but also the frontier of German settlement in central Europe in the Middle Ages and the still earlier frontier of Anglo-Saxon settlement in Great Britain. It is obvious that, even were data available, such a problem would be extremely 'complicated, since it involves not merely different periods of world history but also different areas of the world of radically different character.

In general, the problems which must be handled in systematic studies in history are far too complex, and involve factors too difficult to observe and measure, to permit of the development of generic concepts and principles similar to those developed in systematic geography. There are, to be sure, some students of history-chiefly non-historians-who assume that it is possible to develop scientific laws concerning the rise and fill of states, the causes of revolutions, or the development of particular social movements, but their theses are more notable for the ardor with which they are advanced than for the evidence which has been brought to support them. Most professional historians are sceptical of the possibility of developing a systematic history in which the phenomena important in history may be classified in generic concepts leading to principles; The rather naive belief of some geographers that geography can provide this deficiency in history has not, as yet at least, been substantiated.

This contrast, therefore, between history and geography results from the fact that the interrelations of the phenomena that vary most notably in historic times are far more complex than the interrelations of the phenomena that vary most notably in the earth surface. It does not affect the logically common nature of the two fields, as sciences that attempt to integrate phenomena as they are found in reality.

The simplest form of study in systematic geography is the consideration of the differential character of the earth surface in terms of any single geographic factor. In the past such studies were in large degree limited to the natural factors-the climatic factors, landforms, soils, etc. but, as many students have pointed out, if geography in general is to consider human or cultural features, they must be studied in systematic as well as regional geography. Most students today recognize that in this respect the development of systematic geography to date has been one-sided. An extreme illustration of the contrast is offered, as we noted earlier, in Finch and Trewartha's "Elements of Geography." In contrast to nearly six hundred pages of masterly treatment of the systematic geography of the natural elements, they present hardly a tenth of that amount on the much more complicated problems of systematic cultural geography. This difference, to be sure, hardly does justice to the present development of systematic cultural geography, even in the literature of this country; it is not clear why the authors should appear to disregard a number of excellent studies, including some by Finch himself. Further, a large number of systematic studies have been made in cultural geography by many German writers, including notably Schluter and his students the study of bridges, has been noted for an example of a full detailed survey of systematic human geography.

With the increasing interest, in both Germany and America, in full regional studies, necessarily including cultural as well as natural features, geographic research suffers not only from the lack of adequate foundation in systematic cultural geography but also, as Broek notes, in the lack of training of most geographers in the social studies. If students are to be prepared to make full studies in regional geography and there is fairly general agreement that all geographers should do at least some work in this field it would be logical to require supplementary training in the related social sciences no less than in the related natural sciences. This is the case in few, if any, of our departments of geography. Since, in addition, the academic relations of geographers are commonly closer to geology and other natured sciences than to the social sciences, the individual geographer knows that his work will be subjected to careful criticism so far as it touches natural science. For the same reasons he has been free to indulge in almost any sort of economic theorizing or political speculation that occurred to him, with little or no risk of being called to account.

On the other hand, the relatively recent shift from the emphasis on a geomorphology closely related to geology to the emphasis on human geography, has resulted in a tendency on which Krebs has very recently commented: much material that is essentially economic, historical, or sociological is taken over and presented without digestion in presumably geographic studies. American geographers appear also to be aware of the need for independent research to develop concepts and principles of systematic cultural geography-indeed there is evidently a widespread feeling that this is the single most pressing need in geography today.

It is desirable, therefore, to consider carefully the distinction between systematic studies in geography and the studies made in the related systematic sciences. Especially is it necessary, as Schmidt insists, for those who work on the borders of geography to keep the distinction in mind. Though they must be familiar with the concepts and methods of the neighbouring sciences and may use these in their work, they must use them for purposes dictated by the point of view of geography as distinct from that of the related systematic science. In particular, Schmidt has contributed a very thorough and valuable study of the relations between economic geography and economics, as well as a detailed study of systematic economic geography.

The divisions of systematic geography, as we noted earlier, correspond to the divisions of the systematic sciences and there is inevitably close relationship between each branch of systematic geography and the corresponding systematic science. This relation is not accurately expressed by the phrase "neighbouring sciences," since geography is not a branch of science situated beside the systematic sciences, but represents a point of view in science which cuts through all the systematic sciences. There is, therefore, no line separating systematic geography from the systematic sciences, but there is an essential difference in point of view which must be maintained by the individual geographer who wishes to do geographic work rather than work in some other branch of science.

The distribution of a particular kind of phenomena is significant both in geography and in the systematic science concerned with that kind of phenomenon, with this difference: in geography the focus of attention is concentrated, not on the phenomenon—one of whose aspects is its distribution but on the relation of that distribution to the total areal differentiation of the world.

The Contrast

This contrast in point of view may he illustrated in the case of the production of corn (maize). The totals of production by countries and the resultant effects on national and international markets are presumably of concern in economics, but not in geography. One may be confused here by the fact that economists have generally been willing to leave instruction in the geographic aspects of economics to economic geographers, with the result that geographers have frequently attempted to do research in what is essentially a part of economics. Likewise the relation of annual variations of corn production to annual variations in rainfall is of great concern to the student of agriculture but is not of direct concern in geography, whereas the fact that the variations in rainfall in Nebraska have a greater effect on corn yields, than the same degree of variation in Pennsylvania, is of geographic concern. Geography is concerned with the marked areal variation in corn production, since this represents a part of the total areal differentiation, in which it is associated on the one hand, in its relation to differences in climate, soil, relative location, or cultural conditions and on the other hand, in its relation to differences in the total crop-livestock element-complex, the character of barns, the presence of grain elevators, etc.

In other words, the facts of distribution of corn production are not in themselves "geographic," not even when shown on a map. It is what is studied concerning those facts that is significant for geography. Merely to describe and analyze the facts of distribution of physical and social phenomena found in different areas is to produce a compendium, not a geography, either systematic or regional. The facts concerning the areal differences in these phenomena must be studied in their *areal relations*, that is, their significance to the area as determined by their relations to other phenomena of the same place, and by their spatial connections with phenomena in other areas.

In a study in systematic economic geography, for example, unless the geographic point of view is clearly maintained from the beginning, the work may turn out to be a study in the geographic aspects of economics. The reason for this, of course, is that both of these kinds of studies start with the same step, namely, the establishment of the facts of distribution of the particular phenomena studied. As this first step focusses the attention of the student on the phenomena themselves, it often results in his continuing the work with that point of view, thus producing a study of those phenomena—*i.e.*, a study in a systematic science.

Because the botanist or economist is concerned with the location of his phenomena in only some of his studies, whereas the geographer is always concerned with the location of facts, it is often supposed that the determination (and interpretation) of the "Where" of things is exclusively a function of geography, if not the whole, of geography—i.e., geography as the science of distributions. But it would be both presumptuous and contrary to what actually takes place for us to claim that the zoologist, geologist, or economist concerned with the distribution of his phenomena must look to the science of geography for the answers. Likewise the fact that in such studies the students of other sciences may use the geographic technique of mapping does not make them geographers;

the economist or political scientist may often use the historical method in determining the "When" of past events, but their work does not thereby become history. Similarly all the students of the systematic sciences may use geographic methods in presenting the distribution of their phenomena-whether particular kinds of plants, animals, or factories-without depending upon the science of geography.

In the reverse direction, geographers have, in fact, long been accustomed to looking to certain of the systematic sciences for their knowledge of the distribution of certain kinds of facts. For the location of mineral deposits and different kinds of surface rocks, we depend upon geology; for the occurrence of soils, on soil science; for the distribution of native plants and animals, on botany and zoology. As Hettner, in dependence on Wallace, has insisted, these latter represent geographic studies in botany and zoology, as distinct from studies in plant and animal geography in which the interest is focussed on areas, studied in terms of their plant and animal contents. It is only in those fields in which the systematic sciences concerned have given little attention to the geographic aspects-notably in economics-that the geographer has been forced to do his own spade-work in determining distribution. It is significant, however, that once geographers had introduced their technique of mapping into the study of domestic crops and animals, agricultural economists have taken over this work as an integral part of their field.

The study of the distribution of phenomena presumes a classification of objects into types. In many cases the objects are sufficiently simple so that the classification is both obvious and acceptable to all the sciences concerned *e.g.*, the classification of cultivated plants into different kinds of crops: corn, oats, wheat, etc. If less simple phenomena are involved, however, we noted that the classification will depend on what aspects are selected as most important for the particular study. Consequently two sciences concerned in studying the distribution of the same phenomena may differ, even in the presentation of facts of distribution; though this difference may in itself be slight, it may be of major significance in later stages of study.

One should not forget, however, that economy of effort is an axiomatic *desideratum* in scientific work. In any case where the classification and establishment of facts concerning the distribution of phenomena that have been developed in another science are found to be suitable for the purposes of systematic geography, there is no call for the geographer to do the work over again in a different way. Nevertheless, as we noted earlier, in the consideration of relatively complicated phenomena, such as land forms, the facts established for the purpose of another science have been found not to be the facts needed in geography; consequently the systematic study in geography must begin anew at the first step.

Any presentation of facts in science calls for interpretation. Consequently, geographers have often presumed that, in presenting the facts of distribution of any phenomenon, it was also the function of geography to study the causes of that distribution. But in every branch of science facts are presented and utilized whose interpretation is the function of some other branch, of science. In this case, namely the distribution of any phenomenon, does interpretation of the facts of distribution fit logically into geography or into the systematic science?

This question is not raised here in order to argue over the location of a borderline between sciences, certainly not with the idea of establishing any rules of conduct for geographers. In considering this question we will, I believe, come to a clearer understanding of the whole relation of systematic geography to the systematic sciences.

One point appears clear. Whichever student is to interpret the distribution of a particular phenomenon will study that phenomenon in terms primarily of those aspects which indicate it causal development. If the distribution be measured in terms of other aspects they must first be referred back to the genetic aspects in order to provide interpretation. One of the essential contrasts between geography (and history) on the one hand, and the systematic sciences on the other, is that the former are interested in the integration of phenomena, the latter in analyzing the *processes* of particular kinds of phenomena. The fact that studies of processes involve the time element does not make them history, as Kroeber has emphasized; neither does the fact that distribution involves the element of space make its study a part of geography.

The explanation of the world distribution of a particular kind of phenomenon would appear to be an end resultant of the study of the processes of development of that phenomenon; it is therefore the proper subject of study in a systematic science. In systematic geography, however, it represents the world picture of an element with which one is concerned in its functional relation to the differential character of the areas of the world. In other words, though geography must know where things are, the study of the "Where" is not geography nor an integral part of geography, and it is therefore not the function of geography to explain the "Where"—that is, to give the full explanation of why a phenomenon is found where it is found. Consequently, systematic geography is free to overlook generic concepts based on *genetic* aspects of phenomena in order to develop generic concepts based on aspects that are functionally significant.

Although we conclude that it is not the function of the geographer to explain the distribution of any phenomenon, it is at the same time clear that he may be concerned with such an explanation in order to interpret the relations of that phenomenon to other geographic phenomena. For example, in the geography of soils, the interpretation of the relation of the soil of any area to the character of its climate and bedrock, necessitates an understanding of the whole development of soils; but it is the function of the soil scientists to provide the explanation of soil development in terms of all its factors and processes.

Systematic Geography

The systematic geography of any particular phenomenon depends we conclude, *for* the principles governing its distribution; on the systematic science concerned with that phenomenon. In many cases, however, the geographer may find that the students of the appropriate field have not been interested in developing such principles. In such cases he can hardly be expected to wait indefinitely, but may have to undertake the study himself. If, however, he does that without realizing that he is shifting his point of view, he may later discover that he has definitely passed over into a field in which he may not be adequately prepared.

Since the previous paragraph might seem to be pointed at individual geographers, it may be appropriate to illustrate it from the writer's own experience. Geographers had long recognized that the concentration of iron and steel mills in certain areas was somehow related to the presence of coal and iron mines in the same or other areas, and every text in economic geography attempted to state that relationship. None of these statements, however, were found to be adequate and the reason for this was, no doubt, the failure of students of the economics of industry to study the problem of the distribution of iron and steel mills. The geographer wishing to interpret the character of areas found to have, as one of their major characteristics, intensive development of this industry, must be able to explain the relationship of these factories to other geographic features.

The writer, therefore, undertook to develop the principles governing the location of the iron and steel industry, on the basis of which that industry could be studied as a part of the systematic geography of the United States. The interest developed by the first study led the writer into similar studies of the principles governing the location of other industries, and of industry in general. It has since become clear to me that only in the study of the iron and steel industry in the United States was my attention focussed on the areal significance of the industry as a particular characteristic of certain areas; in the others the centre of attention was the industry as a phenomenon of which one aspect, namely its location, called for explanation. It is not surprising, therefore, that any interest shown in these studies has been confined almost entirely to economists.'

The conclusion which the writer has drawn from this personal experience may have general application. The study of the *"Standorts"* problem-the determination of principles governing location of units of production—not only requires more training in economics than in geography, but also requires a full concentration of interest on the problem for the sake of the problem itself, rather than for the sake of the results; it is the economist, who is interested in the problem, the geographer in the results.

On the other hand it, might be claimed that, regardless of a logical division of problems between the sciences, geographers had in fact, developed this particular subject to such an extent as to justify their retaining it as a part of their field, by right of cultivation, regardless of the logical division of work in science. Undoubtedly geographers have made contributions to the location of economic activities, but we have hardly pursued the problems with sufficient consistency and system to register a valid claim based on thorough and successful cultivation. The world of knowledge as a whole does not look to us to supply the principles governing these phenomena.

Confirmation of this conclusion may be drawn from critical survey of the work of geographers in this field made by the Swedish economist Palander, in the introduction to his exhaustive study of the theory of the *Standorts* problem. It is only fair to add, however, that geographers entered this field only because the results which they needed in their work had not been adequately developed by economists. American economists in particular have shown little interest in this field and at the time the writer was concerned with it were hardly aware of Weber's work, which, in any case, Palander has shown to be impracticable. Geographers, therefore, will welcome the attention which economists are now giving to this problem. Geography will not be merely receptive in this relation, however, for even though we may agree that the problem comes logically under the point of view of economics, it certainly represents a geographic problem in economics, which requires some understanding of the geographic point of view, and for which geography can continuously contribute both positive materials and effective criticism.

Since much has been said of the dependence of geography on the systematic senses, we may appropriately note one or two significant suggestions that geography may contribute to the problem of interpreting the distribution of economic features.

Interpretation of Distribution

The first step toward an interpretation of the distribution of any phenomena is, of course, to portray that distribution. Students of the systematic sciences who have acquired something of the geographic point of view, will realize that the only language in which the location of things on the earth's surface can be portrayed intelligibly is the map, and that reliable interpretations require maps more detailed than cartograms of units as large as our States. Although this proposition is axiomatic in geography and geology, and is now thoroughly recognized in agricultural economics, in other branches of economics it is frequently overlooked. When economists attempt to interpret the location of the steel industry in the United States in terns simply of the amount of development in the States of Pennsylvania, Ohio, Indiana, and Illinois, it is not surprising that they should reach a defeatist conclusion as to the possibility of principles of location.

Even though our census figures by counties are less complete than those by States, they must be used to gain an approximately accurate measure of the development of the steel industry in the areas of South-eastern Pennsylvania and Maryland, the Pittsburgh-Youngstown region, the Lake Erie Ports, and the Calumet District. " Likewise one cannot hope to interpret the contrast between the industrial development of Wisconsin" and that of Minnesota until one has seen the facts portrayed on a detailed map and observed that the concentration of distinctively manufacturing cities is not to be considered in terms of Wisconsin versus Minnesota, but in terms of proximity to the west shore of Lake Michigan.

To the student of economics who has not been trained in geography, even a detailed map of distribution of an economic feature may appear to present a comparatively simple problem in comparative location that is, he is apt to think almost exclusively in terms of relative location, considered purely geometrically, and to ignore other variants of areas. . Thus, in many economic texts the consideration of the distribution of different types of agricultural production has long been dominated by Thunen's simple picture (constructed, we may note, by a writer living in the relatively homogeneous North German Plain) of concentric belts of differential production surrounding a city centre. Whatever validity this analysis may have had in earlier periods has been very largely destroyed by the development of modern commercial facilities that have made relative location a factor of secondary importance in determining land use. O. E. Baker showed, some years ago, that, with such well-known exceptions as market-gardens and fluid-milk harms, the location of different types of agricultural production is far more dependent on climate, relief, soil, and drainage than on relative location; this writer has demonstrated the fact in detail in the agriculture of Europe. On the other hand, this does not mean that the factor of relative location may be entirely ignored in such problems, as is often the case in studies in agricultural geography. Waibel has recently considered Thunen's law in full detail in the light of the radical changes in conditions since it was first stated.

This discussion of the importance of the geographic point of view to the problem of interpreting the distribution of any phenomena may appear to suggest that, in spite of the logic of classification, the geographer is best equipped to handle the problem. Before any geographers accept that conclusion, they should first examine the specific problems treated in Palander's masterly treatise-in particular the enormous amount of economic detail required in the handling of the transportation problem; they should observe the technique of economic analysis developed in a study of the location of manufacturing in the United States by Garver and associates-even though that study provided the examples of lack of map-mindedness to which we referred above; and finally they should consider the studies made by economists who have had some" geographic training, for example, Hoover's " study of the boot and shoe industry.

All that we have shown is that the study of geographic aspects of any field of science, such as economics, requires something of the geographic point of view. That historical problems in economics require something of the historical point of view requires no emphasis, because all economists have no doubt had training, in one way or another, in history; but relatively few have in geography. Every science, Schmidt concludes, "that concerns itself with the study of the areal distribution of its objects on the earth is necessarily led to the geographic method; it must interpret the differences of its objects in relation to area, and so must make use of the method of thinking in geographic comparisons as one of the most important means attaining general concepts in its own field and of penetrating into the character of the scientific objects of its own science. Thus, every research worker in economics must be a geographer in the sense that he must use the geographic method whether he will or not the sooner he wills it and knows it, the better for him and his research".

In brief, we conclude that, both in terms of the logic of the classification of the sciences, and in terms of the professional equipment of the students—in techniques and in knowledge of the literature the problem of principles of distribution of economic phenomena can best be studied by the student who is primarily an economist, but it is necessary also that he be in some degree a student of geography.

To avoid any misunderstanding it may be necessary to add that throughout this discussion-indeed throughout this paper the term "geographer" is to be understood as an abbreviation of "student of geography." Any individual person may presumably be simultaneously a student of geography and a student of economics and may study wherever he finds himself interested and believes himself competent. Undoubtedly individual geographers have made, and may continue to make, important contributions of thought to the work in related fields. Indeed, such personal interconnections between the different fields of science are not to be regarded as merely permissible, but rather, as Penck properly insists in his own defense, are greatly to be desired. If this is true of science in general, it is particularly true of geography, which not only is related to other sciences along border zones that "would be left fallow if scholars always limited themselves to a single science," but in every part of its field intersects the studies of the various systematic sciences. It is fortunate, therefore, that "the boundaries of the sciences are not insurmountable walls," as Penck writes in discussing the same question later, and one might ask him to modify slightly the analogy that he does suggest, of "the boundaries of states, which one can cross if one has the necessary pass, in our case, capacity." Perhaps Professor Penck's own experience on certain international frontiers would persuade him to agree that there should be no border guards along the boundaries of science, but that each student is to be his own judge of his pass, subject always to the ultimate verdict of those qualified to judge in the field into which he crosses. In any case, all will agree that "he who works across the border areas of geography must be able to ride in several saddles". Further, as Penck has elsewhere indicated, the requirement that the student should feel himself competent wherever he works, requires that he should himself know in what field he is working at any time. And in order that geography may maintain clearly its own fundamental point of view, any cross-fertilization should be recognized for what it is, and not accepted as an extension of our field.

On much the same basis we may be permitted to dispose of the difficult problem of the relation of geomorphology to geography without attempting to solve it. This question has long been a matter of controversy, particularly in the English speaking countries. As early as 1908, Chisholm, in agreement with Geikie, expressed the view that if the study of landforms follows genesis it leads to geology, and many others have echoed that view since. At the same time, however, a large part of the work in that field in America, if not in England-has been carried on by geographers-particularly of course as a result of the work and influence of Davis. In Germany, the course set by Richthofen, and followed particularly by Penck and his students, has made geomorphology so definitely a part of the field of geography that few, if any, question its permanent inclusion. In the Netherlands, in contrast, geographers apparently distinguish more clearly between geography and geology in this field. Examining geography from the point of view of knowledge as a whole, the *Erkenntnistheoretiker* Kraft concluded that the study of geomorphology, including the genesis of landforms, disrupts the logical unity of the field, but that, as a result of historical evolution, this field is in fact included in geography—in Germany at least. Its inclusion cannot, therefore, be questioned, he concludes so long as, one may add, the geomorphologists continue also to be geographers. In other words, the geographer in Germany is, by his training, a geomorphologist as well, and geography, therefore, as a division of labour within the sciences includes that special field.

Conclusion based on History

Whether the same conclusion, based on the history rather than the logic of the field, applies in this country, the writer would not attempt to judge. It is important to note, however, that the close association of geomorphology with geography has brought the latter not only undoubted advantages, but also certain disadvantages. If geomorphology is primarily concerned with landforms as objects to be studied in themselves, as the botanist is concerned with plants, then, as Michotte notes, the point of view is that of a systematic science, in contradiction to that of geography as a choreographic science.

A general result of this contradiction is to be found in the difficulty that many geographers who have been trained primarily as geomorphologists have experienced in maintaining consistently the geographic point of view, as Penck himself has observed, not to mention the confusion that many of them have introduced into methodological thought in geography. A more specific result has been suggested in an earlier connection. The study of landforms as objects in themselves, leads logically to a classification of them as individual objects of a systematic science rather than to a classification of the areal character of landforms—"the character of the various morphologic areas of the world," as Michotte puts it. While geographers have felt free to classify climates, natural vegetation, or farm types independent of the classifications of the corresponding systematic sciences, as long as geomorphology was regarded as an integral part of geography, they were inhibited from developing a different classification of landforms suitable for choreographic description. In many cases, to be sure, the types of individual landforms were suitable, but the attempt to make them usable in all cases led, for example, to that paradox of areal terminology, the description of the White Mountains as "a collection of monadnocks."

Whatever conclusion may be drawn with respect to the relation of geomorphology to geography, it is necessary to note that if, as Kraft holds for German science, the facts of historical development make it a part of the field of geography, that conclusion does not provide an argument for the inclusion of logically analogous problems in other parts of geography in which they have not established themselves in the past. Thus, the claim made by Maull, and more recently by East, that the processes of evolution of state-areas are as properly a problem in geography as the study of the evolution of landforms, would be valid only if it could be shown that such studies in the geographic aspects of political history have in fact been developed primarily by geographers, rather than by political scientists or historians.

We may summarize briefly our examination of the relation of systematic geography to the systematic sciences. Ideally, systematic geography receives from other sciences, or from general statistical sources, the necessary data concerning the distribution of any phenomenon; it classifies the various forms of that phenomenon in any way that is suitable for geographic purposes—i.e" in terms of characteristics significant to regional character—whether or not such classification is available from other sciences. Further, ideally, it receives from the systematic sciences the explanation of the distribution of the phenomenon; that is, its genesis. Whether it be landforms, forests, crops, steel mills, or political states, the principles of development and the causes of distribution, as such, are the concern of the appropriate systematic fields. Geography starts with those facts and principle—assuming always, of course, that the systematic sciences concerned' have provided them as frankly borrowed material.

We have dwelt on the preceding question at some length because it is particularly in systematic geography that the student is likely to lose his sense of the geographic point of view, so that he may, as Lehmann has suggested, give a false picture by disproportionate consideration of phenomena that are the objects of the systematic sciences, or he may leave geography and enter entirely into other fields. It is doubtful if this can be prevented merely by drawing boundaries, however sharp and clear they may be. The reader may already have objected that we have drawn no clear boundary between systematic geography and any systematic science. No such attempt has been made, and if we remember that the relationship involved is not the borderland of neighbouring fields but rather the intersection of fields lying in different planes, no such boundaries are needed. The distinction is in the point of view: that of the systematic science is focussed on the particular phenomena, which are studied in terms of distribution; that of systematic geography on the part which that distribution plays in forming areal differentiation. In many studies, the geographer may find it necessary to make excursions in the plane of the related systematic science, away from the common line of intersection. If he has the geographic point of view clearly in mind, he will need no boundary stones to remind him that he is making an excursion out of his field, but will return, as soon as he has established the necessary data or conclusions, to the geographic plane.

To maintain the geographic point of view in systematic geography, it is necessary for the student to refer his work constantly to the field of geography as a chorographic science. Most writers agree that this view is most clearly indicated in regional geography. Consequently, many have agreed with Penck that every geographer, no matter how great his interest in specialized systematic branches of the field, should make regional studies. In any case, it is essential; as Lehmann insists, in making any study in systematic geography, constantly to consider the relation of that study to regional studies. If a systematic study is considered from this point of view it is immediately clear that the interest of the geographer is not in the phenomena themselves, their origins and processed, but in the relations which they have to other geographic features (i.e., features significant in areal differentiation).

On first thought it might be supposed that the conclusions to which we have arrived would result in the elimination of most, if not all, of the work in systematic geography. On the contrary, they free that part of the field for its essential function of providing systematic study of the relation of the differentiation of specific kinds of phenomena to total areal differentiation. The areas of the world differ from each other in terms of a mutually interrelated complex of heterogeneous features, each of which is different in the different areas. The complete interpretation of an individual area requires that, at some level of size, we break it down, mentally, into the component parts formed by the specific categories of phenomena. As Michotte puts it, we must study the vegetative character of the area, its geomorphological character, the character given it by each of the major cultural features, and so on. Further, the comparison of such completed individual studies would not give us a complete understanding of the areal differentiation of the world. It is necessary to know also how these areas, considered solely in terms of their natural vegetation, landforms, or each of various cultural features, differ from, and are related to, each other, Michotte speaks of these comparative studies as "comparative plant geography "comparative morphological geography," etc. Similarly Hettner's title for his several-volume study of systematic geography, *"Vergleichende Landerkunde,"* though unfortunately misleading, is not so inappropriate as might be thought. Likewise, we may add, it is significant that the physiographer, Fenneman, who first presented the chorographic concept to American geography should demonstrate that viewpoint, in systematic geography, in his masterly studies of the regional physiography of the United States.

It is particularly, however, the students whose interest in regional geography has motivated them to make systematic studies who have most clearly indicated the type of work that systematic geography should undertake. Recognizing that the relation of any specific feature to the character of an area is to be measured particularly in terms of its relation to the other factors in that total character, they have perceived that absolute measurements of individual elements are less valuable than relative measurements, or ratios, of elements in reference to each other.

Study in Systematic Geography : One of the most important advances in making studies in systematic geography more geographic in character has been the development of the isoplethic method of mapping ratios. Based on the work of Engelbrecht, this method was developed and effectively presented in this country by Wellington D. Jones and is now in widespread use. Compare, for example, the utility, for a study of agricultural differences in different parts of China, of the maps of crop ratios that Trewartha has recently published with the dot maps showing absolute values upon which we previously had to depend.

That this method may be Carried further, so as to show simultaneously two significant ratios concerning the same phenomenon, is indicated by the writer's isoplethic map of the dairy areas of the United States [the demonstration is very inadequate due to the small scale on which the map is reproduced]. This study (an extension and amplification of Jones's map of a smaller area) actually portrays the differences in areal character of land use resulting from varying degrees of intensity of dairy development, which can be inferred only indirectly, and in many areas incorrectly, from the ordinary census maps showing distribution of dairy cows or milk production. In both of the systems of world division of rural areas discussed in the previous section, the determination of agricultural types and the delimitation of "agricultural regions" depended on the construction of a large number of isopleth maps (not published) showing areal differences in ratios of individual crops to total crops, of cropland to total land, of livestock units to cropland, etc.

The ratio method in systematic geography is not limited to studies in agricultural geography. In addition to ordinary "relief" maps, which actually show directly only elevation, maps of relative relief" have been constructed after Partsch, by various European geographers and, in this country, by Guy–Harold Smith. In "sociological geography," Kniffen has used the method in mapping house-types and the writer has used it to show the areal differences in racial construction of the population of the United States.

Even where the character of the distribution does not permit of isoplethic mapping, as in the treatment of characteristics of cities, the principle of measurement by ratios rather than by absolute figures may be used to bring out the differential character of cities-i.e., that character, other than size, which is most significant in the comparison of cities, and in the comparison of regions in terms of their urban development. This is illustrated by the writer's study of the manufacturing belt of North America, in which the manufacturing functions of cities are measured in relation to their total functions, rather than in absolute values.

A much more complicated technical tool for work in systematic geography has but recently been presented by John K. Wright, under the title of "Some Measures of Distribution". The fact that it involves rather complicated mathematical formulae should not prejudice geographers either for or against its use. The various small examples which Wright offers suggest that we may have here a new technique of great value in enabling systematic geography to arrive at conclusions useful in regional geography that will be far more accurate than those now available. While this possibility is suggested by the examples which he gives, for the writer, at least, they seem inadequate to establish the utility of his technique. It is to be hoped therefore, that some student will be interested in applying the technique to some actual problem to see what results it may produce.

In their simplest form systematic studies in geography are confined to single elements. We have previously noted, however, the importance of the concept of "element-complex" in geography—i.e., an interrelated association of various elements, regardless of kind. If approximately the same element complex is found repeatedly in different areas and its distribution is geographically significant, it may also be studied systematically-over the whole world or any large area. Such studies, interconnecting different branches of systematic geography, may be considered as stepping stones from the study of single elements to the study of the total complex of a particular area in regional geography.

A single element-complex may represent an interrelation of elements at a single point—*e.g.*, rainfall, temperature, slope, soil, drainage, and vegetation in which case we may speak of a vertical complex of indefinite horizontal extent. On the other hand, the elements may be situated at different points so that their interrelation constitutes an areal form of more or less definite horizontal extent. Thus a longitudinal Ushaped Alpine valley in its primeval condition was a natural-complex in which the factors of slope, soil, drainage, and vegetation varied in a definite manner from the mountain shoulder on one side to that on the other. It is an areal form fairly definitely determined in the transverse direction though its limits in the longitudinal direction are indefinite so far as the concept itself is concerned—*i.e.*, are determined only in each specific case. A *polje*, in contrast, is a similar complex areal unit definitely limited in all directions.

Markus's Contribution : The Estonian geographer, Markus, has contributed an interesting and suggestive study of element-complexes *(Naturkomplexe)* confined largely, if not entirely, to combinations of natural elements. He notes that geographers have long recognized certain more obvious cases of element-complexes by such terms as "tundra," "high moor" and "low moor," "grassrich depressions in steppes," etc. Noting that changes in any factor in a complex do not cause immediate change in the others, but rather that these adjust themselves to the new conditions at different rates of speed, he distinguished between "normal complexes," in which all the elements correspond to each other' completely, and "abnormal" in which the adjustment has not yet reached completion. These terms give a clearer description than the distinction between "harmonious" and "inhârmonious" that other writers have suggested in a similar connection. Markus speaks of a "positive shifting" of an element complex where it is pushing into an area of another complex that requires a lesser amount of any particular factor—as in the advance of forest into steppe and negative shifting in the reverse direction. Further, he projects a complete classification of element-complexes in which real complexes are reduced to abstract species or types—by consideration of their essential common characteristics—and in which these are arranged in families, orders, etc.

This ultimate object—establishment of a Linnean classification of types, even of abstract types, of element-complexes-faces essentially the same insurmountable difficulties as we have met in the attempt to arrange regions in a single system of classification. Forested mountains, forested plains, mountain steppes, and plain steppes are four distinct types of element-complexes that cannot logically be arranged in any unilateral system of classification, since we have no method of deciding objectively whether the difference between mountain and plain is more important than the difference between grass and forest. Likewise, we cannot accept Markus' further implication that a geographic region *(Landschaft)* can be expressed as a single type of element-complex; if we consider all the elements involved in the complex of which the region consists, we arrive at the unique case, not the type. Nevertheless, though we cannot accept the more optimistic conclusions that Markus draws, we can expect valuable results from the systematic study of particular types of natural element-complexes, each of which is to be regarded as expressing more of the character of any area than a single element, though not its full character, not even in outline.

Our previous discussion of regional division indicated that we may expect much more useful results from the study of the many element-complexes, extending in many cases over wide areas, that have been produced by the organizing hand of man. These complexes are of a different order from the natural element complexes that we have discussed, in that they are not merely the sum total result of the interaction of forces accidentally placed together any one of which may be understood by itself in its relation to the others. These cultural element complexes have been purposely created by man for the sake of the ultimate result, and the presence of anyone element is to be under stood not in terms of its relations to the others but in its relation to the result. For example the importance of oats in the Cora Belt is to be explained in terms of its significance to the total crop and live stock association that man has organized on Corn Belt farms, or rather, if one will, in terms of his ultimate purpose of securing the highest monetary return with the least expenditure of labour, capital, and land. Consequently, as noted earlier, these cultural element-complexes are, to a considerable degree, organized as unit Wholes, an understanding of which should be a first step in the development of cultural regional geography.

The relatively small units of cultural element-complexes—e.g., farms—involve a much larger number of factors than those commonly found in natural element complexes and include both material and immaterial elements. O. E. Baker has recognized that contrasts in types of farms in the United States involve contrasts in the character of the farm population: farmers represent farm elements just as definitely as do the livestock and crops.

Though the cultural complexes commonly form but small areal units they may be found to be organized together, by man, into looser areal complexes that, individually, cover relatively large areas.

Finally we may recognize complexes involving both cultural and natural elements. Because man, in many cases, has developed the same cultural element-complex in areas of similar natural conditions, we may expect to find a number of complexes consisting of cultural element-complexes in interrelation with certain natural elements. Since man, however, has been far from consistent in his form of adjustment to natural conditions, we must expect these compound element-complexes to have relatively restricted applicability.

Certain Element Complexes

Geography finds that certain element-complexes with which it is concerned have been studied by other sciences. If the complex includes only elements of the same general category—e.g., the natural vegetation as a complex of different plants, or an iron and steel works as a complex of blast-furnaces, steel furnaces, rolling mills, fabricating mills, storage yards, etc.—one of the systematic sciences will presumably be concerned with the study of the complex in itself and in its distribution over the world. The more complex forms however-involving combinations of heterogeneous phenomena—may be of concern to the geographer alone. In either case, the classification of the types of element complexes for use in geography must be adapted to geographic purposes. If economists have produced a classification of farms suitable for geographic purposes, the geographer will utilize that classification; but if not, he is free to develop his own. In this particular case it appears likely that both groups of students working in cooperation may develop a classification suitable from the point of view of both sciences. But if economists' classification of manufacturing industries offers little of value to geography, geographers must develop their own.

Any study in systematic geography, whether of an element or an element-complex, concentrates on one particular kind of phenomena or phenomenon-associations: It naturally leads, therefore, to the establishment of generic concepts; that is, for each element, or element complex, a logical system of types may be established. On this basis the relations of the feature studied to other geographic features, for which types have also been established, may be stated in the form of principles, however limited or inaccurate in application. Whether one considers rainfall, soils, stream deposits or erosion, crop-animal associations, steel mills, or political boundaries-in each case over wide areas if not the whole world-it should be possible to establish principles of relationships between the feature studied and other features geographically significant.

It is by no means possible, however, to express all the findings of systematic geography in generic terms, whether of concepts or principles. In the systematic study of volcanoes, Krakatoa cannot be adequately treated solely as an example of a type. Likewise, in the relations of one geographic feature to another, innumerable cases will be found, each of which is unique. Nevertheless a very large part of the work in systematic geography does deal with universals and leads to the development of principles. Do these provide geography with that precious power that is often regarded as a hall-mark of "science," the power to predict?

The essential characteristics of that form of "knowing" which we call science to use the term suggested by the physicist, John T. Tate1—are not determined by the character either of the knowledge or of the capacities acquired; these are rather resultant products of that manner of pursuing knowledge which is science. The ability to predict in any branch of science represents the attainment of such a degree of certainty of knowing that, by deduction from principles, the future outcome of a combination of present factors is known almost as certainly as it can later be known as an observed fact. The qualification "almost" represents more than a margin of error, of inaccuracy in measurements: in every field there is an ever-present margin of uncertainty, of not-knowing, which cannot be eliminated even in the physical sciences-nor do modern physicists expect that it may ever be eliminated.

The ability of any science to predict is therefore the result and outward evidence of a high degree of attainment of the ideals of accuracy, certainty, universality, and system. It is not the test of a science, but only the test of success in "knowing" in any science. That success is not to be attained by striving directly for its result, the power to predict, but rather by striving for the highest degree of attainment of the fundamental ideals of "knowing."

In plain words, we will not learn to predict in geography by attempting to predict. It is a corollary of scientific principles that we should seek to know to what extent our knowledge is incomplete or uncertain. To attempt predictions in situations for which we know we lack the necessary knowledge is to be unscientific. Science does not require that we be able to predict. The sound demonstration of a low capacity for prediction in any particular field of knowledge is not evidence that field is not a branch of science, but, on the contrary, is a scientific conclusion testifying to the scientific character of that field.

No professional geographer, would claim that research in systematic geography had as yet reached such a high degree of attainment of the requisite ideals as to enable it to make predictions of high degree of certainty. Though a more maturely developed geography should show far higher attainments, we must recognize certain insurmountable difficulties and limitations that will always be present in geography.

Knowledge of Phenomena and their Relation : We know, in the first place, that the nature of most of the phenomena that must be measured in systematic cultural geography, and of many of those in systematic natural geography, will never permit of such accurate and certain measurements as are possible in some of the natural sciences. This difficulty geography shares with many of the systematic sciences, notably, of course, the social sciences.

We know that our knowledge of phenomena and their interrelations, in every branch of systematic geography, can only incompletely be contained in generic concepts and principles, and that there is inevitably, therefore, a margin of uncertainty in prediction. While this margin is present in every field of science, to greater or less extent, the degree to which phenomena are unique is not only greater in geography than in many other sciences, but the unique is of the very first practical importance. This is true not only of geography and the social sciences but likewise the human physiology and psychology-from the point of view of the individual and his family, at least and of certain aspects of meteorology and geology. To predict that the islands of Japan will experience innumerable earthquakes is of little value; who will predict the date and location of the next major earthquake?

We bow further that the-complex interrelations of phenomena that we study in systematic geography cannot be taken into the laboratory where some of the variables may be controlled in experiments, so that we could learn the exact significance of each factor. This handicap, again, geography shares not only with the social sciences but also with human physiology, with most of the branches of geology, and with astronomy. The students in all these fields can work only in the laboratory of reality, can observe only those experiments that reality chooses to perform for them. Actually the same limitation applies to the physicist studying the actions of electrons in the laboratory; he cannot control the individual electrons which he is attempting to study. There is "no hard and fast line between observation and experiment," Cohen concludes.

In the performances that reality presents us as substitutes for laboratory experiments, we know that geography is handicapped in two ways. Whereas some fields are presented with thousands or millions of repetitions of nearly similar cases, in geography, as well as in geology and astronomy, and in parts of all the social sciences, there may be only hundreds of similar cases, or only a handful, or often but a single case. Where the number of factors involved in the relationships is relatively small, as in astronomy, and the clearness and exactness of observation may be at least as fine as in the laboratory, a few cases may suffice to provide an adequate basis upon which to develop scientific laws of high degree of certainty; but in geography, geology, and the social sciences, one or both of those conditions are lacking.

One further difficulty remains in any branch of science that must deal with extremely complex functions of a large number of more or less independent variables. If some divine power should present the scientist with complete statements of everyone of the interrelations involved, expressed (if that were possible) in mathematical equations of the greatest complexity, and if then, in any particular situation, divine power should also provide complete and accurate knowledge of the individual factors, the complexity of the problem which must then be solved to arrive at certain knowledge of the outcome would be beyond the ability of finite minds.

In conclusion, therefore, geography is, by its nature, one of the branches of science from which we are to expect relatively little knowledge of the future of such a degree of certainty as to justify the word "prediction." Undoubtedly one could postulate many cases where such certainty was possible—even in our present state of development. If large deposits of high-grade iron ore should be discovered in West Virginia we could not only predict the mining development that would result, but we could no doubt predict, in a general way, notable changes in the iron and steel industry in the Pittsburgh and Calumet Districts. The reader will observe that this is not only an extreme case but one in which the relationships are unusually simple, since but three variables are of major importance. For the most part, the knowledge of the future that systematic geography can provide is limited to that lesser degree of certainty that we express by such terms as "trends" or "likelihoods," and must further be qualified by many uncertain factors due to the more or less arbitrary action of individual men or groups of men.

In sum, we may' justifiably predict that a mature geography developed to the maximum, cannot attain more than a very restricted capacity for prediction.

While we may dismiss the question of ability to predict as not fundamentally relevant, the pursuit of universals, generic concepts, and principles, must be regarded as of major importance for the development of any science. In geography, the greatest opportunity to develop generic concepts is in systematic geography. A large part of the work in each section of systematic geography is concerned with phenomena and relations between phenomena that repeat themselves in similar specimens in different parts of the world, so that it is possible to express them in universals and thereby to develop principles. Consequently, for those among the ranks of geographers who by reason of temperament, ability, or training, prefer to study the generic, with the opportunity to develop scientific principles or laws, there is plenty of work to be done in geography. Since such work is not merely an integral part of geography, but forms the necessary base for the studies of regional geography, no geographer need berate these students as deserters from the field. Those who use Fenneman's picture of the field of geography and speak of regional geography as the centre, should not overlook his qualification: "There is no intention of assigning more dignity to one part of the field than to another, nor of asking any man to turn aside from that which interests him to something else. There is no more inherent worth in a centre than in a border".

Our examination of the character of systematic geography emphasizes the inescapable comprehensiveness of geography, a condition that would not be reduced in the slightest if one were to omit regional geography; on the contrary, the one method by which the diversity of interests is brought into unified study would be lost. Even if one should attempt to reduce systematic geography to the study of natural, non-human, elements, it would still be concerned with a heterogeneity of phenomena as great as that of all the systematic natural sciences put together, and the elimination of human factors would make it impossible to unify this diversity in the study of actual regions. Any attempt to arrive at a unified field by further whittling can change the situation only relatively if one throws out plant and animal geography, one still has subjects as different as the study of climates and of landforms. As these are both physical sciences, and are both concerned with the earth, they can logically be combined, either from the point of view of physics, or of the earth. From the point of view of physics they are widely separated fields that are not brought into logical combination by the incidental fact that they both concern the earth. If the earth forms the unifying framework they are combined only in the earth surface, as broadly conceived, where they are inextricably intermixed with the elements studied in plant, animal, and human geography. Only in the study of all earth surface features in their actual interrelated combinations in regions, can the heterogeneity of systematic geography be unified into one science. "We need not be frightened away by the fullness and breadth of the problems," Richthofen concluded. "The field is great. But the work can be divided among many. No one today can do research in all the parts of geography. But he who devotes himself seriously to geography, can master it sufficiently to follow advances in all branches and he who, through modest limitation, is fortunate enough to investigate productively in one part, should always strive to comprehend the relation of that part to the rest and never to lose sight of the interconnection of the whole".

At the same time, Richthofen felt that the individual geographer who wished to contribute research to the advance of geography, "the higher he sets his goal, the more should he concentrate his preparation on one part of (systematic) geography and the particular systematic science that forms its foundation, without neglecting instruction in the other parts. It was natural for Richthofen to emphasize geology as "the surest foundation," since that had been his own, but both Oberhummer and Gradmann are drawing the logical conclusion from his general principle when they state that individual geographers may just as properly select some other science as their principal supplementary field-whether meteorology, botany, economics, or some other.

An individual geographer who specializes in a branch of systematic geography, and is adequately equipped in the corresponding systematic science, will no doubt have occasion to make studies in that other field as well as in systematic geography. Just as it has always been regarded as appropriate for individual geographers who were adequately equipped therefor to do research in geology, it may similarly be appropriate for individual geographers under the same condition—to do research in anthropology, economics, or political science. Inasmuch, however, as geographers are not capable of judging the research in other fields, it seems logical that such research should be presented, not to geographers, but to the workers in those other fields.

Need it be added that such transfers of point of view may equally well be made in the opposite direction? The student of a systematic science, interested in the geographic aspects of his field, will frequently be able to contribute to the field of geography, and one trusts that in this exchange there need be no grumbling of trespassing on either side.

Regional Geography

The development of geography during the past thirty-odd years has been marked by an increasing interest in regional geography. Under the leadership of Vidal in France, of Hettner, Penck, Gradmann, Passarge, and many others in Germany, European geographers gradually shifted away from the concentration on systematic geography, which had been a natural result of the emphasis on universals in all science. Likewise, in this country, the programmatic papers of Barrows and Sauer, however divergent in other respects, agreed in the emphasis on regional studies as the core of geography. Though Pfeifer is correct in noting the similarity of these, the two most influential methodological statements in current American geography, he overestimates their importance in determining the course of current thought in American geography by failing to note the major degree to which, like the earlier methodological pronouncements of the presidents of this association, they simply "mirror geographic opinion in America". As Platt has pointed out, the roots of the current movement, in particular of the tendency for detailed studies of small areas, reach back to geological field courses before the World War and military mapping during the War. It is neither possible nor necessary to determine even approximately what forces or what individuals have been responsible for this development. Mention should certainly be made of the influence that Bow-man, as Director of the American Geographical Society, exerted towards intensive regional studies. Possibly most important of all has been the personal influence exerted by the group of Midwestern geographers whose annual field conferences, in the years 1923 and following, concentrated the attention of a much larger number of workers on the problems of regional mapping [note, for example, the report of the joint conclusions of this group.

If geography, in America as well as in Europe, may be said to have returned, in a certain sense, to the point of view that was common with Humboldt and Ritter, its long period of concentration on systematic studies has enabled it to return far better equipped with generic concepts and principles with which to interpret the findings of regional geography—though unfortunately this equipment is relatively deficient in respect to human or cultural features, both in geographic literature and in the training of most of its students.

Many geographers who have accepted this shift in emphasis evidently have done so under the provisional assumption that regional geography is to be made as "scientific" as systematic geography has been, that somehow it must be raised to the plane on which scientific principles may be constructed. We have noted a number of difficulties into which this ambition has led. In our final consideration of regional geography it is necessary to understand clearly certain limitations imposed upon the student that are not found in systematic geography.

After a number of unsuccessful attempts to express the special nature of regional study in words, we find it can be most clearly presented if we may use mathematical symbols, though we shall not, of course, find it possible to express such complicated problems in any real mathematical formulae or equations.

Varying Features

Any particular geographic feature, *z*, varying throughout a region, might theoretically be represented as a function, f(x, *y*), *x* and *y* representing coordinates of location. As a function of two variables, any such feature that we are able to measure mathematically such as slope, rainfall, or crop yield can be represented concretely by an irregular surface. Such a surface would then present the actual character of that feature for the whole region; it would, theoretically, be correct for every point, and for every small district. Furthermore, if the function involved were not too complicated, the theory of integral calculus would permit us to integrate the total of that feature for any limited section, as well as for any individual point. In a sense, part of our work in systematic geography corresponds to this form of presentation.

Likewise, the relation of any two or three geographic factors to each other within a region—e.g., the relation of crop yield to rainfall and humus content of soil-might be represented as a functional equation involving that many variables: *Z3* = f′ *(Z1, Z2)*. The concrete representation of this relation would require again a surface form. More commonly, in systematic geography, we consider only the relation of one factor to but one other, which we may then represent as a curve on a plane surface. Each of these factors, *z*, is of course a different function, *f(x, y)*, and the more complex equation, $Z_3 = f' (Z_1{}' Z_2)$ holds true only if Z_3 is unaffected by other Z factors, or if those which affect it are constant throughout the region under consideration. Neither of these conditions is strictly true: almost any geographic element we may consider is affected by more than two of the natural elements, and may also be affected by incommensurable, or quite unknown, human factors; and all of the factors considered vary to some extent no matter how small the area considered. Consequently, we have introduced a degree of distortion of reality even at this step in systematic geography.

We may introduce a further step by establishing element-complexes, *u*, each representing functions of many Z elements, varying, by more or less regular rules, with the variations in a smaller number of those elements. Thus, given certain conditions of soil, slope, temperature, and rainfall we may presume within a wide margin of both inaccuracy and uncertainty, certain conditions of natural vegetation and wild animal life, and we may express the total of all these *z* elements by one *u* element-complex. If it were conceivable that we could express this feature, *u*, arithmetically, its character over an area would likewise form an irregular surface that would indicate its character for any limited part. From the nature of these element-complexes, however, it is obvious that any such representation would have a high degree of unreliability.

In regional geography, however, we are concerned with a vastly more complicated function of the location coordinates. It cannot be expressed as the function of anyone element or element-complex, but rather of various semi-independent element-complexes, *u*, and of additional semi-independent elements, *z'*. Thus, the total geography, *w*, at any point, might be expressed by the function, $F(u_1, u_2 .... u_{n'}, z'_1, z'_2 ... z'_{n'})$. If we could have accurate and complete information concerning the form of the function, *F*, and every one of the element-complexes; '*u*—each as a function of various *z* elements-and of the semi-independent elements *z'*, the function would be so complicated that we could not hope to represent it by any concrete form, even in terms of n-dimensional space. We would have a function that could be solved only for each point, *x*, y, in the region, but could not be correctly expressed for any small part larger than a point. In other words, we could study the geography of the area only from the study of the geography of the infinite number of points within it. This task, being infinite, is impossible. The problem of regional geography, as distinct from a geography of points, is how to study and present the geography of finite areas, within each of which the total complex function involved depends on so many complex functions, complexly interrelated, as to permit of no solution by any theory of integration.

Consequently we are forced to consider, not the infinite number of points at each of which w is in some degree different, but a finite number of small, but finite, areal divisions of the region, within each of which we must assume that all the factors are constant. In order, then, to cover an entire region we will need but a finite number of resultants, *w*, each representing the geography of a small unit of area rather than of a point. This method is legitimate only if one remembers that it inevitably distorts reality. The distortion can be diminished by taking ever smaller unit areas, but it cannot be eliminated entirely; no matter how small the unit, we know that the factors which we assume to be constant within it are in fact variable. In practice, the smallest units that we can commonly take time to consider are sufficiently large to permit of a marked degree of variation, and therefore of a significant distortion of reality in our results.

To express our conclusion in more common terms, in any finite area, however small, the geographer is faced with an interrelated complex of factors, including many semi-independent factors, all of which vary from point to point in the area with variations only partially dependent on each other. He cannot integrate these together except by arbitrarily ignoring variations within small units of area, *i.e.*, by assuming uniform conditions throughout each small, but finite unit. He may then hope to comprehend, by analysis and synthesis, the interrelated phenomena within each particular unit area.

Although the studies of all the unit areas added together will constitute an examination of the entire region, this does not complete the regional study. As Penck has emphasized, it is not sufficient to study individual "chores" (approximately homogeneous districts) and to establish types of chores. "Above all geography must consider the manner in which these are fitted together to form larger units, just as the chemist does not limit himself merely to studying the atoms, but investigates also the manner of their situation beside each other in individual combinations. The comprehension of geographic forms *(Gestalten)* has scarcely been taken into consideration by the new geography." Just as a mosaic cannot be comprehended, Penck continues, by classifying and studying the individual stones of which it is made, but requires also that we see the arrangement and grouping of the individual pieces, so the study of the arrangement of the "chores" will present different structural forms of significance.

Our second step-in a theoretical approach to regional geography is to relate the unit areas to each other to discover the structural and functional formation of the larger region. Since all the factors concerned, and therefore the resultants, have been made arbitrarily constant for each small unit, it may be permissible to speak of functional relations between one factor in one unit and another in another unit, as though these were functional relations between the units themselves-provided that we understand that this is not strictly true. Further, the regional structure produced by this method will have the character of a mosaic of individual pieces each of which is homogeneous throughout, many of them so nearly alike that in any actual method of presentation they will appear as repetitions in different parts of the region. But we are not to be deceived into regarding this mosaic which we have made as a correct reproduction of reality. It is simply the device by which finite minds can comprehend the infinitely variable function of many semi-independent variable factors. The fiction involved is threefold: we have arbitrarily assumed each small unit area to be uniform throughout; we have delimited it from its neighbours arbitrarily, as a distinct unit (individual); and we have arbitrarily called very similar units identical in character.

Fundamental Limitations

There are certain other fundamental limitations that must be insisted upon if we are to compare the face of the earth, even in the more or less distorted form in which the geographer must present it, to a mosaic. We may say that there is a similarity in the detail of technique but, unless we are to return to some teleological principle, we cannot liken the face of the earth to any work of art, for we cannot assume that it is the organized product of a single mind. On the contrary, if we may transfer Hettner's analogy of a building built by several architects working independently to Huntington's picture of "The Terrestrial Canvas," we may say that the face of the earth has been produced by the interrelated combination of different colour designs each applied by different artists working more of less independently, and each changing his plan as he proceeded. In systematic geography one might say, we attempt to separate each of the individual designs in order to understand its form and its relation to the others and, thereby, to the total picture. Since the total pictures were not produced simply by superimposing different colour plates in printing, but are, to some extent, causally related to each other, this separation involves the analysis of the causal and functional relations of each design to the others.

In regional geography we first reduce the subtle gradations which the different artists of nature have applied and intermixed on the face of the earth, to the stiff and arbitrary form of the mosaic technique. When we then survey the formation in the mosaic pieces, we are not to expect some unified organized pattern such as every work of art must have. On the other hand, neither need we expect mere chaos, or a kaleidoscope; for we]mow, from our studies in systematic geography, that there were principles involved in the individual designs, and if our determination of the unit areas of homogeneity has not been purely arbitrary, but has been based on the combination of careful measurement and good judgement, we may expect the combinations of these designs to show more or less orderly, though complex patterns. Further, whatever the explanation of these patterns may be, their form is significant to each of the parts, since the development in each unit part is affected by that in the others.

The last thought leads us finally to another major respect in which any analogy of the earth surface to a work of art is inadequate, namely, the fact that, while the latter is static, consisting of motionless forms, the face of the earth includes moving objects that are constantly connecting its various parts. To attempt to introduce the artist's special use of such terms as "lines of force," "movement," "opposite forces," etc., would merely add to confusion here. In other words the geographer must consider function as well as form. In establishing our arbitrary small unit areas we not only assume that each is uniform throughout in character, but also in function. Likewise, in combining these units into larger regional divisions our problem is complicated by the fact that we must consider the functional relations of the units to one another as wen as their form. For example, if two neighbouring areal units are so similar that we have painted them as much alike as two pieces of mosaic of the same colour, but one of them is functionally related to a city centre in one region, the other to a city centre in another, are we to include them in the different regions, or, if in the same region, in which? Any answer to this question can only be more or less intelligent: there can be no one "correct answer."

Just as it is necessary to know the arrangement of unit areas in a region; it is likewise necessary to understand the arrangement of regions to each other. Both Penck and Grano who follows a similar line of thought would carry the process on to larger units; the size of the areas concerned is immaterial. Regional geography, therefore, studies the manner in which districts are grouped and connected in larger areas, the manner in which these larger areas are related in areas of greater scale, and so on, until one reaches the final unit, the only real unit area, the world.

There is, however, one important difference at the different levels of integration. Bath Penck and Grano appear to ignore the fact that the small, but fundamental, element of fiction in the assumption of hamogeneity of the smallest units of area increases progressively as one advances to larger divisions. Consequently, the determination of these larger divisions requires increasingly arbitrary distortions of fact.

Assuming the first step, the establishment of "homogeneous units" of area, we may proceed to the second by enclosing in a continuous area which we call a region, the greatest possible number of "homogeneous units" that we judge to be nearly similar, together with the smallest number of dissimilar units. Our Judgement of similarity will involve subjective Judgement as to which characteristics of the homogeneous units are of greater importance than others, so that, at best, the determination of the region is in a sense arbitrary.

Furthermore we seldom find in reality such a simple solution as that described. Though same geographic features vary but gradually from place to place, the irregular and steep variations of others—such as sails, slopes in mountainous areas, urban settlement, and all the features of essentially linear form, rivers, roads, and railroads-will farce us to include in any region, "units" of quite different character. It is necessary therefore to determine which kinds of units are, either in actual interrelation are merely in juxtaposition, characteristic of the region as approximately considered, and then so determine it as to include the greatest number of those several kinds of similar units, with the smallest number of units of other kinds.

In considering any large area in which we have first recognized "homogeneous units" and are attempting to farm them into regions, which we can briefly characterize in terms of similarities or relations among some of those units, we may find the task relatively simple in parts of the area, where perhaps the great majority of the units are notably similar. But it may be extremely difficult in parts between these, which may be characterized by units that are, in some respects, similar to units an one side of them, in other respects, to units on another side. Further we will find areas containing such a variety of different kinds of units that we cannot see where to include them. In some cases, to be sure, we may recognize such areas as transition zones, but that merely postpones the fundamental problem without salving it. Likewise, to call them "characterless areas," or areas of "general" or "mixed" types is simply to dodge the problem entirely.

The individual student, no doubt, would gladly wipe such troublesome areas off the map, but he is not 'granted that privilege. Neither is a science which seeks to know what the world is like permitted to ignore mare difficult areas and confine itself to those easier to organize into its body of knowledge. Since these doubtful areas are commonly not merely narrow borders of transition, but areas of wide extent, perhaps as great or greater than those mare clearly classified, there is no basis for assuming that they are of less importance in the total picture of the larger area, or of the world, than the areas whose character we can more readily describe. Fenneman's statement with reference to the different parts of geography applies even mare literally to parts of an area—"there is no more inherent worth in a centre than in a border."

Consequently, when we divide any given area into parts which we call regions, so determined that those characteristics that we have judged to be mast important may be most economically stated for each region, we cannot avoid many decisions based on Judgement rather than on measurement. We must, therefore, acknowledge that our regions are merely "fragments of land" whose determination involves a considerable degree of arbitrary Judgement. On the other hand, if all possible adjective measures have been used, and the arbitrary decisions are based on the student's best Judgement, we may properly regard his regions as having mare validity than is expressed by the bare phrase "arbitrarily selected." On the other hand, the view of various writers previously noted, that geographers could be expected to come to approximate agreement on the specific limits of regions—or even on their central cores—appears, in view of all the difficulties listed, overly optimistic.

No basis for Regional Division

It hardly needs to be added that the conclusion that geography cannot establish any precise objective basis far regional division does not permit it to shirk the task of organizing regional knowledge into areal divisions determined by the best judgement possible. In order to utilize the generic concepts and principles developed in systematic geography to interpret the findings of regional geography, the latter must be organized into parts that are as significant as is possible. In the present state of development of the field—if not indefinitely we do not have what would be the simplest solution, namely, a single standardized and universally accepted division and subdivision of the world into regions. Therefore, each student of regional geography has imposed upon him the task of standardizing his own system of regional division-unless he can utilize that of some colleague. "Standardized" is used here to indicate that the regional system is based on certain standards specifically stated, so that other students may know precisely what the organization is.

The complete organization of regional knowledge in geography requires whether as a final or as a primary step the division of the whole world. In whichever direction the process is carried on and we noted that it requires consideration in both directions—the completed system must provide a regional division of the world in which our knowledge of each small part may be logically placed. For' this extremely difficult problem we found two different methods of solution. Geographical knowledge may be logically arranged in systems of areas classified according to certain characteristics of the areas. Though this method has distinct utility for comparative purposes, it does not permit organizing all regional knowledge into one system, but requires several independent systems. Furthermore, it does not present the actual relations of areas as parts of larger areas. These relations can be included only in a realistic division of the world into a system of specific regions, in which all regional knowledge may be incorporated in a single logical system. Such a system unfortunately is not provided the geographer by any natural division present in reality, nor by anything corresponding to the simple division of organic forms. It must be developed and constantly modified by geographers as a result of research, at the same time that it is being used, always in tentative form, as the organizing structure of regional research.

We have suggested, in very general terms, the manner in which the problem of delimiting regions may be met, in order that geographic knowledge may be organized intelligently in regional units. What kind of knowledge is to be included within the regional study itself? So far as the nature of the material is concerned, we have previously indicated that a complete geography of a region includes all the kinds of phenomena that are included in systematic geography-insofar as they may be present in the particular region. The only field of geography that is not included in regional geography, as well as in systematic geography, is historical geography. As there was a different geography in every past period, there may be any number of independent historical geographies, each including its own systematic and regional divisions.

The kinds of phenomena present in regions, the particular manner in which they are present, and the nature of their interrelations, both within each unit area and across unit divisions, determine the particular forms and the functions of the area. Though most students agree in theory that these are of coordinate importance, much of the work of recent decades tends to emphasize the study of forms to the neglect of functions. We found this to be particularly pronounced in the work of the "landscape purists". On the other hand, Grano finds that many students, like Spethmann in particular, conceive of an area as "the field of forces, as a dynamic complex." Geography, Grano insists, is not the study of forces, of interrelations, but the study of things in interrelation in areas. Judging by the major example which he has presented in German, Grano himself tends to emphasize physiognomy and gives but little attention to the functions of areas.

When we speak of the functions of areas, we are not to forget that in reality the area is not a thing that functions, it is only certain things within it that have functional relations to things in other areas. If our fiction of the small homogeneous areal unit, uniform in both form and function, permits us to speak figuratively of the unit area as having a functional relation to other unit areas, we are not to ignore the fictive character of this concept by attempting to consider areas as having, in themselves, functional relationships.

Breakdown of Concept

In particular, it is necessary to note that the concept of the small areal unit breaks down when we attempt to study "the genesis of an area." When we study the previous historical stages in the geography of the area, we find that anyone of our small unit areas of homogeneity may not have had in the past even that incomplete degree of validity that we may grant it today. That is to say, since areas, no matter how small, do not grow as units, but change only as a result of the differential change of different things within them, the unit area of today was probably not a unit area in an earlier stage, and will probably not be in a future stage. The very concept of mosaic is incompatible with the concept of gradual and differential change. Consequently, the study of genesis in geography can only be undertaken in the form of systematic studies: the study of "the genesis of an area" can only be broken down into studies of the genesis of each of the various objects contained within it. These are therefore studies in systematic geography; to what extent they may be desirable for an understanding of the geography of any region is a controversial question which we touched on earlier, and need not reconsider here.

We should now be in a position to answer the question that is of greatest importance in contemplating the possible development of regional geography may we hope to progress in this branch of our field to the construction of universals, of generic concepts, and scientific laws or principles?

One form of generalization used in regional geography we have already described: the construction of regions from small unit areas. The philosopher, Kries, has distinguished such generalizations of heterogeneous and semi-independent parts as a third type of scientific description, together with type concepts and the description of the unique. The importance of the distinction lies in the fact that this form of generalization offers no basis for establishing general principles; for that we must have type concepts.

It is obvious that any universal principles that we might attempt to construct on the basis of the fictive areal units set up for the purposes of description, could have no more validity than the units themselves. Unless these are taken as extremely small units, the margin of error introduced by our personal Judgement would lead, in any principles we might set up, to a degree of error so great as to render them-of very doubtful value.

Regardless of that essential difficulty, however, we found that even these arbitrary units, each involving a complex combination of associated forms, cannot be classified into a system of types based on the sum totals of its varied and semi-independent factors. Though in anyone region we find unit areas so similar that we may, with but a minor degree of error, call them alike, we do not find unit areas of that kind of similarity in other regions of the world. A small district somewhere in the Upper Rhine Plain may be very much like many other such districts in the same region, but no matter how small a district we take, it is fundamentally different from any unit area in any other world region.

We arrive, therefore, at a conclusion similar to that which Kroeber has stated for history: "the uniqueness of all historical phenomena is both taken for granted and vindicated. No laws or near-laws are discovered". The same conclusion applies to the particular combination of phenomena at a particular place.

One is not to suppose, however, that regional geography is studied without the use of generic concepts and principles. On the contrary, the interpretation of the interrelations of phenomena within each region depends upon die type concepts and principles developed in systematic geography. In other words, for the individual items included in regional geography, and the simpler relations between them, we depend constantly on universal concepts supplied from the systematic studies, but the total interrelated combination of each areal unit represents an essentially unique case for which we can have no universals.

An objection may be made that one form of study used by many geographers in the consideration of regions represents an approach to the construction of scientific laws namely what has been called "comparative regional geography," the comparison of regions of notable similarity. As a current example we may cite Maull's effective comparison of the Amazon, Congo, and Insulindia areas. The fact that in other sciences "comparative studies" have marked an adolescent period preceding the flowering of a nomothetic science, has led many to suppose that regional geography might be expected to grow out of its youth by progressing from comparisons to scientific principles.

The essential idea involved is nothing new in geography. Introduced by Humboldt-if not by earlier writers it was used, according to Hettner, by Brehm, Nehring, and particularly by Richthofen. Plewe found, however, that these represented merely occasional examples, that our literature contained no comparative regional geography as a branch of the field. Such occasional comparisons, he noted, are used in all sciences, citing as an example, Th. Litt's comparative study of Kant and Herder. Historians, we may add, frequently find it valuable to compare the developments of any two or more periods that are significantly similar in certain respects. These examples should make us sceptical of the likelihood of cur discovering anything that could be called' laws, or near-laws, of regional geography.

Passarge recognized the limitations that prevent a comparative *Landerkunde* from developing universal concepts, but still (in 1936) believes that these can be avoided or overcome in a comparative *Landschaftskunde*. In order to discover the laws of regions it is necessary, he says, to have a *tertium comparationis* and this, he believes, is provided by his system of abstract types. As we saw in our previous discussion he has in part, merely reduced the difficulties, by reducing the size of areas concerned, and for the rest he has simply dodged the limitations ,by setting up types that are not even in outline complete abstractions of real areas. The difference between the real *Land* and the real *Landschaft* (as area) is only a difference in size; a *tertium comparationis* is equally impossible in both cases. We may go on comparing areas of whatever size forever with no hope of discovering regional laws.

Plewe concludes, therefore, that the comparative study of regions is neither a preparatory step to a nomothetic regional geography nor an independent branch of geography. Ritter's introduction of the concept, over a century ago, represented a transfer from a quite different kind of science; he never clearly defined his concept, and others who have taken it up have used it in many different ways but without leading to any important development.

Nevertheless the use of this method, as a supplementary device, appears to offer certain distinct advantages. If widely separated regions are in many respects similar, so that, in respect to certain elements or element-complexes, they may be classified as of the same type, the comparison of their similarities, and particularly of their differences, may well serve as a check on the interpretations we place upon the relation of phenomena within each one of them.

Even more useful is the employment of this method in the comparison of localities within a major region, where there may be a much larger number of element complexes of the same type. By selecting those localities that are alike with respect to the greatest number of features, and comparing them with those that are like them in many, but not all, of these features, we may have a key to the significance of specific features for the area as a whole.

To take a well-known example: the consideration of the major characteristics of the Cotton Belt as a whole might lead one to suppose that—taking certain cultural conditions for granted the importance of cotton in the area was to be explained simply in terms of climatic conditions. We have learned, however, by contrasting the localities in which cotton is the all-important crop, with those where cotton is of minor importance though the climatic conditions are the same, that the cotton crop of the South as a whole is not to be understood without considering the character of the soil.

Likewise, American geographers, at least, have long realized what is not so clearly recognized in popular thought—or even by many European geographers that the climatic conditions of the South do not directly explain that feature which is of greatest importance in the contrast between North and South-namely the high proportion of Negro population. By the same method of comparison of localities one finds that this element and all the cultural elements associated with it—cannot be understood without considering the combination of climatic and soil conditions that are necessary for cotton. This conclusion, however, is incomplete: in the cotton district of greatest importance today, in central Texas, the proportion of Negro population is low. The complete explanation can be reached only when one also compares the localities which were developed for plantation crops-including tobacco as well as cotton before the end of the slavery period, with those localities developed for the same crops since that time.

This method of comparing localities within the same larger region might appear to lead to generic principles. But it can lead to conclusions that are applicable only to the single larger region concerned. If we should add to the districts of the Cotton Belt a district in the Yangtse Valley and a district in the Bombay province, we could not include them all under any generic concepts of districts. In the comparison limited to districts in the Cotton Belt, we are not, as we noted in an earlier connection, comparing separate units, but only similar parts of a single larger region, parts whose similarity is simply the result of the fact that they are parts of the same region. Valuable as the device may be for checking our interpretations, it leads to no universal concepts or principles.

Regional geography, we conclude, is literally what its title expresses: the description of the earth by portions of its surface. Like history, in the more common sense of periodic history, it is essentially a descriptive science concerned with the description and interpretation of unique cases, from which no scientific laws can be evolved. Though this is undoubtedly a disadvantage, making the interpretation of findings far more difficult than in those fields that are able to develop general laws to explain individual cases, it does not mean that regional geography lacks any scientific goal.

As previously noted, the construction of scientific laws is not the purpose of science, but a means toward its purpose, the understanding of reality. To any "who find the title 'earth description' *(Erdbeschreibung*, or *geographia)* insufficiently learned and scientific." Heiderich has answered, "description is the last and highest goal of scientific work, to be sure, not a mere outward external description that remains on the surface of the object, but a description that aims. . . to comprehend synthetically all that has been learned analytically from the characteristics of the object". All that science requires is that, in order that the interpretive description may have a maximum degree of accuracy and certainty, universals shall be constructed and used wherever possible. Regional geography utilizes all the appropriate generic concepts and principles developed both in the systematic sciences that study particular kinds of phenomena, and in systematic geography which studies their relations to each other over the earth.

The conclusion to which we have arrived concerning the nature of regional geography may enable us to answer one or two questions that have been raised by a number of students in very recent years. The course of thought among American geographers concerning regional studies has been discussed in two articles published in Germany during the past year, by Broek and by Pfeifer. From these surveys, and the critical articles to which they refer, the reader might suppose that after a period of enthusiastic concentration on regional studies that was introduced by the methodological papers of Barrows and, more particularly of Sauer, American geographers had now begun to doubt whether much was to be expected from regional geography after all. It may be that the testimony has been exaggerated in the echoes back and forth across the Atlantic; possibly we are presented with a revolt within a single university department that has reverberated among its present and former members here and in Germany Undoubtedly, however, other American geographers in oral discussions have expressed a note of skepticism concerning the results to be expected from regional studies.

In a number of cases, the sceptics have spoken, or written, as though after a long and earnest attempt to advance geography by regional studies, we had discovered that the works produced did not add up to, or yield, significant general results. It is difficult to believe that this argument is meant seriously. American geography has concentrated its efforts on regional studies for scarcely twenty years and never completely. During that time, perhaps a score or so of research students have each made one, two, or three regional studies in areas scattered from the Peace River Country to Sao Paulo, from Europe to China. Since neither of the two principal American promoters of the regional concept in theory has presented a concrete example of a full study in (present) regional geography, each of the individual research students has had to work, out more or less independently his own methods of determining his region, of selecting the phenomena for consideration in it, and of presenting his results. Would anyone seriously consider that we have had a fair test of the possibilities of developing general results from regional studies? Even if all the work had been carried out under standardized procedures, such a small number of cases scattered over more than half the world could hardly be expected "to add up" to any general results, or to provide the basis for generalizations.

It seems more likely that many students have begun to suspect, for other reasons, that no matter how many regions are studied, no matter by what methods, no scientific laws will be forthcoming. This conclusion we have found can be thoroughly demonstrated in theory, so that we can agree that any who have made regional ,studies with that ultimate purpose in mind have been following a will-o'-the-wisp; the sooner it is abandoned the better for all concerned.

If, however, the purpose of geography is to gain a knowledge of the world in terms of the differential development of its different areas, the task of studying regions as areal divisions of the world, is not subject to question in geography. Neither need the workers in any science feel discouraged if the efforts of a relatively small number of workers over a period of less than twenty years have produced less than enthusiasts led them to expect. Though the object of geography, the world, is large, it is limited in size, and we must assume that geography has a long life ahead of it. No doubt the group efforts of American geographers would show more productive results if they could concentrate the attention of all or most of their members on ,some limited part of the world as French geographers have done on their own country. But the many factors that persuade students to travel far afield are not to be restrained, even if it were desirable. One can only hope for a larger total number of workers, and possibly, for increasing concentration within this country of the work of particular groups—as at Wisconsin—on the regions of a relatively limited area. In particular, as Finch notes, we should not expect results of far-reaching scientific value from the practice of "skimming the cream of the more clearly given from a region and its abandonment for another area". The value that studies of this kind may have for teaching purposes may justify the time and effort spent on them-provided that the areas concerned are significant for class instruction. Lasting progress in research in regional geography will require a much greater amount of concentration of the individual's time—whether or not one would go as far as Finch and consider one region as sufficient for one student's life work.

On the other hand, these considerations do inevitably raise the question of what size of area should be considered worthy of research in regional geography. The regional studies launched under Vidal's leadership in France formerly examined areas the size of a province, but increasingly smaller areas have been selected. Demangeon feels that the extreme limit of "microscopic" study has been reached by Allix, whose examination of "L'Oisans," a part of an Alpine valley in the Dauphine smaller than an *arrondissment*, requires 915 pages, with a bibliography of 861 works. This averages, Demangeon reckons, a little over a page per square kilometer or for 12 inhabitants. American geographers, by comparison, hardly seem justified in applying to their work the word "microscopic."

The question raised admits of no simple answer. Historians welcome extremely detailed studies of very short periods, in addition to less intensive studies of an extensive series of periods. The criterion, in either field, is the same—namely the significance of the study—but that is a criterion for which we have no objective measure. We have previously suggested two major considerations-namely, the significance of the area in itself, and its possible significance as representative of a large area, or a large number of similar small areas. Outside of the proper interest of the citizens of L'Oisans itself in the geography of their own district, we may assume that the world of knowledge in general has little need for such an exhaustive study of this small and unimportant district. On the other hand, if we had but little knowledge of the valleys of the French Alps, and reconnaissance had shown that this particular district was in large degree representative of hundreds of others, such a study might provide us with an approximate view of the regional geography of the entire area—or of a large part of it.

Presumably, however, such a study would be limited by the desire to express primarily those characteristics that were representative, and one might question whether that" would require a thousand pages. Demangeon finds much of the study superfluous because' it merely duplicates findings that Blanchard and others have presented in works on similar districts. Insofar as Allix's work has served to corroborate that of his predecessors, that fact might have been more briefly presented. On the other hand, another competent critic finds that Allix has contributed a much more thorough treatment of the problems representative of the French Alps than have any of his predecessors.

It is particularly against such "micro-geographic" studies to use Platt's term-that criticism of regional geography in America has been directed. Recognizing that it would be impracticable, in any reasonable length of time, to cover the whole land area of the world by the total addition of such small studies and that the total might be indigestible if it could be attained—critics have feared that we would have but a miscellaneous collection of scattered pieces selected at random. More particularly, however, the critics have asked what general principles can we expect to derive from such minute and scattered studies. Even some who have made such micro-geographic studies, like James, have given expression to a later feeling that "the more detailed and specific is the study the more insignificant are the results".

Platt's Views : To these attacks, Platt, in particular has replied vigorously, both in two published papers and in unpublished statements read before this Association. Microscopic geography, he observed in one of the latter, developed "as a rational and timely drive against the limitations of armchair compilation from promiscuous data of subjective impressions from casual travel, and of environmental theory not founded on data." To attain these purposes, geographers took to the field and "in the field all geographers are microscopic." There "they face the geographer's dilemma in trying to comprehend large regions while seeing at once, only a small area." They do not, he insists, plunge into detailed studies of minute areas because the methodological conclusions of others have led them to believe that thereby something will ultimately be gained for geography. On the contrary, their own efforts to comprehend areas of larger extent has led them to the reasoned conclusion that, in addition to general reconnaissance studies and detailed systematic studies covering large areas, accurate generalizations for larger regions require an examination of the total fundamental complex of inter, related features that can be examined, in detail, only in the small area.

Platt's defence of micro-geography, however, is based less on theoretical discussion than on the actual work that he has been carrying on for some years in Hispanic America, which forms the most significant series of micro-geographic regional studies in American geography. Of the sceptical questions that have been raised concerning the value of such a series of studies, many appear irrelevant to its purpose. This, we take it, is simply to increase our organized, objective and reliable knowledge of the lands south of the Rio Grande. That such knowledge of the different parts of the world is desirable and requires the research of trained workers is, we repeat, the fundamental justification for the field of geography. That our present knowledge of the Hispanic American area is inadequate is obvious to anyone who has attempted to gather the materials necessary even for an elementary course concerned with that part of the world. Consequently, we are not to test the value of such a series of detailed studies of scattered districts by asking whether they can yield us any "scientific principles," or whether they will aid us in drawing conclusions concerning "the larger relationships" of which Pfeifer speaks. So long as Platt does not claim that all geography should consist of such "micro-geographic" studies, or of regional studies in general, these questions are irrelevant. The relevant question is, granted that we want more adequate knowledge of the geography of South America, is his method of study appropriate to produce such knowledge?

Few will question the inadequacy of the general surveys of South America now available. In his most recent study, on coastal plantations in British Guiana, Platt has noted that the best available, generalized maps of the continent give erroneous impressions of the soils, vegetation, and population density of the specific districts he studied. Even if we had accurate detailed information on the climates, land forms, soils, crops, races, and commerce of South America, these would not add up to the geography-the areal differentiation—of the different parts of that continent. In studies limited even to provincial scale, the American student is frequently baffled because he lacks the detailed knowledge of the cultural element-complexes that are basic to the cultural geography of the region. For areas in United States or Europe, he may have acquired that knowledge unconsciously, whether as a by-product of field work, or merely from his general knowledge. These essential features must be studied first in relatively small areas-particularly in a world area where there is lack of cultural homogeneity. If, then, one has acquired an understanding of a particular ranch in Panama and may be permitted to assume somewhat similar features scattered through a large area, one has a more correct picture of the geography of the larger area concerned than can be 'acquired by any small-scale measures.

Essential Assumption

The essential assumption in this proposition, of course, is that the minute district studied is in fact representative of others; as Finch notes, it can hardly be typical in any full sense. If it is representative, however, it presumably will be typical in certain limited respects, and it is important that we know in what respects it is approximately typical. In areas that are adequately covered by census and climatological data, geological, topographic, and soil surveys, it may be possible to give approximate answers to these questions from the study of such data. The utility of element-ratios and isopleth maps in this connection has been suggested previously. In other areas, one can only depend on the student's Judgement formed from reconnaissance.

Though such Judgement can only give answers that are far removed from scientific certainty, they are better than no answers at all, and should therefore be provided-even at the risk of being shown erroneous by later work by the same or other students. Perhaps only in his most recent study has Platt clearly demonstrated the relation of these detailed studies of small districts to the reconnaissance study of large areas. Though the micro-geographic area which he studied in detail is not, in this case, "typical of broad regional types," it is shown to be "a normal feature of a coherent plantation district, which in turn has a consistent place in the intricate geographic pattern of South America". No doubt the significance, to broader regional knowledge, of his previous studies in small and widely separated districts, apparently chosen at random, will be made clear in the ultimate publication of his "reconnaissance study of Hispanic America," of which these detailed' unit studies are to form integral parts.

In sum, the student who presents a study of a small area of no special importance in itself, needs to keep in mind that the purpose is not to present the area in itself, but to provide an accurate illustration of the representative character of a larger region, too large to permit of such intensive study. So long as he keeps this broader purpose in mind, there are no grounds apparent on which we can prescribe the minimum size of area that may be studied.

Geography's Integrated Dualism : The final question raised by our examination of the nature of geography, as presented to us both by its historical evolution and by the logical consideration of its position among the sciences, is the same question that has provoked so much controversy throughout almost the entire history of modern geography—certainly ever since Bucher raised the issue in 1827. If geography studies the areas of the world according to the differential character of their phenomenal contents, either according to a systematic point of view, category by category, or, on the other hand, according to an areal point of view, each area in terms of all its heterogeneous phenomena, how can these two points of view be related to each other in a unified field of geography?

Our historical survey showed that, while modern geography from its beginnings has included both of these points of view-in theory even in Varenius's outline-it has experienced notable shifts in emphasis from one to the other. Whereas the work of Humboldt combined both points of view, under the influence of Ritter, systematic studies were placed in a subordinate position and easily lost sight of. Though the protests of Bucher and Frobel were of no avail at the time, a later generation following Peschel, and motivated by scientific standards developed in such fields as geology, swung the centre of interest the other way. As late as 1919 Hettner found that in Germany systematic geography was generally regarded as "something higher, more distinguished" than regional geography.

He therefore repeated the arguments that he had presented at various times during nearly a quarter of a century to show that the two parts of the field were scientifically on the same level. Less than a decade later, however, he found it necessary to present opposite arguments to urge the same conclusion; for "youth, which is given to exaggeration, has turned far too much away from systematic geography" (we can hardly suppose that Hettner was unaware of the fact that some of those concerned were not much younger them he). The reaction that had taken place earlier in France, under Vidal, swept German geography in the post-War 'years toward an increasing emphasis on regional geography, as the real goal of geographic work. Thus Obst, believing that one could develop a science of *"Landerkundliche Typologie"* wished to shift the time-honoured term of "general geography" to the study of regional types as the goal of geography, and placed systematic geography (as *allgemeine Erdkunde)* in the subordinate position of a necessary propaedeutic; somewhat similar views have been expressed by Braun.

Likewise in this country, the emphasis that Barrows, and particularly Sauer, had placed upon the study of regions (in the latter case, "landscapes") led some to regard systematic studies as necessary only for instructional purposes but inappropriate for geographical research.

On the other hand, such veterans as Hettner and Penck have never wavered in their insistence that both points of view were of equal importance in geography [Hettner has said as much in almost every methodological treatment he has written. The very fact that geography has experienced these successive shifts in emphasis from one side to the other is in itself indirect evidence that both are of coordinate importance in the field.

In his critical investigation of geography as a single, unified field of science, Kraft finds that, while one could dismiss the charge of dualism of content natural and human features as invalid, the inclusion of the systematic and the regional points of view was an unquestionable form of dualism. He agrees with Hettner, however, that this dualism cannot be expressed simply as the combination of a nomothetic and an idiographic science; systematic geography must include the study of unique cases, and regional geography must use generic concepts and principles. In any case, neither construction of laws nor the description of the unique represents the purpose of geography, or of any other science. The purpose of geography is the same in both branches, the comprehension of the areal differentiation of the earth, and this purpose cannot be solved either by systematic studies alone nor by regional studies alone, but requires both approaches. Consequently, he concludes, this dualism in approach is justified as necessary for the single aim which makes geography a unified science.

This view, we may add, is further supported by the fact, stressed by Hettner, that it is frequently difficult to classify particular studies under one heading or the other. The difference is not in the substance, but in the point of view, and in certain kinds of studies these may be combined. For example, the systems of land use classification previously discussed are intended to provide backgrounds for agricultural regional geography and they involve, in outline, a major part of the regional study of any area. At the same rime however, they represent systematic studies of particular element complexes in their world distribution, so that it is by no means clear whether they belong more in the one or the other of our two major divisions.

Finally, if one agrees that both regional and systematic studies are included as essential parts of geography, we may perhaps dismiss any question of relative importance as irrelevant. For systematic geography, regional studies provide, not merely a source of detailed factual information that otherwise would hardly be available, but they also indicate problems of relationships that might easily be overlooked in systematic geography, and they provide the final testing ground for the generic concepts and principles of systematic geography. On the other hand, it is even more obvious that progress in interpretation of the interrelated phenomena of regional geography is constantly dependent on the development of such universals by systematic studies. Any assumption that these studies can be left to the systematic sciences concerned with each particular category of phenomena has been shown by experience to be unwarranted. The aspects of these phenomena with which geography is concerned their relation to other earth phenomena in different parts of the world—are not of direct concern to those systematic sciences and are more commonly left unstudied, unless geographers study them, as Lehmann has shown. Systematic geography, he therefore concludes, is not to be thought of as a border area of geography, or merely as a propaedeutic, but represents "organs vital to the growth of geography, without which its regional crowning can as little exist as a real tree without its roots".

Further, Lehmann suggests, the point of view developed in systematic geography is different from the general point of view in regional geography but at the same time of such value to it that every regional geographer should work productively in some systematic branch or branches (he recommends two or more). On the other hand, Penck, whose most notable contribution has no doubt been the systematic study of landforms, urges that "the cultivation regional studies is indispensable for the geographer; they form for him the touchstone of his whole concept of geography, of his geographic system".

The mutual dependence of the two interconnected points of view in geography has been consistently maintained by Hettner from his earliest methodological treatment of more than forty years ago to the present time. The development of sound universal concepts in systematic geography is the essential basis for progress-in regional geography, but since systematic geography is in method similar to the systematic sciences, "the geographer who *works* only in it and does not cultivate regional geography runs the risk of leaving the ground of geography entirely. He who does not understand regional geography is no true geographer. While regional geography alone, without systematic geography, is incomplete, it remains geographic; systematic geography without regional geography cannot fulfil the full function of geography and easily falls out of geography".

We may assume, therefore, that there is plenty of work to be done in the field of geography by both methods of approach. It is not for any student specializing in either approach to speak with scorn or condescension of those who are working in the other. "Differences of approach," as Kroeber suggests, "are probably at bottom largely dependent on differences of interest in individuals". Paraphrasing his statement further, we may conclude that it is perfectly legitimate to confine one's interest to the specific approach of systematic geography, or to the integrating approach of regional geography, or to use alternately one or the other according to occasion. But sympathetic tolerance is intrinsically desirable and certainly advantageous to understanding to *scientia*.

Our examination of the great variety of different ideas that have been suggested for geography has repeatedly led us into sidtracks that proved to be blind alleys or routes leading outside of geography. No doubt also we have lingered at other points along the way to investigate in detail certain important problems within the field. It may be well therefore to summarize briefly the positive conclusions to which we have arrived concerning the nature of geography.

In its historical development geography has occupied a logically defensible position among the sciences as one of the chorographical studies, which, like the historical studies, attempt to consider not particular kinds of objects and phenomena in reality but actual sections of reality; which attempt to analyze and synthesize not processes of phenomena, but the associations of phenomena as related in sections of reality.

Geography in Historical Development : Whereas the historical studies consider temporal sections of reality, the chorographical studies consider spatial sections; geography, in particular, studies the spatial sections of the earth's surface, of the world. Geography is therefore true to its name; it studies the world, seeking to describe, and to interpret, the differences among its different parts, as seen at anyone time, commonly the present time. This field it shares with no other branch of science; rather it brings together in this field parts of many other sciences. These parts, however, it does not merely add together in some convenient organization. The heterogeneous phenomena which these other sciences study by classes are not merely mixed together in terms of physical juxtaposition in the earth surface, but are causally interrelated in complex areal combinations. Geography must integrate the materials that other sciences study separately, in terms of the actual integrations which the heterogeneous phenomena form in different parts of the world. As Humboldt most effectively established, in practice as well as in theory, though any phenomenon studied in geography may at the same time be an object of study in some systematic field, geography is not an agglomeration of pieces of the systematic sciences: it integrates these phenomena according to its distinctive chorographic point of view.

Since geography cuts a section through all the systematic sciences, there is an intimate and mutual relation between it and each of those fields. On the one hand, geography takes from the systematic sciences all knowledge that *it* can effectively utilize in making its descriptions of phenomena and interpretations of their interrelations as accurate and certain as possible. This borrowed knowledge may include generic concepts or type classifications, developed in the systematic sciences; but, where these are found unsuitable for geographic purposes, geography must develop its own generic concepts and systems of classification.

In return, geography has contributed, and continues to contribute, much to the systematic sciences In its naive examination of the interrelation of phenomena in the real world it discovers phenomena which the sophisticated academic view of the systematic sciences may not have observed, shows them to be worthy of study in themselves and thus adds to the field of the systematic studies. Further, geography constantly emphasizes one aspect of phenomena which is frequently lost sight of in the more theoretical approach of the systematic fields, namely, the geographic aspect. It serves, therefore, as a realistic critic whose function it is constantly to remind the systematic sciences that they cannot completely understand their phenomena by considering them only in terms of their common characteristics and processes. They must also note that differences in those phenomena that result from their actual location in different areas of the world. In order to interpret these differences correctly, and to interpret the resultant world distribution of their phenomena, the systematic sciences take from geography something of the particular techniques which its point of view has required it to develop—notably the techniques of maps and map interpretation.

Geography, like history, is essential to the full understanding of reality. The naked, schematic study of the systematic sciences divides up reality into academic compartments, and thereby necessarily destroys something of its essential character:

"Ach, von ihrem lebenwarmen Bilde
Blieb de" Schatten nur zuruck."

Geography adds, as Vidal said, "the aptitude of comprehending the correspondence and correlation of facts, be they in the terrestrial milieu which includes them all, be they in the regional milieu in which they are localized".

It is a corollary of this proposition that, in the application of science to society, as Finch observes, the chorological science of geography can function directly, since many of the problems of society-notably those concerned with the most efficient organization of land use—are, in fact, regional problems. But that statement does not mean—and I assume that Finch did not intend it to mean that the chorological point of view requires the justification of utility. On the contrary, whatever value geography has in relating science to the problems of society, merely confirms the fact that in pure science itself the pursuit of knowledge for the sake of gaining more knowledge—there is need for a science that interprets the realities of areal differentiation of the world as they are found, not only in terms of the differences in certain things from place to place, but also in terms of the total combination of phenomena in each place, different from those at every other place.

Geography, like history, is so comprehensive in character, that the ideally complete geographer, like the ideally complete historian, would have to know all about every science that has to do with the world; both of nature and of man. The converse of this proposition, however, is that every student of a systematic science is somewhat at home in some part of geography. Furthermore, both geography and history endeavour to describe and interpret actual sections of reality as they exist, and in these sections they observe phenomena by methods that, in a general way are available to the common man. Consequently geography, like history, is a field apparently open for layman to enter. Whereas the study of history, other than current history, at least requires the degree of learning sufficient to utilize the records of the past, geography may be studied by anyone who has the opportunity to travel and the ability to describe what he sees. Consequently, geography was in fact studied by laymen long before any organized subject of geography was constructed, and countless non-professional travellers since have contributed more or less useful data to its literature. This characteristic, likewise, it shares, for good or ill, with history.

In consequence, Richthofen has noted, "many have the delusion that geography is a field in which one can reap without sowing. Because a great part of that which the serious research students have won in it is easily understood, one thinks that he can work, successfully in it without preparatory training, and can win laurels by the easy means of describing fleeting travel observations or by uncritical compilations. An endless flood of superficial literature, which, in spite of its deficiencies, may not be denied the service of popularization, has been able to obscure the Judgement of a great part even of the educated public concerning the scientific content of geography. But, just as with history, the apparent ease with which a great part of the facts secured can readily be understood, stands in contrast to the difficulty of sound research". Allen Johnson, among many others, has discussed the importance of the same contrast in history.

Since the vulnerability of both geography and history to occasional trespass by wandering laymen is a result of the fundamental character of the field in each case, little would be gained by attempting to set up barbed wire fences in the form of erudite technical terms designed to bar trespass. Few geographers, presumably, will wish to have their subject strive for prestige by hiding its knowledge behind smoke-screens. On the contrary, in a subject in which the field includes vast areas that few professionals will have the opportunity to explore, the assistance of the interested amateur may heartily be welcomed. The sole provision that we might like to suggest is that the amateur, as in any activity of life, should recognize his need of securing as much knowledge and training from professionals as is possible for him, so that his efforts may produce results of greater accuracy and interest in themselves and of more lasting value for the science of geography.

Geography and History, Alike : Geography and history are alike in that they are integrating sciences concerned with studying the world. There is, therefore, a universal and mutual relation between them, even though their bases of integration are in a sense opposite-geography in terms of earth spaces, history in terms of periods of time. The interpretation of present geographic features requires some knowledge of their historical development; in this case history is the means to a geographic end. Likewise the interpretation of historical events requires some knowledge of their geographic background; in this case geography is the means to an historical end. Such combinations of the two opposite points of view are possible if the major emphasis is dearly and continuously maintained on one point of view. To combine them coordinately involves difficulties which, as yet at least, appear to be beyond the limitations of human thought. Possibly one approach to such a combination can be made in geography by the lantern-slide method of successive views of historical geographies of the same place. An attempt to develop a motion picture would produce a continuous variation with respect to both time and space which would, of course, represent reality in its completeness, but which appears to be beyond our capacity even to visualize, not to say, to interpret.

Though the point of view under which geography attempts to acquire knowledge of reality is distinct, the fundamental ideals which govern its pursuit of knowledge are the same as those of all parts of that total field of knowledge for which we have no ether name than science.

Geography seeks to acquire a complete knowledge of the areal differentiation of the world, and therefore discriminates among the phenomena that vary in different parts of the world only in terms of their geographic significance—i.e., their relation to the total differentiation of areas. Phenomena significant to areal differentiation have areal expression not necessarily in terms of physical extent over the ground, but as a characteristic of an area of more or less definite extent. Consequently, in studying the interrelation of these phenomena, geography depends first and fundamentally on the comparison of maps depicting the areal expression of individual phenomena, or of interrelated phenomena. In terms of scientific techniques, geography is represented in the world of knowledge primarily by its techniques of map use.

There are no set rules for determining which phenomena are, in general, of geographic significance. That must be determined, in any particular case, on the basis of the direct importance of the phenomenon to areal differentiation, and of its indirect importance through its causal relation to other phenomena. In order to determine his findings as accurately as possible, the individual student, in any particular case, must depend upon those among the significant phenomena for which he is able to secure some sort of measured data. Non-measurable, but geographically significant, phenomena must be studied indirectly, by whatever measurable effects they have produced.

These general principles lead to no general exclusion of any kind of phenomena, nor of any aspect of the field. In any particular study in systematic geography or in any partial study of a region, particular kinds of phenomena may logically be excluded only if they are not significant to the interrelations of those that are being studied. Finally, the ideal of completeness requires geography to consider not only those features and relationships that can be expressed in generic concepts but a great number of features and relationships that are essentially unique.

In order to make its knowledge of interrelated phenomena as accurate and as certain as possible, geography considers all kinds of facts involved in such relations and utilizes all possible means of determining the facts, so that results obtained from one set of facts, or by one method of observation, may be checked by those secured from other facts or from other observations.

With the same ends in view, geography accepts the universal scientific standards of precise logical reasoning based on specifically defined, if not standardized, concepts. It seeks to organize its field so that scholarly procedures of investigation and presentation may make possible, not an accumulation of unrelated fragments of individual evidence, but rather the organic growth of repeatedly checked and constantly reproductive research.

In order that the vast detail of the knowledge of the world may be simplified, geography seeks to establish generalized pictures of combinations of dissimilar parts of areas that will nevertheless be as nearly correct as the limitations of a generalization permit, and to establish generic concepts of common characteristics of phenomena, or phenomenon-complexes that shall describe with certainty the common characteristics that these features actually possess. On the basis of such generic concepts, geography seeks to establish principles of relationships between the phenomena that are Arielle related in the same or different areas, in order that it may correctly interpret the interrelations of such phenomena in any particular area.

Finally, geography seeks to organize its knowledge of the world into interconnected systems, in order that any particular fragment of knowledge may be related to all others that bear upon it. The areal differentiation of the world involves the integration, for all points on the earth's surface, of the resultant of many interrelated, but in part independent, variables. The simultaneous integration allover the world of the resultant of all these variables cannot be organized into a single system.

In systematic geography each particular element, or element-complex, that is geographically significant, is studied in terms of its relation to the total differentiation of areas, as it varies from place to place over the world, or any part of it. This is in no sense the complete study of that particular phenomenon, such as would be made in the appropriate systematic science, but the study of it solely in its geographic significance-namely in its own areal connections, and in the relations of its variations to those of other features that determine the character of areas. Although the study of any single earth feature is thus organized into a complete system in systematic geography, it is clear that at every point on the earth it is connected with the coordinate systems concerned with the other features.

In regional geography all the knowledge of the interrelations of all features at given places-obtained in part from the different systems of systematic geography-is integrated, in terms of the interrelations which those features have to each other, to provide the total geography of those places. The areal integration of an infinite number of place-integrations of factors varying somewhat independently in relation to place, is possible only by the arbitrary device of ignoring variations within small unit areas so that these finite areal units, each arbitrarily distorted into a homogeneous unit, may be studied in their relations to each other as parts of larger areas. These larger areas are themselves but parts of still larger divisions ultimately divisions of the world.

Problem of Dividing the World : The problem of dividing the world, or any part of it, into sub-divisions in which to focus the study of areas, is the most difficult problem of organization in regional geography. It is a task that involves a complete division of the world in a logical system, or systems, of division and subdivision, down to, ultimately, the approximately homogeneous units of areas. Difficult though the task may be, the principles of completeness and organization demand that geography seek the best possible solution.

One method of providing such an organization represents perhaps an intermediate step between systematic and regional geography. On the basis of anyone element or element-complex-which latter may represent a great number and variety of closely related elements-we may construct a logical system of division and subdivision of the world according to types. Each of these systems of division determined on the basis of generic concepts of element-complexes, may be carried through by objective decisions based on measurement. Possibly as few as three such systems-each based on a cultural complex of many elements-may be adequate to provide outlines into which to organize most of our regional knowledge of the world. In each case, however, we are, organizing separately different aspects of the geography of regions, we are not organizing the complete geography of regions.

A single system in which to organize the complete geography of the regions of the world must be based on the total character of areas, including their location as parts of larger units. Such a system of specific regions requires the consideration of all features significant in geography, some more significant in some areas, others in others. The determination of the divisions at any level involves, therefore, subjective Judgement as to which features are more, which less, inportant in determining similarities and dissimilarities, and in determining the relative closeness of regional interrelations. At any level therefore, the regions are fragments of the land, so determined that we may most economically describe the character of each region, that is, that in each region we will have a minimum number of different generalized descriptions of approximately similar units, each" description involving the maximum number of nearly common characteristics and applicable to the maximum number of similar units.

Although all the fundamental ideals of science apply equally in all parts of geography, there are differences in the degree to which they can be attained in the different parts. These differences among the special divisions of geography-physical, economic, political, etc.—are differences in degree corresponding to the similar differences in degree to which the various systematic sciences are able to attain those ideals.

The greatest differences in character within geography are found between the two major methods of organizing geographic knowledge—systematic geography and regional geography-each of which includes its appropriate part of all the special fields. In addition to the difference in form of organization in the two parts, there is a radical difference in the extent to which knowledge may be expressed in universals, whether generic concepts or principles of relationships.

Systematic geography is organized in tends of particular phenomena of general geographic significance, each of which is studied in tends of the relations of its areal differentiation to that of the others. Its descriptive form is therefore similar to that of the systematic sciences. Like them, it seeks to establish generic concepts of the phenomena studied and universal principles of their relationships, but only in terms of significance to areal differentiation. No more than in the systematic sciences, however, can systematic geography hope to express all its knowledge in tends of universals much must be expressed and studied as unique.

While there are no logical limitations to the development of generic concepts and principles in systematic geography, the nature of the phenomena and the relations between them that are studied in geography present many difficulties preventing the establishment of precise principles. These difficulties are of the same kind as are found, in differing degree, in all parts of science. In many of the systematic sciences, both natural and social, the degree of difficulty is as great, or greater, than in geography. In that field which is most "nearly the counterpart of geography, namely history, the difficulties are in almost every case far greater. Systematic geography is therefore far more able to develop universals than is "systematic history."

Nevertheless the degree of completeness, accuracy, and certainty, both of the principles established and of the facts known in regard to any particular situation, seldom permit definite predictions in geography. This characteristic, geography shares not only with history, but also with many other sciences, both natural and social.

Regional geography organizes the knowledge of all interrelated forms of areal differentiation in individual units of area, which it must organize into a system of division and subdivision of the total earth surface. Its form of description involves two steps. It must first express, by analysis and synthesis, the integration of all interrelated features at individual unit places, and must then express, by analysis and synthesis, the integration of all such unit places within a given area. In order to make this possible, it must distort reality to the extent of considering small but finite areas as homogeneous units which can be compared with each other and added together in areal patterns of larger units. These larger, likewise arbitrary units, are so determined as to make possible a minimum of generalized description of each unit "region," that will involve a minimum of inaccuracy and incompleteness.

Since the units with which it deals are neither real phenomena nor real units but, at any level of division represent distortions of reality, regional geography itself cannot develop either generic concepts or principles of reality. For the interpretation of its findings it depends upon generic concepts and principles developed in systematic geography. Furthermore, by comparing different units of area that are in pan similar, it can test and correct the universals developed in systematic geography.

The direct subject of regional geography is the uniquely varying character of the earth's surface-a single unit which can only be divided arbitrarily into parts that, at any level of division, are, like the temporal parts of history, unique in total character. Consequently the findings of regional geography, though they include interpretations of details, are in large part descriptive. The discovery, analysis and synthesis of the unique is not to be dismissed as "mere description"; on the contrary, it represents an essential function of science, and the only function that it can perform in studying the unique. To know and understand fully the character of the unique is to know it completely; no universals need be evolved, other than the general law of geography that all its areas are unique.

Science Required Systematic and Integrating Field : In the same way that science as a whole requires both the systematic fields that study particular kinds of phenomena and the integrating fields that study the ways in which those phenomena are actually related as they are found in reality, so geography requires both its systematic and its regional methods of study of phenomena and organization of knowledge. Systematic geography is essential to an understanding of the areal differences in each kind of phenomena and the principles governing their relations to each other. This alone, however, cannot provide a comprehension of the individual earth units, but rather divests them of the fullness of their colour and life. To comprehend the full character of each area in comparison with others, we must examine the totality of related features as that is found in different units of area-i.e., regional geography. Though each of these methods represents a different point of view, both are essential to the single purpose of geography and therefore are properly included in the unified field. Further, the two methods are intimately related and essential to each other. The ultimate purpose of geography, the study of areal differentiation of the world, is most clearly expressed in regional geography; only by constantly maintaining its relation to regional geography can systematic geography hold to the purpose of geography and not disappear into other sciences. On the other hand, regional geography in itself is sterile; without the continuous fertilization of generic concepts and principles from systematic geography, it could not advance to higher degrees of accuracy and certainty in interpretation of its findings.

It would be an error to interpret the current interest in methodological discussion as a sign that American geography had entered a period of unusual dissension. In large part, no doubt, it represents the crystallization in print of disagreements hitherto held in the more liquid solution of oral discussion, but which have been suddenly precipitated by the first challenge to fundamental principles in geography to appear in print in more than a decade. Although the basic position of geography as a chorographic science has been questioned, the challenge does not appear to have produced dissension. On the contrary, it has revealed that those who were accustomed to find themselves in opposite camps in methodological discussions, were actually in opposition only on secondary questions and were fundamentally at one on the major function of geography among the sciences.

More than at any previous time in the development of American geography, there is notable agreement, in practice as well as in theory, on the importance of studies in regional geography, while at the same time there is a continued drive to develop the various aspects of systematic geography. Further, the apparent gulf between these two aspects of the field is being narrowed: students of regional geography depend increasingly on studies in systematic geography and those making systematic studies have recognized that their value to geography as a whole depends on the extent to which they are correlated with the viewpoint of regional geography.

If American geography is approaching that major degree of common understanding on the fundamental nature of its field that was attained in Germany two or three decades earlier, and likewise underlies—even though less definitely expressed a great part of the work in French geography, we may hope that the immediate future in this country will show a period of correspondingly rich production along a wine, but common front. Agreement on methodological questions, as on any others, can be attained, by those who are free to think for themselves, only by thorough examination of the problems involved, with adequate and fair consideration of the divergent views expressed by other students, past as well as present. By directing on current methodological problems in our field a critical review organized out of the rich literature of more than a century of geographic thought, we hope to have contributed to a more general understanding of our fundamental purposes and problems.

University of Minnesota.

June, 1939.

2

Planetary Movement

Movement of the Earth

The study of forces affecting the crust of the earth or of geological processes is of paramount significance because these forces and resultant movements are involved in the creation, destruction, recreation and maintenance of geo-materials and numerous types of relief features of varying magnitudes. These forces very often affect and change the earth's surface. In fact, the change is law of nature. The geological changes are generally of two types e.g. (i) long period changes and (ii) short-period changes. Long-period changes occur so slowly that man is unable to notice such changes during his life-period. On the other hand, short-period changes take place so suddenly that these are noticed within few seconds to few hours, e.g., seismic events, volcanic eruptions etc., The forces, which affect the crust of the earth, are divided into two broad categories on the basis of their sources of origin e.g. (i) endogenetic forces and (ii) exogenetic forces.

Endogenetic Forces

The forces coming from within the earth are called as endogenetic forces which cause two types of movements in the earth viz. (i) horizontal movements and (ii) vertical movements. These movements motored by the endogenetic forces introduce various types of vertical irregularities which give birth to numerous varieties of relief features on the earth's surface (e.g. mountains, plateaus, plains, lakes, faults, folds etc.). Volcanic eruptions and seismic events are also the expressions of endogenetic forces. Such movements are called sudden movements and the forces responsible for their origin are called sudden forces. We do not know precisely the mode of origin of the endogenetic forces and movement because these are related to the interior of the earth about which our scientific knowledge is still limited. On an average, the origin of endogenetic forces is related to thermal conditions of the interior of the earth. Generally, the endogenetic forces and related horizontal and vertical movements are caused due to contraction and expansion of rocks because of varying thermal conditions and temperature changes inside the earth. The displacement and readjustment of geo-materials some times take place so rapidly that earth movements are caused below the crust. The endogenetic forces and movements are divided, on the basis of intensity, into two major categories viz. (1) diastrophic forces and (2) sudden forces.

Sudden Forces and Movements : Sudden movements, caused by sudden endogenetic forces coming from deep within the earth, cause such sudden and rapid events that these cause massive destructions at and below the earth's surfaces. Such events, like volcanic eruptions and earthquakes, are called 'extreme events' and become disastrous hazards when they occur in densely populated localities. These forces work very quickly and their results are seen within minutes. 'It is important to note that these forces are the result of long-period preparation deep within the earth. Only their cumulative effects on the earth's surface are quick and sudden' (Savindra Singh, 1991, Environmental Geography, p. 68). Geologically, these sudden forces are termed as 'constructive forces' because these create certain relief features on the earth's surface. For example, volcanic eruptions result in the formation of volcanic cones and mountains while fissure flows of lavas form extensive lava plateaus (e.g. Deccan plateau of India, Columbian plateau of the USA etc.) and lava plains. Earthquakes create faults, fractures, lakes etc.

Diastrophic Forces and Movements : Diastrophic forces include both vertical and horizontal movements which are caused due to forces deep within the earth. These diastrophic forces operate very slowly and their effects become discernible after thousands and millions of years. These forces, also termed as constructive forces, affect larger areas of the globe and produce meso-level reliefs (e.g.) mountains, plateaus, plains, lakes, big faults etc.). These diastrophic forces and movements are further subdivided into two groups viz. (i) epeirogenetic movements and (ii) orogenetic movements.

(i) *Epeirogenetic Movements* : Epeirogenetic word consists of two words viz. 'epiros' (meaning thereby continent) and 'genesis' (meaning thereby origin). Epeirogenetic movement causes upliftment and subsidence of continental masses through upward and downward movements respectively. Both the movements are, in fact, vertical movements. These forces and resultant movements affect larger parts of the continents. These are further divided into two types viz. (i) upward movement and (ii) downward movement. Upward movement causes upliftment of continental masses in two ways e.g.(a) the upliftment of whole continent or part thereof and (b) the upliftment of coastal land of the continents. Such type of upliftment is called emergence.

Downward movement causes subsidence of continental masses in two ways viz. (i) subsidence of land area. Such type of downward movement is called as subsidence. (ii) Alternatively, the land area near the sea coast is moved downward or is subsided below sea-level and is thus submerged under sea water. Such type of downward movement is called as submergence.

(ii) *Orogenetic Movement* : The word orogenetic has been derived from two Greek words, 'oros' (meaning thereby mountain) and 'genesis' (meaning thereby origin or formation). Orogenetic movement is caused due to endogenetic forces working in horizontal manner. Horizontal forces and movements are also called as 'tangential forces.' Orogenetic or horizontal forces work in two ways viz. (i) in opposite directions and (ii) towards each other. This is called 'tensional force' when it operates in opposite directions. Such types of force and movement are also called as divergent forces and movements. Thus, tensional forces create rupture, cracks, fracture and faults in the crustal parts of the earth. The force, when operates face to face, is called compressional force or convergent force. Compressional force causes crustal bending leading to the formation of folds or crustal warping leading to local rise or subsidence of crustal parts.

Crustal Bending

When horizontal forces work face to face the crustal rocks are bent due to resultant compressional and tangential force. In other words, when crustal parts move towards each other under the influence of horizontal or convergent forces and movements, the crustal rocks undergo the process of 'crustal bending' in two ways e.g. (i) warping and (ii) folding. The process of crustal warping affects larger areas of the crust wherein the crustal parts are either warped (raised) upward or downward. The upward rise of the crustal part due to compressive force resulting from convergent horizontal movement is called upwarping while the bending of the crustal part downward in the form of a basin or depression is called downwarping. When the processes of upwarping or downwarping of crustal rocks affect larger areas, the resultant mechanism is called broad warping. When the compressive horizontal forces or convergent forces and resultant movements cause buckling and squeezing of crustal rocks, the resultant mechanism is called folding which causes several types of folds.

The Folds

Wave-like bends are formed in the crustal rocks due to tangential compressive force resulting from horizontal movement caused by the endogenetic force originating deep within the earth. Such bends are called 'folds' wherein some parts are bent up and some parts are bent down. The upfolded rock strata in arch-like form are called 'anticlines' while the down folded structure forming trough-like feature is called 'syncline'. In fact, folds are minor forms of broad warping. The two sides of a fold are called limbs of the fold. The limb which is shared between an anticline and its companion syncline is called middle limb. The plane which bisects the angle between the two limbs of the anticline or middle limb of the syncline is called the axis of fold or axial plane. On the basis of anticline and syncline these axial planes are called as axis of anticline and axis of syncline respectively.

It is desirable to explain the characteristics of 'dip' and 'strike' as it becomes absolutely necessary to understand them in order to understand the structural form. The inclination of rock beds with respect to horizontal plane is termed as 'dip'. It is apparent that we derive two information about the dip e.g. (i) the direction of maximum slope down a bedding plane and (ii) the angle between the maximum slope and the horizontal plane. The direction of dip is measured by its true bearing in relation to east or west of north. e.g. 60°N.E. ; while the angle of dip is measured with an instrument called clinometer. For example, if any rock bed is inclined at the angle of 60° with respect to horizontal plane and the direction of slope is N, then the dip would be expressed as 60° N. 'The strike of an inclined bed is the direction of any horizontal line along a bedding plane' (A. Holmes and D.L. Holmes). The direction of dip is always at right angle to the strike.

Anticlines : The upfolded rock beds are called anticlines. In simple fold the rock strata of both the limbs dip in opposite directions. Some times, folding becomes so acute that the dip angle of the anticline is accentuated and the fold becomes almost vertical. When the slopes of both the limbs or sides of an anticline are uniform, the anticline is called as 'symmetrical anticline' but when the slopes are unequal, the anticline is called as 'asymmetrical anticline'. Anticlines are divided into two types on the basis of dip angle e.g. (i) gentle anticline when the dip angle is less than 40°, some times 1 ° or 2° and (ii) steep anticline when the dip angle ranges between 40° and 90°.

Synclines : Downfolded rock beds due to compressive forces caused by horizontal tangential forces are called synclines. These are, in fact, trough like form in which beds on either side 'incline together' towards the middle part. If folded intensely, the syncline assumes the form of a canoe.

Anticlinorium : Anticlinorium refers to those folded structures in the regions of folded mountains where there are a series of minor anticlines and synclines within one extensive anticline. Anticlinorium is formed when the horizontal compressive tangential forces do not work regularly. Consequently, due to difference in the intensity of compressive forces such structures are formed. Such type of folded structure is also called as fan fold.

Synclinorium : Synclinorium represents such a folded structure which includes an extensive syncline having numerous minor anticlines and synclines. Such structure is formed due to irregular folding consequent upon irregular compressive forces.

The Types : The nature of folds depends on several factors e.g., the nature of rocks, the nature and intensity of compressive forces, duration of the operation of compressive forces etc. The elasticity of rocks largely affects the nature and the magnitude of folding process. The softer and more elastic rocks are subjected to intense folding while rigid and less elastic rocks are only moderately folded. The difference in the intensity and magnitude of compressive forces also causes variations in the characteristics of folds. Normally, both the limbs of a simple fold are more or less of equal inclination but in most of the cases of different folds the inclinations of both the limbs are different. Thus, based on the inclination of the limbs, folds are divided into 5 types.

(1) Symmetrical folds are simple folds, the limbs (both) of which incline uniformly. These folds are an example of open fold. Symmetrical folds are formed when compressive forces work regularly but with moderate intensity. In fact, symmetrical folds are very rarely found in the field.

(2) Asymmetrical folds are characterized by unequal and irregular limbs. Both the limbs incline at different angles. One limb is relatively larger and the inclination is moderate and regular while the other limb is relatively shorter with steep inclination. Thus, both the limbs are asymmetrical in terms of inclination and length.

(3) Monoclinal folds are those in which one limb inclines moderately with regular slope while the other limb linclines steeply at right angle and the slope is almost vertical. It may be pointed out that vertical force and movement are held responsible for the formation of monoclinal folds. There is every possibility for the splitting of the limbs of such folds because of intense folding. Splitting of limbs gives birth to the formation of faults. It is also opined that monoclinal folds are also formed due to unequal horizontal compressive forces coming from both the sides.

(4) Isoclinal folds are formed when the compressive forces are so strong that both the limbs of the fold become parallel but not horizontal.

(5) Recumbent folds are formed when the compressive forces are so strong that both the limbs of the fold become parallel as well as horizontal.

(6) Overturned folds are those folds in which one limb of the fold is thrust upon another fold due to intense compressive forces. Limbs are seldom horizontal.

(7) Plunge folds are formed when the axis of the fold instead of being parallel to the horizontal plane becomes tilted and forms plunge angle which is the angle between the axis and the horizontal plane.

(8) Fan folds represent an extensive and broad fold consisting of several minor anticlines and synclines. Such fold resembles a fan. Such feature is also called as anticlinorium or synclinorium.

(9) Open folds are those in which the angle between the two limbs of the fold is more than 90° but less than 180° (i.e. obtuse angle between the limbs of a fold). Such open folds are formed due to wave-like folding because of moderate nature of compressive force.

(10) Closed folds are those folds in which the angle between the two limbs of a fold is acute angle. Such folds are formed because of intense compressive force.

Nappes : Nappes are the result of complex folding mechanism caused by intense horizontal movement and resultant compressive force. Both the limbs of a recumbent fold are parallel and horizontal. Due to further increase in the continued compressive force one limb of the recumbent folds slides forward and overrides the other fold. This process is called 'thrust' and the plane along which one part of the fold is thrust is called 'thrust plane'. The upthrust part of the fold is called 'overthrust fold'. When the compressive force becomes so acute that it crosses the limit of the elasticity of the rock beds, the limbs of the fold are so acutely folded that these break at the axis of the fold and the lower rock beds come upward. Thus, the resultant structure becomes reverse to the normal structure. Due to continued horizontal movement and compressive force the broken limb of the fold is thrown several kilometres away from its original place and overrides the rock beds of the distant place. Such type of structure becomes unconformal to the original structure of the place where the broken limb of the fold of the other place overrides the rock beds. Such broken limb of the fold is called 'napple'.

Several examples of nappe are traceable in the present folded mountains. The nappes of the Alps have been more systematically studied. Four major nappes have been identified in the Alps mountains. The structure has become very much complex because of superimposition of one nappe upon another nappe. The four major groups of Alpine nappes from below upward are (i) Helvetic nappe, (ii) Pennine nappe, (iii) Austride nappe and (iv) Dinaride nappe. In fact, these nappes are located like a series of earthwaves. In most of the localities the overriding nappes have been eroded away because of dynamic wheels of denudational processes and thus burried basic structure has been exposed. When the portion of lower nappe is seen because of denudation of overriding nappe, the resultant open structure is called 'structural window'. Several examples of 'complete window' have been discovered in the eastern Alps.

A few examples of nappes have also been traced out in the Himalayas. The existence of nappes has been discovered by Wadia from Kashmir Himalaya, by Pilgrim from Simla Himalaya, by Auden from Garhwal Himalaya and by Heim and Gansser from Kumaun

Himalaya. It is desirable to mention some facts about nappe structure. When the broken limb of a fold overrides the other fold near to the broken fold, the resultant nappe is called autochthonous nappe. On the other hand, when the limb of a fold, after being broken, overrides the other fold at a distant place (several kilometres away), the resultant nappe is called exotic nappe.

Crustal Fracture

Crustal fracture refers to displacement of rocks along a plane due to tensional and compressional forces acting either horizontally or vertically or some times even in both ways. Crustal fracture depends on the strength of rocks and intensity of tensional forces. The crustal rocks suffer only cracks when the tensional force is moderate but when the rocks are subjected to intense tensional force, the rock beds are subjected to dislocation and displacement resulting into the formation of faults. Generally, fractures are divided into (i) joints and (ii) faults. A joint is defined as a fracture in the crustal rocks wherein no appreciable movement of rock takes place, whereas a fracture becomes fault when there is appreciable displacement of the rocks on both sides of a fracture and parallel to it.

The Faults

A fault is a fracture in the crustal rocks wherein the rocks are displaced along a plane called as fault plane. In other words, when the crustal rocks are displaced, due to tensional movement caused by the endogenetic forces, along a plane, the resultant structure is called a fault. The plane along which the rock blocks are displaced is called fault plane. In fact, there is real movement along the fault plane due to which a fault is formed. A fault plane may be vertical, or inclined, or horizontal, or curved or of any type and form. The movement responsible for the formation of a fault may operate in vertical or horizontal or in any direction. During the formation of a fault the vertical displacement of rock blocks may occur upto several hundred metres and horizontally the rock blocks may be displaced upto several kilometres but it does not mean that the total displacement occurs at a single time. In fact, fault movement or the displacement of rocks occurs only upto a few metres only at a time. Fault, in fact, represents weaker zones of the earth where crustal movements become operative for longer duration. A few terms regarding an ideal fault should be understood before going into the details of the mode of formation of various types of faults.

(1) Fault plane is that plane along which the rock blocks are displaced by tensional and compressional forces acting vertically and horizontally to form a fault. A fault plane may be vertical, inclined, horizontal, curved or of any other form.

(2) Fault dip is the angle between the fault plane and horizontal plane.

(3) Upthrown side represents the uppermost block of a fault.

(4) Downthrown side represents the lowermost block of a fault. Some times, it becomes difficult to find out, which block has really moved along the fault plane ?

(5) Hanging wall is the upper wall of a fault.

(6) Foot wall represents the lower wall of a fault.

(7) Fault scarp is the steep wall-like slope caused by faulting of the crustal rocks. Some times, the fault scarp is so steep that it resembles a cliff. It may be pointed out that scarps are not always formed due to faulting alone, rather these are also formed due to erosion, but whenever these are formed by faulting (tectonic forces), these are called 'faultscarpts. Types of Faults-The different types of faulting of the crustal rocks are determined by the direction of motion along the fracture plane. Generally, the relative movement or displacement of the rock blocks or the slip of the rock blocks occurs approximately in two directions viz. (i) either to the direction of the dip or (ii) to the direction of the strike of the fault plane. Thus, the displacement or movement of rock blocks may be distinguished as (a) dip slip movements and (b) strike-slip movements. Thus, on the basis of the direction of slip or displacement faults are divided into (i) dip-slip faults and (ii) strike-slip faults. Again, the displacement of rock blocks mainly upper blocks may be either down the direction of the dip (then the resultant fault is called normal fault) or up the dip (the resultant fault becomes reverse or thrust fault). In the case of strike-slip movement and fault, the relative displacement of the rock blocks may be either to the right (then the resultant fault will be right-lateral or dextral fault) or to the left side (the resultant fault becomes left-lateral or sinistral fault). Strike slip faults are also called as wrench faults, tear faults or transcurrent faults. The combinations of normal and wrench faults or reverse and wrench faults are called as oblique slip faults.

(i) Normal faults are formed due to the displacement of both the rock blocks in opposite directions due to fracture consequent upon greatest stress. The fault plane is usually between 45° and the vertical. The steep scarp resulting from normal faults is called fault-scarp or fault-line scarp the height of which ranges between a few metres to hundreds of metres. It may be mentioned that it becomes very difficult to find out the exact height of the faultscarps in the field because the height is remarkably reduced due to continued denudation.

(ii) Reverse faults are formed due to the movement of both the fractured rock blocks towards each other. The fault plane, in a reverse fault, is usually inclined at an angle between 40 degree and the horizontal (0 degree). The vertical stress is minimum while the horizontal stress is maximum. It may be mentioned that in a reverse fault the rock beds on the upper side are displaced up the fault plane relative to the rock beds below. It is apparent that reverse faulting results in the shortening of the faulted area while normal faults cause extension of the faulted area. It is, thus, also obvious that some sort of compression is also involved in the formation of reverse faults. Reverse faults are also called as thrust faults. Since the reverse fault is formed due to compressive force resulting from horizontal movement and hence this is also called as compressional fault. When the compressive force exceeds the strength of the rocks, one block of the fault overrides the other block and the resultant fault is called as overthrust fault wherein the fault plane becomes almost horizontal.

(iii) Lateral or strike-slip faults are formed when the rock blocks are displaced horizontally along the fault plane due to horizontal movement. These are called left-lateral or sinistral faults when the displacement of the rock blocks occurs to the left on the far side of the fault and right-lateral or dextral faults when the displacement of rock blocks takes place to the right on the far side of the fault. In majority of the cases there are no scarps in such faults, if they occur at all, they are very low in height.

(iv) Step faults- When a series of faults occur in any area in such a way that the slopes of all the fault planes of all the faults are in the same direction the resultant faults are called as step faults. It is a prerequisite condition for the formation of step faults that the downward displacement of all the downthrown blocks must occur in the same direction.

Graben or Rift Valley

Rift valley is a major relief feature resulting from faulting activities. Rift valley represents a trough, depression or basin between two crustal parts. In fact, rift valleys are long and narrow troughs bounded by one or more parallel normal faults caused by horizontal and vertical movements motored by endogenetic forces. Rift valleys are actually formed due to displacement of crustal parts and subsidence of middle portion between two normal faults. Rift valleys are generally also called as 'graben' which is a German word which means a trough-like depression. These two terms are synonymously used in various parts of the world. 'Tensional crustal forces, literally puling the crust apart, are responsible for these down dropped fault blocks' (F. Press and R. Siever, 1974). A few scientists have attempted to differentiate a graben from a rift valley on the basis of size and dimension. They believe that a graben is relatively smaller in size than a rift valley but this minor difference of size is not acceptable to others. Thus, both the terms, graben and rift valley should always be considered as synonym.

A rift valley may be formed in two ways viz. (i) when the middle portion of the crust between two normal faults is dropped downward while the two blocks on either side of the down dropped block remain stable or (ii) when the middle portion between two normal faults remains stable and the two side blocks on either side of the middle portion are raised upward.

Normally, a rift valley is long, narrow but very deep. Rhine rift valley is the best example of a well defined rift valley. It stretches for a distance of 320 km having an average width between the cities of Basal and Bingen. The one side of this great rift valley is bounded by Vosges and Hardt mountains (block mountains-horst) and the other side is bordered by Black Forest and Odenwald mountains. The example of the longest rift valley is the valley that runs from the Jordon river valley through Red Sea basin to Zambezi valley for a distance of 4,800 km. A few of the rift valleys are so deep that their bottom/floor is below the sea-level. Death Valley of the southern California (USA) is a good example of such graben. Dead Sea of Asia presents an ideal example of typical rift valley. The floor of the Dead Sea is about 867 m below sea-level. The floors of the Jordon rift valley and Death Valley are also 433 m below sea-level. The Narmada valley, the Damodar valley and some stretches of the Son Valley, the Tapi valley etc., are considered to be examples of rift valleys but this view is still controversial and is not acceptable to all geologists.

It may be mentioned that the rift valleys are not only confined to the continental crustal surfaces but they are also found on sea-floor. In fact, the deepest grabens are found in the form of 'ocean deeps' and trenches. The Bortlet Trough located to the south of Cuba is 4.8 km deep while Java Deep is 6.4 km deep from the sea-floor. The central plain of Scottland, Spencer Bay of south Australia etc., are examples of rift valleys.

The Origin

The riddle of the problem of the origin of the rift valleys and grabens, typical topographic expressions of faulting, still remains a mystery. Though many scientists have propounded their views regarding the origin of the rift valleys based on their studies of respective rift valleys but their concepts and theories are still controversial and no commonly acceptable theory could be propounded as yet. The hypotheses regarding the origin of the rift valleys are generally grouped in two categories e.g. (1) tensional hypothesis and (2) compressional hypothesis.

(1) *Tensional Hypothesis* : The earlier hypothesis of the origin of the rift valleys was based on the basic concept of the 'dropped keystone of the arch' of a building.

According to this concept the rift valleys were related to the hollow space created by the dropping of the keystone of an arch of a building downward. In other words, an open space is formed at the middle portion of an arch of a building when the keystone or keybrick falls downward due to cracks developed in the arch. Similarly, when two parallel cracks develop in the crustal surface due to tensional forces and when the bounding side blocks on either side of the two cracks or fractures are pulled apart due to tensional forces, the middle portion between two parallel normal faults moves downward and thus an open space is formed. This open space becomes a rift valley.

This 'key stone hypothesis' was severely criticized because it was based on erroneous concepts and beliefs. For example, there is wide open space below the arch of a building and hence the keystone or keybrick, after the arch develops cracks, can easily fall down but there is no open space beneath the crustal rocks and thus there would be difficulty for the middle block between the two parallel normal faults to slip downward. The faulted middle block can only be slipped downward when it would be able to displace the magma lying below the crustal blocks. If this process is accepted then the formation of the rift valley must be followed by volcanic activities because the displaced magma would try to ascend through the faults. Some times, the mechanism may be so sudden that there may be sudden violent volcanic eruption, but the observations of several deep rift valleys denote the fact that rift valley formation is not necessarily always associated with volcanic eruptions. The observations and several experiments have revealed the fact that already existing volcanic activities and active volcanoes ceased to operate at the time of the formation of rift valleys. It might have become possible only when the exit of the ascent of magma would have been plugged due to faulting activity. This explanation is also refuted on the ground that if we accept the mode of formation of a rift valley due to horizontal tensional forces and resultant pulling of bounding, faulted side blocks of two normal faults apart, then the upwelling of magma in the form of lava cannot be stopped, rather the pouring of lava can be stopped due to compressive forces. Thus, the tensional hypothesis of the origin of the rift valleys is rejected on this ground.

(2) *Compressional Hypothesis* : In order to remove the difficulties of the tensional hypothesis of the origin of the rift valleys compressional hypothesis was postulated by a number of scientists e.g. Wayland, Baily Willis, Waren D. Smith, E.C. Bullard etc. Wayland through his studies of Lake Albert and Ruwenzori section and Baily Willis based on his studies of Dead Sea have postulated the concept that the rift valleys are not formed by tensional forces but are formed due to compressional forces at greater depth. Due to intense compression the side blocks are thrown

up along the thrust faults in the form of horsts. These upthrown blocks are called overthrusting rift blocks. The middle portion is forced to slip downward because of the pressure resulting from the rising side blocks. Thus, the downward slipping middle portion between two faults is called as rift block which is narrow upward but broader downward. In other words, the rift block gradually broadens out downward. Thus, the rift valleys are formed due to slipping of middle block or rift block downward between two rising side blocks caused by thrust faulting under the impact of convergent compressional forces.

(3) *Hypothesis of E.C. Bullard* : E.C. Bullard, while conducting the gravity survey, postulated his new concept of the origin of the rift valleys in 1933-34. According to him the rift block cannot slip downward under the impact of gravity, like a keystone of an arch of a building. Thus, the rift valley can be formed only due to compression coming from two sides. According to Bullard the formation of a rift valley is not completed during a single phase but is completed through a series of sequential phases. First Stage, there is compression in the crustal rock beds of the rigid part of a plateau due to active horizontal movement. The horizontal compressive forces work face to face from both the sides of the land. This lateral compression causes buckling of the crustal rocks. As the compressive forces continue to increase, the buckling and squeezing of the crustal rocks also continue to increase. When the compression becomes so enormous that it exceeds the strength of the rocks, a crack is developed at a place in the crustal rocks. This crack is gradually enlarged due to continuous increase in the compressive force.

Second Stage, due to the formation of a crack, one portion overrides the other portion. This process is called as 'thrusting.' On the other hand, the second part is thrown downward relative to the first part. This process is called 'downthrusting'.

According to E.C. Bullard the width of the rift valley (A-B) depends upon the elasticity of the rocks, depth of the rift valley and the density of the substratum. If the density of the substratum is taken to be 3.3, then the width of the rift valley would be 40 km if the depth of the valley is 20 km. Similarly, for a 40-km deep valley the width would be 65 km.

It may be concluded that neither the tensional hypothesis nor the compressional hypothesis could be able to solve many of the intricate problems of the origin of the rift valleys.

Exogenetic Forces

The exogenetic forces or processes, also called as denudational processes, or 'destructional forces or processes' are originated from the atmosphere. These forces are continuously engaged in the destruction of the relief features created by the endogenetic forces through their weathering, erosional and depositional activities. Exogenetic processes are, therefore, planation processes. Denudation includes both weathering and erosion where weathering being a static process includes the disintegration and decomposition of rocks in situ whereas erosion is dynamic process which includes both, removal of materials and their transportation to different destinations. Weathering is basically of three types viz. (i) physical or mechanical weathering, (ii) chemical weathering and (iii) biological weathering. Weathering is very important for the biospheric ecosystem because weathering of parent rocks results in the formation of soils which are very essential for the sustenance of the biotic lives in the biosphere. The erosional processes include running water or river, groundwater, sea-waves, glaciers, periglacial processes and wind. These erosional processes erode the rocks, transport the eroded materials (except periglacial processes) and deposit them in suitable places and thus form several types of erosional and depositional landforms of different magnitudes and dimensions.

3

Continental Drift

Continents and Ocean Basins

Continents and ocean basins being fundamental relief features of the globe are considered as 'relief features of the first order'. It is, therefore, desirable to inquire into their mode of possible origin and evolution. Different views, concepts, hypotheses and theories regarding the origin of the continents and ocean basins have been put forth by the scientists from time to time. Before examining these views about their origin we should know the characteristic features of the distributional patterns and arrangement of the continents and ocean basins as seen at present. About 70.8 per cent of the total surface area of the globe is represented by the oceans whereas remaining 29.2 per cent is represented by the continents. Even the distribution of different continents and oceans in both the hemispheres is not uniform. The following characteristic features of the distributional pattern of the continents and the ocean basins may be highlighted :

(1) There is overhwelming dominance of land areas in the northern hemisphere. More than 75 per cent of the total land area of the globe is situated to the north of the equator (i.e. in the northern hemisphere). Contrary to this water bodies dominate in the southern hemisphere. If we divide the globe in two such hemispheres where the north pole stands located in the English Channel and the south pole near New Zealand, then the northern hemisphere would be 'land hemisphere' while the southern hemisphere as 'water hemisphere'. Thus, the land hemisphere would represent 83 per cent of the total land area of the globe while the water hemisphere would carry 90.6 per cent of the total oceanic areas of the globe.

(2) Continents are arranged in roughly triangular shape. Most of the continents have their bases (of triangle) in the north while their apices are pointed towards south.

If we take North and South Americas together, they represent equilateral triangles, the base of which would be along the arctic sea while the apex would be represented by Cape Horn. If we take these two continents separately, again they form two separate triangles. Similarly, Eurasia also assumes the form of a triangle the base of which is along the arctic sea while its apex is near East Indies. The base of African triangle is towards north while its apex is the Cape of Good Hope. Australia and Antarctica are the exceptions to this rule.

(3) Roughly, the oceans are also triangular in shape. Contrary to the continents the bases of oceans are in the south while their apices are in the north. The base of the Atlantic Ocean extends between Cape Horn and Cape of Good Hope while its apex is located to the east of Greenland. The base of the Indian Ocean is in the south but its two apices are located in the Bay of Bengal and the Arabian Sea. The apex of the Pacific Ocean is near Aleutian Islands while its base lies in the south.

(4) The north pole is surrounded by oceanic water while south pole is surrounded by land area (of the Antarctic continent).

(5) There is antipodal arrangement (situation) of the continents and oceans. Only 44.6 per cent oceans are situated opposite to oceans and 1.4 per cent of the total land area of the globe is opposite to land area. More than 95 per cent of the total land area is situated diametrically opposite to water bodies. There are only two cases of exceptions to this general rule e.g. (i) Patagonia is situated diametrically opposite to a part of north China and (ii) New Zealand is situated opposite to Portugal and Spain (the Iberian Peninsula).

(6) The great Pacific Ocean basin occupies almost one-third of the entire surface area of the globe.

The validity and authenticity of any hypothesis or theory dealing with the origin and evolution of the continents and the ocean basins would be determined in the light of aforesaid characteristics of the distributional pattern of the continents and ocean basins. The presence of the great Pacific Ocean basin and island arcs and festoons of the Pacific Ocean are teething problems before scientists who venture in the precarious field of the postulation of the relevant theory of the origin of the continents and ocean basins. Keeping the above facts in mind Lowthian Green postulated his 'tetrahedral hypothesis' to explain the intricate problems of the origin of the continents and oceans and characteristic features of their distributional pattern. Besides, Lord Kelvin, Sollas, Love etc. have also attempted to explain the origin of the continents and ocean basins but their views are not discussed here because they are based on discarded and obsolete arguments and assumptions. In fact, all the previous hypotheses and theories dealing with the origin of the continents and ocean basins have faded away after the postulation of plate tectonic theory. We will examine here only the concepts of Lowthian Green, F.B. Taylor, A.G. Wegener and of course plate tectonic theory.

Tetrahedral Hypothesis

A few scientists have attempted to solve the problems of the origin of the continents and ocean basins on the basis of fundamental principles of geometry. The patagonal dodecahedral hypothesis (dodeca is a Greek word which means twelve) of Elie de Beaumont is considered to be the first attempt in this field but the tetrahedral hypothesis of Lowthian Green is most significant of all the hypotheses based on geometrical principles. 'An attractive hypothesis which has enjoyed a considerable vogue was initiated by Lowthian Green in 1875' (S.W. Wooldridge and R.S. Morgan, 1959). His hypothesis is based on the characteristics of a tetrahedron which is a solid body having four equal plane surfaces, each of which is an equilateral triangle.

Lowthian Green postulated his hypothesis after considering the characteristics of the distributional pattern of land and water over the globe. Barring a few drawbacks and defects the tetrahedral hypothesis successfully explains the following characteristics of the continents and ocean basins.

(1) Dominance of land areas in the northern hemisphere and water areas in the southern hemisphere;

(2) triangular shape of the continents and oceans ;

(3) situation of continuous ring of land around north polar sea and location of south pole in land area (Antarctica) surrounded by water from all sides;

(4) antipodal arrangement of the continents and oceans ;

(5) largest extent of the Pacific Ocean covering one third area of the globe and

(6) location of chain of folded mountains around the Pacific Ocean.

The hypothesis of Lowthian Green propounded in the year 1875 is based on the common characteristics of a tetrahedron. He based his hypothesis on the following two basic principles of geometry.

(1) 'A sphere is that body which contains the largest volume with respect to its surface area'.

(2) 'A tetrahedron is that body which contains the least volume with respect to its surface area'.

After many experiments Lowthian Green opined that a sphere if subjected to uniform pressure on all its sides would be transformed into the shape of a tetrahedron. He applied this principle in the case of the earth. According to him when the earth was originated it was in the form of a sphere. In the beginning the earth was very hot but it gradually began to cool down due to loss of heat. First, the outer part of the earth cooled down and thus was formed the crust but inner part of the earth continued to cool down. Consequently, the inner part of the earth was subjected to more contraction due to continued cooling and thus there was marked reduction in the volume of the inner part of the earth. Since the upper part, the crust, was already cooled and solidified and hence it could not be subjected to further contraction. This resulted into possible gap between the upper and inner parts of the earth. Consequently, the upper part collapsed on the inner part and ultimately the earth began to assume the shape of a tetrahedron. Lowthian Green has further maintained that the earth has not been as yet changed into a complete tetrahedron rather as it is being cooled, it is proceeding towards attaining the true shape of a tetrahedron. He has further opined that the earth cannot be in the shape of a real tetrahedron because of its structural variations and thus it is natural that there may be some deviations from a true tetrahedron.

In a tetrahedron a plane face remains always opposite to an apex or coign. The apex or coign is more sharpened in the case of a real tetrahedron. In the case of the earth the oceans represent the plane faces of the tetrahedron and land masses represent the apices or coigns but in the case of the earth the coigns are not much sharpened, rather they are flat and convex. According to Lowthian Green oceans were created on the plane faces of the terrestrial tetrahedron whereas the coigns became continental masses.

Four oceans (e.g. the Pacific Ocean, the Atlantic Ocean, the Indian Ocean and the Arctic ocean) were created on the four plane faces of the terrestrial tetrahedron. These plane faces could retain water because of the fact that these were lower than the level of the apices or coigns of the terrestrial tetrahedron. Continents were formed along the apices or coigns of the tetrahedron. This fact may also be proved on the basis of an experiment. If we submerge a tetrahedron in a hemisphere of water, the flat surface of the tetrahedron would retain water while the edges or apices or coigns will project above the water. Lowthian Green claimed to see a tetrahedral arrangement in the distribution of the continents and oceans in such a way that the earth was linked to a tetrahedron having four flat faces and standing on one point. The upper flat face represents the Arctic Ocean while the remaining three faces represent the Pacific Ocean, the Atlantic Ocean and the Indian Ocean. Similarly, three vertical meridional edges represent North and South America, Europe and Africa and Asia while the lower point is represented by Antarctica. Thus, the presence of water around north pole and the location of south pole in land area (Antarctic continent) are very well explained on the basis of tetrahedral hypothesis. Three coigns out of four coigns of four equilateral triangles are located in the northern hemisphere. Only the fourth coign is located in the southern hemisphere. These three coigns present the oldest rigid masses around which the present continents have grown. These three ancient shields are the Laurentian or Canadian Shield, Baltic

Shield and Siberian Shield. The fourth coign or the pivot of the tetrahedron represents the Antarctic shield. The present continents have grown out of these four ancient shields represented by four coigns of the tetrahedron. All the continents developed along the edges of the tetrahedron taper southward and thus triangular shape of the continents is proved. The location of the oceans along four plane faces and the continents along the edges or coigns of the plane faces of the tetrahedron proves antipodal position of land and water.

Though Gregory accepted the tetrahedral hypothesis of Lowthian Green but he suggested certain modifications. According to Gregory due to shrink age of the earth because of contraction on cooling 'the portion of the vertical tetrahedral edges should be fairly constant, but three edges around the polar depression might develop sometimes in the northern and at others in the southern hemisphere'.

Criticism : Though the tetrahedral hypothesis throws light on the problems of the continents and ocean basins and to major extent it successfully explains the characteristic features of the distributional pattern of the present-day continents and ocean basins but because of certain basic defects and errors the hypothesis is not acceptable to the modern scientific community. It is argued that the balance of the earth in the form of a tetrahedron while rotating on an apex cannot be maintained. Secondly, the earth is rotating so rapidly on its axis that the spherical earth cannot be converted into a tetrahedron while contracting on cooling. Thirdly, this hypothesis believes more or less in the permanency of continents and ocean basins while the plate tectonic theory has validated the concept of continental drift.

Theory of Continental Drift

F. B. Taylor postulated his concept of 'horizontal displacement of the continents' in the year of 1908 but it could be published only in the year 1910. The main purpose of his hypothesis was to explain the problems of the origin of the folded mountains of Tertiary period. In fact, F.B. Taylor wanted to solve the peculiar problem of the distributional pattern of Tertiary folded mountains. The north-south arrangement of the Rockies and the Andes of the western margins of the North and South Americas and west-east extent of the Alpine mountains (Alps, Caucasus, Himalayas etc.) posed a serious problem before Taylor which needed careful explanation. He could not find any help from the 'contraction theory' to explain the peculiar distribution of Tertiary folded mountains and hence he propounded his 'drift or displacement theory. The concept of Taylor, thus, is considered to be first attempt in the field of continental drift though Antonio Snider presented his views about 'drift' in the year 1858 in France. Main purpose behind the postulation of 'drift hypothesis' of Snider was to explain the similarity of the fossils of the coal seams of Carboniferous period in North America and Europe.

Taylor started from Cretaceous period. According to him there were two land masses during Cretaceous period. Lauratia and Gondwanaland were located near the north and south poles respectively. He further assumed that the continents were made of sial which was practically absent in the oceanic crust. According to Taylor continents moved towards the equator. The main driving force of the continental drift was tidal force. According to Taylor continents were displaced in two ways e.g. (1) equatorward movement and (ii) westward movement but the driving force responsible for both types of movement was tidal force of the moon.

Lauratia started moving away from the north pole because of enormous tidal force of the moon towards the equator in a radial manner. This movement of landmass resulted into tensional force near the north pole which caused stretching, splitting and rupture in the landmass. Consequently, Baffin Bay, Labrador Sea and Davis Strait were formed. Similarly, the displacement of the Gondwanaland from the south pole towards the equator caused splitting and disruption and hence the Gondwanaland was split into several parts. Consequently, Great Australian Bight and Ross Sea were formed around Antarctic continent. Arctic sea was formed between Greenland and Siberia due to equatorward movement of Lauratia. Atlantic and Indian oceans were supposed to have been formed because of filling of gaps between the drifting continents with water. Taylor assumed that the landmasses began to move in lobe form while drifting through the zones of lesser resistance. Thus, mountains and island arcs were formed in the frontal part of the moving lobes. The Himalayas, Caucasus and Alps are considered to have been formed during equatorward movement of the Lauratia and Gondwanaland from the north and south poles respectively while the Rockies and Andes were formed due to westward movement of the landmasses.

Criticisms : Since F.B. Talor's main aim was to explain the origin of the Tertiary folded mountains and hence he made the continents to move at a very large scale. In fact, some sort of horizontal movement of the landmasses was essential for the origin of mountains but the displacement of landmasses upto 32-64 km would have been sufficient enough for the purpose. Contrary to this, Taylor has described the displacement of the landmasses for thousands of kilometres. Secondly, the mode of drift as suggested by Taylor has also been erroneous. If the tidal force of the moon was so enormous during Cretaceous period that it could displace the landmasses for thousands of kilometres apart then it might have also put a break on the rotatory motion of the earth and thus the rotation of the earth might have stopped within a year. According to A. Holmes neither tidal force nor any external force can drift the continents apart and can help in the formation of mountains. The responsible force must come from within the earth. Though the concept of F.B. Taylor is not acceptable but his hypothesis is considered to be significant on the ground that Taylor raised his voice very forcefully through deductive postulation against the prevalent concept of the permanency of the continents and ocean basins and forcefully objected to the 'contraction theory' and showed a new direction to solve the problem of the origin of the continents and ocean basins. A. Holmes has rightly remarked, 'but Taylor must be given credit for making an independent and slightly an earlier start in this precarious field'.

Wegener's Theory

Aim of the Theory : Professor Alfred Wegener of Germany was primarily a meteorologist. He propounded his concept on continental drift in the year 1912 but it could not come in light till 1922 when he elaborated his concept in a book entitled 'Die Entstehung der Kontinente and Ozeane' which was translated in English in 1924. Wegener's displacement hypothesis was based on the works and findings of a host of scientists such as geologists, palaeoclimatologists, palaeontologists, geophysicists and others. The main problem before Wegener, which needed explanation, was related to climatic changes. It may be pointed out that there are ample evidences which indicate widespread climatic changes throughout the past history of the earth. In fact, the continental drift theory of Wegener 'grew out of the need of explaining the major variations of climate in the past'. The climatic changes which have occurred on the globe may be explained in two ways.

(1) If the continents remained stationary at their places throughout geological history of the earth, the climatic zones might have shifted from one region to another region and thus a particular region might have experienced varying climatic conditions from time to time.

(2) If the climatic zones remained stationary, the landmasses might have been displaced and drifted. Wegener opted for the second alternative as he rejected the view of the permanency of continents and ocean basins. Thus, the main objective of Wegener behind his 'displacement hypothesis' was to explain the global climatic changes which are reported to have taken place during the past earth history.

Basic Premise of the Theory : Following Edward Suess, Wegener believed in three layers system of the earth e.g. outer layer of 'sial', intermediate layer of 'sima' and the lower layer of 'nife'. According to Wegener sial was considered to be limited to the continental masses alone whereas the ocean crust was represented by the upper part of sima. Continents or sialic masses were floating on sima without any resistance offered by sima. He assumed, on the basis of evidences of palaeo-climatology, palaeontology, palaeobotany, geology and geophysics, that all the landmasses were united together in the form of one landmass, which he named Pangaea, in Carboniferous period. There were several smaller inland seas scattered over the Pangaea which was surrounded by a huge water body, which was named by Wegener as 'Panthalasa', representing primaeval Pacific Ocean. Lauratia consisting of present North America, Europe and Asia formed northern part of the Pangaea while Gondwanaland consisting of South America, Africa,

Madagascar (now Malagasy), Peninsular India, Australia and Antarctica represented the southern part of the Pangaea. South pole was located near present Durban (near Natal in southern Africa) during Carboniferous period. Thus, Wegener's theory of continental drift begins from Carboniferous period, he does not describe the conditions during pre-Carboniferous times 'but the postulation of a Carboniferous Pangaea does not mean that he disbelieves in pre-Carboniferous drift; events before this time are known with much less certainty, and the distribution of plants and animals can largely be explained by movements which have taken place since the Carboniferous' (J.A. Steers, 1961, p. 160). The Pangaea was disrupted during subsequent periods and broken landmasses drifted away from each other and thus the present position of the continents and ocean basins became possible.

Supporting Evidences : Wegener has successfully attempted to prove the unification of all landmasses in the form of a single landmass, the Pangaea, during Carboniferous period. On the basis of evidences gathered from geological, climatic and floral records, he claimed that all the present-day continents could be joined to form Pangaea. The following evidences support the concept of the existence of Pangaea during Carboniferous period.

(1) According to Wegener there is geographical similarity along both the coasts of the Atlantic Ocean. Both the opposing coasts of the Atlantic can be fitted together in the same way as two cut off pieces of wood can be refitted (jig-saw fit).

(2) Geological evidences denote that the Caledonian and Hercynian mountain systems of the western and eastern coastal areas of the Atlantic are similar and identical. The Applachians of the north-eastern regions of North America are compatible with the mountain systems of Ireland, Wales and north-western Europe.

(3) Geologically, both the coasts of the Atlantic are also identical. Du Toit, after detailed study of the eastern coasts of South America and western coast of Africa, has said that the geological structures of both the coasts are more or less similar. According to Du Toit both the landmasses (i.e. South America and Africa) cannot be actually brought together but near to each other because a gap of 400-800 km would separate them due to the existence of continental shelves and slopes of these two landmasses.

(4) There is marked similarity in the fossils and vegetation remains found on the eastern coast of South America and the western coast of Africa.

(5) It has been reported from geodetic evidences that Greenland is drifting westward at the rate of 20 cm per year. The evidences of sea floor spreading after 1960 have confirmed the movement of landmasses with respect to each other.

(6) The lemmings (small sized animals) of the northern part of Scandinavia have a tendency to run westward when their population is enormously in creased but

they are foundered in the sea water due to absence of any land beyond Norwagian coast. This behaviour of lemmings proves the fact that the landmasses were united in the ancient times and the animals used to migrate to far off places in the western direction.

(7) The distribution of glossopteris flora in India, South Africa, Australia, Antarctica, Falkland islands etc. proves the fact that all the landmasses were previously united and contiguous in the form of Pangaea.

(8) The evidences of Carboniferous glaciation of Brazil, Falkland, South Africa, Peninsular India, Australia and Antarctica further prove the unification of all landmasses in one landmass (Pangaea) during Carboniferous period.

The Theoretical Process : As stated earlier the main aim of Wegener behind the postulation of his 'drift theory' was to explain major climatic changes which are reported to have taken place in the past geological history of the earth, such as Carboniferous glaciation of major parts of the Gondwanaland. Besides, Wegener also attempted to solve other problems of the earth e.g. origin of mountains, island arcs and festoons, origin an evolution of continents and ocean basins etc.

(1) *Force Responsible for the Drift :* According to Wegener the continents after breaking away from the Panagaea moved (drifted) in two directions e.g. (i) equatorward movement and (ii) westward movement. The equatorward movement of sialic blocks (continental blocks) was caused by gravitational differential force and force of buoyancy. As already stated the continental blocks, according to Wegener, were formed of lighter sialic materials (silica and aluminium) and were floating without any friction on relatively denser 'sima'. Thus, the equatorward movement of the sialic blocks (continental blocks) would depend on the relation of the centre of gravity and the centre of buoyancy of the floating continental mass. Generally, these two types of forces operate in opposite directions. 'But because of the ellipsoidal form of the earth, these forces are not in direct opposition, but are so related that, if the buoyancy point lies under the centre of gravity, the resultant (force) is directed towards the equator' (J.A., Steers, 1961, p. 164).

The westward movement of the continents was caused by the tidal force of the sun and the moon. According to Wegener the attractional force of the sun and the moon, which was maximum when the moon was nearest to the earth, dragged the outer sialic crust (continental blocks) over the interior of the earth, towards the west. It may be pointed out that in any drift theory the weakest point and the most difficult problem is related to the competent force responsible for the movement of the continents. 'Such a force (tidal force/ attractional force of the sun and the moon) is extraordinarily small, but, as in the case of other forces, the question of time is all important; given sufficient time, it is claimed that even these very small forces are able to cause movements' (J.A. Steers, 1961, p. 164).

(2) *Actual Drifting of the Continents :* The disruption, rifting and ultimately drifting of the continental blocks began in Carboniferous period. The movement of the continental blocks away from the poles was dramatically called by Wegener as 'the flight from the poles'. Pangaea was broken into two parts due to differential gravitational force and the force of buoyancy. The northern part became Lauratia (Angaraland) while the southern part was called by Wegener as Gondwanaland. The intervening space between these two giant continental blocks was filled up with water and the resultant water body was called Tethys Sea. This phase of the disruption of Pangaea is called 'Opening of Tethys'. Gondwanaland was disrupted during Cretaceous period and Indian peninsula, Madagascar, Australia and Antarctica broke away from Pangaea and drifted apart under the impact of tidal force of the sun and the moon. North America broke away from Angaraland and drifted westward due to tidal force. Similarly, South America broke away from Africa and moved westward under the impact of tidal force. Due to northward movement of Indian Peninsula Indian Ocean was formed while the Atlantic Ocean was formed due to westward movement of two Americas. It may be mentioned that North and South Americas were drifting westward at different rates and hence 'S' shape of the Atlantic Ocean could be possible. Arctic and North Sea were formed due to flight of the continental blocks from north pole. The size of the Panthalasa (primitive Pacific Ocean) was remarkably reduced because of the movement of continental blocks from all sides towards Panthalasa. Thus, the remaining portion of Panthalasa became the Pacific Ocean. It may be mentioned that disruption, rifting and displacement (drifting)of continental blocks continued from Carboniferous period to Pliocene period when the present pattern and arrangement of the continents and ocean basins was attained. There have been frequent changes in the positions of the equator and the poles as given in table.

Table : Shifting of the Positions of the Poles

Period	*North Pole*	*South Pole*
Silurian	14°N latitude 124°W longitude	to the north-west of Madagascar
Carboniferous	16°N latitude 147°W longitude	near Durban in Natal
Tertiary	51°N latitude 153°W longitude	near 53°S latitude to the south of Africa

Equator was located at the most northerly location during Silurian period as it passed north of Norway. It passed through London during Carbon iferous period and through present locations of the European Alpine mountains during Tertiary period. 'The south Pole and Equator obviously moved into accordant positions. The prevailing westward and equatorward movements must be referred to these positions' (J.A. Steers, 1961, p. 166).

(3) *Mountain Building :* A.G. Wegeneralso attempted to solve the problem of the origin of folded mountains of Tertiary period on the basis of his continental drift theory. The frontal edges of westward drifting continental blocks of North and South Americas were crumpled and folded against the resistance of the rocks of the sea-floor (sima) and thus the western cordilleras of the two Americas (e.g. Rockies and Andes and other mountain chains associated with them) were formed. Similarly, the Alpine ranges of Eurasia were folded due to equatorward movement of Eruasia and Africa together with Pennisular India (equator was passing thorough Tethys sea at that time). Here, Wegener postulated contrasting view points. According to Wegener sial (continental blocks) was floating upon sima without any friction and resistance but during the later part of his theory he pointed out that mountains were formed at the frontal edges of floating and drifting continental blocks (sialic crust) due to friction and resistance offered by sima. How could it be possible ? The question remains unanswered. Inspite of this serious flaw in the continental drift theory of Wegener, S.W. Wooldridge and R.S. Morgan have remarked, 'certainly the problem of mountain building is one in which the hypothesis of continental drift solves more difficulties than it creates'.

(4) *Origin of Island Arcs :* Wegener has related the process of the origin of island arcs and festoons (of eastern Asia, West Indies and the arc of the southern Antilles between Tierra del Fugo and Antarctica) to the differential rates of continental drift. When the Asiatic block (part of Angaraland) was moving westward, the eastern margin of this block could not keep pace with the westward moving major landmass, rather lagged behind, consequently the island arcs and festoons consisting of Sakhalin, Kurile, Japan, Phillippines etc. were formed. Similarly, some portions of North and South Americas, while they were moving westward, were left behind and the island arcs of West Indies and southern Antilles were formed.

(5) *Carboniferous Glacitation :* There are ample evidences to demonstrate that there was large scale glaciation during Carboniferous period when Brazil, Falkland, Southern Africa, Peninsular India, Australia, Antarctica etc. were extensively glaciated. According to Wegener all the continental blocks were united together in the form of one landmass called as Pangaea. South Pole was located near the present position of Durban in Natal. Thus, south pole was located in the middle of

Pangaea. Consequently, ice sheets might have spread from south pole outward at the time of glaciation and the aforesaid land areas, which were closer to south pole, might have been covered with thick ice sheets. At much later date, these land areas might have parted away due to disruption of Pangaea and related continental drift. Glossopteris flora might have also been distributed over the aforesaid areas when these were united together.

The Evaluation : It may be pointed out that Wegener's continental drift theory widely departed from the contemporary orthodox geological ideas of the nineteenth century and the time-honoured thermal contraction theory of the mountain building and thus it was obvious that the believers of contraction theory should not only criticize the new theory of horizontal displacement of the continents but should also discard it. 'It is now widely agreed that he (Wegener) handled his case as an advocate rather than as an impartial scientific observer, appearing to ignore evidences unfavourable to his ideas and distort other evidences in harmony with the theory' (S.W. Wooldridge and R.S. Morgan, 1959, p. 40). The critics of Wegener's continental drift theory fall in two broad categoreis e.g. (i) the critics and writers who always attempted to search errors and discrepancies in Wegener's original synthesis and (ii) the scientists who attempted to modify, enlarge and correct the original theory of Wegener while retaining its basic tenet. The following flaws and defects have been pointed out by different scientists in Wegener's theory of continental drift.

(1) The forces applied by Wegener (differential gravitational force and the force of buoyancy and tidal force of the sun and the moon) are not sufficient enough to drift the continents so apart, 'The tidal force as invoked by Wegener to account for the supposed westerly drift of the continents would need to be 10,000 million times as powerful as it is at present to produce the required effects, and, if it had such a value, it would stop the earth's rotation completely in a year' (S. W. Wooldridge and R.S. Morgan, 1959, p. 40). Similarly, the differential gravitational force and the force of buoyancy are also not adequate to cause equatorward movement of the continents, instead the force, if so enormous, might have caused the concentration of the continents near the equator.

(2) Wegener has described several contrasting view points. Initially, sialic masses (continents) were considered by Wegener as freely floating over 'sima' without any friction offered by 'sima' but in later part of his theory he has described forceful resistance offered by 'sima' in the free movement of sialic continents to explain the origin of mountains along the frontal edges of floating continents. Moreover, 'it is difficult to show how the sial blocks, in their passage through the sima, would crumple at their frontal edges and produce mountains' (J.A. Steers, 1961, p. 195). According to Wills no compression could be possible to form the

Rockies and the Andes if the 'sima' is more rigid than the 'sial'. Bowie has maintained that sima has no strength to crumple sial to form mountains.

(3) Both the coasts of the Atlantic Ocean cannot be completely refitted. Thus, the concept of juxtaposition' or 'jig-saw fit' cannot be validated.

(4) Wegener has not elaborated the direction and chronological sequence of the displacement of the continents. He did not describe the situations of pre-Carboniferous times. Many questions remain unanswered such as, What kept Pangaea together till its disruption in Mesozoic era?' Why did the process of continental drift not start before Mesozoic era ? etc. Some writers argue that 'it is not a fair criticism to say that any pre-Carboniferous mountain building cannot be explained on Wegener's hypothesis merely because he does not develop his scheme in earlier geological times' (J.A. Steers, 1961, pp. 161-161).

It may be concluded that 'even if all the matter of his theory is wrong, geologists and others can but remember that it is largely to him that we owe our more recent views on world tectonics' (J.A. Steers, 1961, p. 174). Though most point of Wegener's theory was rejected but its central theme of horizontal displacement was retained. In fact, the postulation of plate tectonic theory after 1960 is the result of this continental drift theory of Wegener. Wegener is, thus, given credit to have started thinking in this precarious field.

Theory of Plate Tectonic

The rigid lithospheric slabs or rigid and solid crustal layers are tectonically called 'plates'. The whole mechanism of the evolution, nature and motion of plates and resultant reactions is called 'plate tectonics'. In other words, the whole process of plate motions is referred to as plate tectonics. 'Moving over the weak asthenosphere, individual lithospheric plates glide slowly over the surface of the globe; much as a pack of ice of the Arctic Ocean drifts under the dragging force of currents and winds' (A.N. Strahler and A.H. Strahler, 1978, p. 373). Plate tectonic theory, a great scientific achievement of the decade of 1960s, is based on two major scientific concepts e.g. (i) the concept of continental drift and (ii) the concept of sea-floor spreading. Lithosphere is internally made of rigid plates. Six major and 20 minor plates have been identified so far (Eurasian plate, Indian-Australian plate, American plate, Pacific Plate, African plate and Antarctic plate.

It may be mentioned that the term 'plate' was first used by Canadian geophysicist J.T. Wilson in 1965. Mckenzie and Parker discussed in detail the mechanism of plate motions on the basis of Euler's geometrical theorem in 1967. They postulated 'a paving stone' hypothesis wherein the oceanic crust was considered to be newly formed at mid-oceanic ridges and destroyed at the trenches. Isacks and Sykes confirmed the 'paving stone hypothesis' in 1967. W.J. Morgan and Le Pichon elaborated the various aspects of plate tectonics in 1968. Now the continental drift and displacement are considered a reality on the basis of plate tectonics.

It may be highlighted that tectonically plate boundaries or plate margins are most important because all tectonic activities occur along the palte margins e.g. seismic events, vulcanicity, mountain buiiding, faulting etc. Thus, the detailed study of plate margins is not only desirable but is also necessary. Plate margins are generally divided into three groups, as follows:

(1) *Constructive Plate Margins :* These are also called as 'divergent plate margins' or 'accreting plate margins'. Constructive plate margins (boundaries) represent zones of divergence where there is continuous upwelling of molten material (lava) and thus new oceanic crust is continuously formed. In fact, oceanic plates split apart along the mid-oceanic ridges and move in opposite directions.

(2) *Destructive Plate Margins :* These are also called as 'consuming plate margins' or 'convergent plate margins' because two plates move towards each other or two plates converge along a line and leading edge of one plate overrides the other plate and the overridden plate is subducted or thrust into the mantle and thus part of the crust (plate) is lost in the mantle.

(3) Conservative Plate Margins are also called as shear plate margins. Here, two plates pass or slide past one another along transform faults and thus crust is neither created nor destroyed.

H. Hess postulated the concept of 'plate tectonics' in 1960 in support of continental drift. The continents and oceans move with the movement of these plates. The present shape and arrangement of the continents and ocean basins could be attained because of continuous relative movement of different plates of the second Pangaea since Carboniferous period. Plate tectonic theory is based on the evidences of (1) sea-floor spreading and (ii) palaeomagnetism.

Sea-Floor Spreading : The concept of sea floor spreading was first propounded by professor Hary Hess of the Princeton University in the year 1960. His concept was based on the research findings of numerous marine geologists, geochemists and geophysicists. Mason of the Scripps Institute of Oceanography obtained significant information about the magnetism of the rocks of sea-floor of the Pacific Ocean with the help of magnetometer. Later on he surveyed a long stretch of the sea-floor of the Pacific Ocean from Mexico to British Columbia along the western coast of North America. When the data of magnetic anomalies obtained during the aforesaid survey were displayed on a chart, there emerged well defined patterns of stripes. Based on these information Hary Hess propounded that the mid-oceanic ridges were situated on the rising thermal convection currents coming up from the mantle. The oceanic crust moves in opposite directions from mid-oceanic ridges. These molten lavas cool down and solidify to form new crust along the trailing ends of divergent plates (oceanic crust). Thus, there is continuous creation of new crust along the mid-oceanic ridges and the expanding crusts (plates) are destroyed along the oceanic trenches. These facts prove that the continents and ocean basins are in constant motion.

W.G. Vine and Mattheus conducted the magnetic survey of the central part of Carlsberg Ridge in the Indian Ocean in 1963 and computed the magnetic profiles on the basis of general magnetism. When he compared the computed magnetic profiles with the profiles of magnetic anomalies plotted on the basis of actual data obtained during the survey, he found sizeable difference between the two profiles. When he plotted the magnetic profiles on the basis of alternate bands of normal and reverse magnetism in separate stripes of 20 km width on either side of the ridge, he found complete parallelism between the computed profiles and observed profiles.

Vine and Mattheus have opined on the basis of the evidences of temporal reversal in the geomagnetic field and the concept of sea-floor spreading as propounded by Deitz and Hess that when molten hot lavas come up with the rising thermal convection current along the mid-oceanic ridges and get cooled and solidified, these (lavas) also get magnetized, at the same time, in accordance with the then geomagnetic field and thus alternate bands or stripes of magnetic anomalies are formed on either side of the mid-oceanic ridge. In other words, when molten lavas are upwelled along the mid-oceanic ridges, these divide the earlier basaltic layer into two equal halves and these basaltic layers slide horizontally on either side of the mid-oceanic ridges. The findings of Cox, Doell and Dalrympal (1964), Opdyke (1966) and Heritzler (1966) have validated the following facts - (i) there is reversal in the main magnetic field of the earth (known as geocentric dipole magnetic field), (ii) normal and reverse magnetic amomalies are found in alternate manner on either side of the mid-oceanic ridges, (iii) there is complete parallelism in the magnetic anomalies on either side of the mid-oceanic ridges and (iv) there is parallelism in the time sequence of palaeomagnetic epochs and events calculated for 4.5 million years on the basis of magnetism of basaltic rocks or sedimentary rocks.

It may be concluded, on the basis of above discussion, that there is continuous spreading of seafloor. New basaltic crust is continuously formed along the mid-oceanic ridges. The newly formed basaltic layer is divided into two equal halves and is thus displaced away from the mid-oceanic ridge. Alternate stripes of positive and negative magnetic anomalies are found on either side of the mid-oceanic ridges. Such magnetic anomalies (positive and negative) 'are formed because of temporal reversal in the geomagnetic field. The rocks formed during reverse polarity (reversed geomagnetic field) denote negative magnetic anomalies'.

The age of magnetic stripes, the rate of sea-floor spreading and the time of drifting of different continents are calculated on the basis of above facts. The dating of the magnetic stripes formed upto 4.5 million years before present has been completed on the basis of information obtained from the survey of palaeomagnetism of the sea-floors of different oceans.

The rate of sea-floor spreading is calculated on two bases e.g. (i) on the basis of the age of isochrons (isochrons are those lines which join the points of equal dates of the magnetic stripes plotted on the map) and (ii) on the basis of distance between two isochrons. Thus, the rates of spreading (drifting) of different oceans have been determined on the basis of above principles. The Atlantic and Indian Oceans are spreading (expanding) very sluggishly i.e. at the rate of 1.0 to 1.5 cm per year while the Pacific Ocean is expanding at the rate of 6.0 cm per year. It may be pointed out that the rate of seafloor spreading always means the rate of expansion only on one side of the mid-oceanic ridge. For example, if the rate of sea-floor is reported to be 1.0 cm per year, the total spreading of the concerned ocean would be 1 + 1 = 2 cm per year. The recent studies have shown that (i) the maximum spreading of the Pacific Ocean is 6 to 9 cm per year (total expansion 12 to 18 cm/year) along the eastern Pacific ridge between equator and 30°S latitude, (ii) the southern Atlantic Ocean is spreading along the southern Atlantic ridge at the rate of 2 cm per year (total expansion 4 cm/year) and (iii) the Indian Ocean is expanding at the rate of 1.5 to 3 cm per year (total expansion being 3 to 6 cm/ year).

Continental Displacement and Plate Tectonics : On the basis of the evidences of palaeomagnetism and sea-floor spreading it has been now validated that the continents and ocean basins have never been stationary or permanent at their places rather these have always been mobile throughout the geological history of the earth and they are still moving in relation to each other. The scientists have discovered ample evidences to demonstrate the opening and closing of ocean basins. For example, the Mediterranean Sea is the residual of once very vast ocean (Tethys Sea) and the Pacific Ocean is continuously contracting because of gradual subduction of American Plate along its ridge. On the other hand, the Atlantic Ocean is continuously expanding for the last 200 million years, Red Sea has started to open (to expand). It may be mentioned that continental masses come closer to each other when the oceans begin to close while continents are displaced away when the oceans begin to open (expand).

Though the sequence of events of continental displacement based on the evidences of palaeomagnetism and sea-floor spreading is available only for the last 200 million years but on the basis of general mechanism of plate tectonics and the evidences from the continents the sequence of earlier events may be reconstructed. Valentine and Moors (1970) and Hallam (1972) have attempted to reconstruct the chronological sequence of the continents and ocean basins from the beginning to the present time. About 700 million years ago all the landmasses were united together in the form of one single giant landmass known as 'Pangaea I'. About 600-500 million years before present first Pangaea was broken because of thermal convective currents coming from within the earth, most probably from the mantle and different landmasses drifted apart. These landmasses were again united together due to plate motions in one land mass known as 'Pangaea II' about 300-200 million years before present. According to A. Hallam Second Pangaea began to break during early Jurassic period and N. W. Africa broke away from N. America and drifted away. The zone of sea-floor spreading continued to extend towards north and south. The separation of South America and Africa was accomplished during middle Cretaceous period, and North America and Europe began to move away from each other.

The opening of North Atlantic was accomplished in many phases. After the separation of North America from Africa, Europe and Greenland broke away from Labrador during late Cretaceous period (about 80 million years before present) and thus Labrador sea was formed. This newly formed sea continued to remain for some time as northern extension of the Atlantic Ocean. Rockall plateau was separated from Greenland during Tertiary period (about 60 million years before present). Labrador Sea and North Atlantic continued to expand between Europe and Greenland upto middle Miocene period because the European and American plates continued to move eastward and westward respectively. The spreading of Labrador Sea stopped by middle Miocene period (about 47 million years before present) but North Atlantic continued to expand.

Indian Ocean did not exist before Cretaceous period. Indian plate began to move towards Asiatic plate through 'Tethys Sea' and Australian-Antarctic plates after breaking away from African plate began to move southward during Cretaceous period. Dan Mackenzie and John Sclater have presented the chronological sequence of the evolution of Indian Ocean on the basis of the study of magnetic anomalies. According to them Indian plate began to move northward at the rate of 18 cm per year during early Tertiary period but the movement stopped during Eocene period. At the same time Antarctica broke away from Australia. Thus, the Pacific Ocean began to shrink in size because of expansion of the Atlantic and Indian Oceans. The Atlantic Ocean began to open about 700 million years before present because of breaking of First Pangaea when the American and Africa-European plates began to move in divergent directions and thus the Atlantic continued to expand till 400 million years before present when the Atlantic again began to close. Because of the closing of the Atlantic Ocean Applachian mountains of North America were formed. The Atlantic Ocean again began to open up about 150 million years before present when Second Pangaea was broken into several landmasses and it still continues to expand because of the movement of American and European plates in opposite directions. It may be pointed out that the Atlantic Ocean is continuously expanding for the past 200 million years but the Pacific Ocean is contracting in size because of westward movement of the America.

The following examples demonstrate the trends and patterns of continental displacement, sea-floor spreading and contraction in the size of the oceans.

Red Sea and the Gulf of Aden-Red Sea is an example of axial trough which is located between Africa and Arabian peninsula. The surveyed magnetic anomalies in this area show, as observed by A.W. Girdler, the pattern of stripe and these are similar to the magnetic anomalies of the ocean basins. F.J. Vine calculated the rate of the spreading of the Red Sea on the basis of the data of magnetic anomalies in the year 1966. According to him the Red Sea is spreading at the rate of one centimetre per year (total spreading 2 cm/year) since the past 3-4 million years. Alen and Morelli calculated the spreading rate in 1969 as 1.1 cm/year (total spreading 2.2 cm/year). Similarly, the rate of spreading of the Gulf of Aden has been calculated on the basis of stripped magnetic anomalies as 0.9 to 1.1 cm/year (total spreading 1.8 to 2.2 cm/year). The Red Sea and the Gulf of Aden are located at the junction of three plates viz. Nubian plate, Somali plate and Arabian plate. Nubian and Somali plates are separated by Ethiopian fault.

The Gulf of California : The Pacific Ocean is a waning ocean because it is continuously being contracted in its size because of gradual encroachment of westward moving American plates. It is believed that like mid-Atlantic ridge there might have been a mid-oceanic ridge in the Pacific Ocean but it has now been remarkably deformed due to plate movement. The magnetic survey of the Gulf of California revealed the presence of stripped magnetic anomaly. This situation validates two facts viz. (i) East Pacific Rise (ridge) is also located in the Gulf of California and there has been continuous spreading of the gulf along the ridge since the past four million years and (ii) Baja, the Californian peninsula, was previously united with the mainland of North America but later on it broke away from the continent due to spreading of sea floor.

The Evaluation : It is commonly agreed by the majority of the scientists that plate tectonics has validated the concept of continental drift, rather continental drift has now become a reality. The only point of argument and question is related to the competent force responsible for the drifting of the continents. Most of the scientists still rely on the thermal convective currents coming from the mantle as the probable adequate force to move the plates (continents) in different directions.

4

Plate Boundaries

Plate Tectonics

New concepts and theories based on evidences and interpretation of sea-floor spreading and palaeomagnetic field have been advanced after 1960 in the field of geology, geophysics and geomorphology, of which theory of plate tectonics is most significant. The present chapter deals with various aspects of plate tectonics viz. meaning and concept, palaeomagnetism, sea-floor spreading, plate margins, pate movements and resultant geologic expression such as continental drift, vulcanicity, seismic activity, mountain building etc.

Concept and Meaning

The rigid lithospheric slabs or rigid and solid crustal layers are technically called plates. The study of whole mechanism of evolution, nature and motions of plates, deformation within plates and interactions of plate margins with each other is collectively called as plate tectonics. In other words, the whole process of plate motions and resultant deformations is referred to as plate tectonics. 'Moving over the weak asthenosphere, individual lithospheric plates glide slowly over the surface of the globe; much as a pack of ice of the Arctic Ocean drifts under the dragging force of currents and winds' (A.N. Strahler and A.H. Strahler, 1978).

Plate tectonic theory, a significant scientific advancement of the decade 1960's, is based on two major scientific concepts e.g. (1) the continental drift and (2) the concept of sea-floor spreading. Lithosphere is internally made of rigid plates. Six major and 20 minor plates have been identified so far (Eurasian plate, Indian-Australian plate, American plate, Pacific plate, African plate, and Antarctic plate). It may be mentioned that the term 'plate' was first used by Canadian geophysicist J. Tuzo Wilson in 1965. Mackenzie and Parkar discussed in detail the mechanism of plate motions on the basis of Euler's geometrical theorem in 1967. They postulated a 'paving stone hypothesis' wherein the oceanic crust was considered to be newly formed at mid-oceanic ridges and destroyed at the trenches. Isacks and sykes confirmed the 'paving stone hypothesis' in 1967. W.J. Morgan and Le Pichon elaborated the various aspects of plate tectonics in 1968. It may, thus, be pointed out that the theory of plate tectonics is not related to any individual scientist rather a host of scientists of various scientific disciplines and research groups and expeditions have contributed in the development of this valuable concept of the second half of the 20th century. Now the continental drift and displacement is considered a reality on the basis of plate tectonics.

Plates are classified into 3 types viz. oceanic plates (having oceanic crust), continental plates (having continental crust) and continental-oceanic plates.

Plate Margins

It may be highlighted that tectonically plate boundaries or plate margins are most significant because all tectonic activities occur along the plate margins e.g., seismic events, vulcanicity, mountain building, faulting etc. Thus, the detailed study of plate boundaries is not only desirable but is also necessary. Plate margins are generally divided into three categories as follows.

(1) Constructive plate margins are also called as 'divergent plate boundaries' or 'accreting plate boundaries'. It may be mentioned that a distinction may be drawn between plate margins and plate boundaries e.g., plate margin represents marginal part of the plate whereas plate boundary represents 'surface trace of the zone of motion between two plates.' Constructive plate boundaries represent zones of divergence where there is continuous upwelling of molten material (lava) and thus new oceanic crust is continuously formed. Oceanic plates split apart along the mid-oceanic ridges and move in opposite directions. Divergent plate margins are constructive in the sense that there is continuous formation of new crust along these margins because of cooling and solidification of basaltic lava which comes up as magma due to rifting of plates along the mid-oceanic ridges. Divergent movement of plates (i.e. movement of two plates in opposite directions) results in (i) volcanic activity of fissure flow of basaltic magma, (ii) creation of new oceanic crusts, (iii) formation of submarine mountain ridges and rises, (iv) creation of transform faults, (v) occurrence of shallow focus earthquakes, (vi) drifting of oceanic plates etc.

(2) Destructive plate margins are also termed as 'convergent plate boundaries' or 'consuming plate margins' because two plates move towards each other (face to face) or two plates converge along a line and collide wherein leading edge of one plate (of relatively lighter material) overrides the other plate (of relatively denser material) and the overridden plate is subducted or thrust into upper mantle and thus a part of the crust (plate) is lost in the mantle, this is why convergent plate margins are called destructive margins. The zone of collision of convergent plates is also called as 'collision zone', 'subduction zone' and 'Benioff zone' (after the scientist Hugo Benioff). Convergence, collision and resultant subduction of heavier plate margin under lighter plate margin results in (i) occurrence of explosive type of volcanic eruptions, (ii) deep focii earthquakes, (iii) formation of folded mountains, island arcs and festoons, oceanic trenches etc.

Plate collisions are of three types viz. (i) ocean-ocean collission (collision of two oceanic plates), (ii) continent-continent collision (collision of two continental plates) and (iii) ocean-continent collision (collision of oceanic and continental plates). Ocean-ocean collision involves collision of two convergent plates having oceanic crusts where one oceanic crust having relatively denser material is subducted into upper mantle. Such collision and subduction occurs along east Asia and the resultant tectonic expression of plate collision and subduction includes deformation in crustal area, vulcanism, metamorphism, formation of oceanic trenches, island arcs and festoons etc., and occurrence of earthquakes. Ocean-continent collision involves collision of one oceanic plate having oceanic crust and other one of continental plate having continental crust along Benioff zone (subduction zone) and the resultant tectonic expressions are deformation of crustal rocks, metamorphism, volcanic eruptions, formation of folded mountains and occurrence of deep-focus earthquakes. Collision of American and Pacific plates is a typical example of this category and formation of majestic western cordillera of N. America and Andes of S. America is significant resultant tectonic expression of such situation. It may be mentioned that one of the manifestations of continent-oceanic plate collision is the exposure of deep ocean rocks through their thrusting in resultant mountain masses. This process is called obduction which is opposite to subduction as the former implies thrusting up while the latter means thrusting down.

Continent-continent collision involves collision of two continental plates along Benioff zone and is responsible for creation of folded mountains and occurrences of earthquakes of varying magnitudes. The collision of Asiatic-Indian plates, and European-African plates is typical example of such situation and the formations of Alpine and Himalayan mountainous chains are major manifestations

(3) Conservative plate margins are also called as shear plate margins and parallel/

transform fault boundaries where two plates pass or slide past each other along transform faults. These are called conservative because crust is neither created nor destroyed. The significant tectonic expression of such situation is the creation of transform faults which move, on an average, parallel to the direction of plate motion. Transform faults off-set mid-oceanic ridges. Besides oceanic transform faults, there are also continental transform faults e.g. San Andreas fault (California, USA), Alpine fault (Africa) etc. It may be mentioned that San Andreas fault 'is ridge to ridge transform fault.' The other manifestations of conservative plate margins include no volcanic activity, seismic events, creation of ridge and valley, fracture zone etc.

Palaeomagnetism

Palaeomagnetism refers to the preservation of magnetic properties in the older rocks of the earth. It may be mentioned that when any rock, whether sedimentary or igneous, is formed it gets magnetised depending on the presence of iron content in the rock and is preserved (frozen at temperature below Curie point, which is generally 600°C). It was the year 1600 A.D. when William Gilbert, the physician of Queen Elizabeth, postulated that the earth behaved like a giant magnet and magnetism of the earth was produced in the inner part of the earth. The magnetic field of the earth is like a giant bar magnet of dipoles, located in the centre (core) of the earth and is aligned approximately along the axis of rotation of the earth. When the long axis of dipole bar magnet is extended it intersects the earth's surface at two centres which are called north and south magnetic poles. It may be pointed out that magnetic south pole of the earth is near its (earth's) geographical north pole and vice-versa (i.e. magnetic north pole is located near geographical south pole). If an ordinary small magnet is freely suspended at the earth's surface then the earth's south magnetic pole attracts north pole of small magnet and earth's north magnetic pole attracts south pole of small magnet. It may be clarified that as per general rule when two magnets are brought together, then their similar poles repel each other but opposite poles attract each other.

A freely suspended magnet on the earth's surface does not indicate geographical north and south perfectly because the axis of magnetic north and south poles is not perfectly aligned along the axis of geographical north and south poles. This causes angular inclination between the magnetic and geographical axes. This angular inclination is called magnetic declination which, in fact, denotes angular inclination between the direction of freely suspended magnet at any part of the earth's surface and the direction of earth's geographical north-south pole axis. On the other hand, angular inclination between freely suspended magnetic needle and horizontal plane of the earth's surface is called magnetic inclination or magnetic dip. If a magnetic needle is freely suspended at the north pole of the earth, the north pole of the magnet being closer to the south magnetic pole of the earth (which is, in fact, near geographical north pole) would be attracted more and magnetic needle becomes perpendicular. Consequently, north pole of the suspended magnetic needle dips downward vertically. The situation is reversed in the southern hemisphere. Thus, magnetic dip becomes 90° on geographical north and south poles of the earth. Magnetic dip becomes zero wherever freely suspended magnetic needle becomes horizontal at the earth's surface. The imaginary line joining places of zero magnetic dip angle is called magnetic equator. The magnetic dip angle increases poleward. It may be pointed out that there may be spatial and temporal variation in the intensity of simple dipole magnetic field.

Geomagnetic Field—Origin and Sources : The origin of geomagnetic field is in no case related to mantle rather it is related to the outer core of the earth because of the fact that there is gradual westward migration of geomagnetic field at the rate of 0.18° per year which proves that the rotation of geomagnetic field is slower than the rotation of the earth. This indirectly proves that the core of the earth rotates at slower rate than the overlying mantle. It may be stated that 'the magnetic field cannot be a permanent property of the material of the core...... must therefore be continuously produced and maintained' (A. and Doris L. Holmes, 1978). If permanent geomagnetic field is not possible then the continuous production and maintenance of geomagnetic field may be possible only when there would be presence of materials of high electrical conductivity in the core so that electrical currents may be generated. It is further pointed out that the generation of electrical currents is possible only in metallic liquid materials and such situation is found in the outer core of the earth which functions as self exciting dynamo. Thus, the energy coming out of the core is transformed into electrical currents which in association with metallic liquid substances produce geocentric dipole magnetic field.

The required energy to maintain geomagnetic field is believed to come from three possible sources:

(1) heat energy released from the disintegration of radioactive elements of the core of the earth. It is argued that this source of energy for the generation of convective currents (electrical currents) may not be possible because if we accept this proposition then difficulty arises in the process of cooling of the crust of the earth because such situation (generation of heat energy from radioactive elements) would also prevail in the mantle and hence the crust cannot cool because there would be constant supply of heat energy from below (from the mantle).

(2) The downward transfer of ferromagnesian materials from mantle into core results in the release of gravity force in the core which in turn produces energy.

(3) The movement of materials from inner core to the outer core results in the heating of outer core through heat energy released from inner core.

Irregular Magnetic Field : The geocentric axial dipole magnetic field represents 95 per cent of the earth's total magnetism. The remaining portion is represented by irregular, scattered and weak magnetic fields. It may be pointed out that there is no such giant bar magnet inside the earth but there is more concentration of magnetism in the rocks of the core of the earth in the shape of a bar magnet. The hot and liquid lava and magma with high ferromagnesian contents, when cooled and solidified to form igneous rocks, get magnetised, the records of which are preserved in the rocks. Such magnetism preserved (frozen) in the rocks are called remanent or palaeomagnetism. It is to be remembered that the newly formed rocks are magnetised in the direction of existing geomagnetic field, and thus the magnetic inclination/dip of newly formed rocks is the same as that of the geomagnetic field at the time of the formation of said igneous rocks. Thus, it is evident that the orientation and magnetic inclination of palaeomagnetism preserved in the rocks is always in accordance with the prevailing magnetic inclination of geomagnetic field. The intensity of such palaeomagnetism/remanent magnetism depends on the composition of minerals of lava and magma at the time of cooling and solidification and on the intensity of geomagnetic field of that period (when the concerned igneous rocks were formed). Similarly, sedimentary rocks, at the time of their formation, are also magnetised, the intensity of which depends on the amount of ferromagnesian minerals present therein. Sometimes, the magnetism (weak) of sedimentary rocks is destroyed due to chemical changes. Remanent magnetism preserved in the rocks is recorded with the help of galvanometer.

Reconstruction

The reconstruction of palaeomagnetism involves the collection of rock samples of the same age from different places and determination and recording of their orientation. It may be pointed out that some changes may take place in the original orientation of magnetism due to tectonic events. Any way, after the determination of orientation of palaeomagnetism, the magnitude, declination and inclination of local force are measured with the help of magnetometer. It is assumed that generally at the time of magnetisation of rocks (palaeomagnetism) the geomagnetic field is dipolar in shape and there is approximate coincidence between average geomagnetic field (average, because it varies temporally) and contemporary geographical poles. Based on this assumption average palaeomagnetic inclination/dip of rocks of a certain place and of a certain time is determined, on the basis of which the latitude of that place existing at that time is determined on the basis of the following equation

$$\tan I = 2 \tan \lambda$$

when I = magnetic inclination

λ = latitude

Thus, the latitude, so determined helps in determining the distance of poles and the direction of poles is determined on the basis of palaeomagnetic declination (D). On the basis of distance and direction of geographical poles from the selected place (from where the rock samples are collected) the position of poles of the globe, at the time of the formation of the sample rocks, is determined. There may be some errors in the aforesaid process of determination of the position of the globe. viz. (i) at the time of palaeomagnetic reconstruction the impact of only geomagnetic field is considered while minor magnetic fields are ignored ; (ii) sampled rocks might have experienced magnetic changes; (iii) some errors may crop up at the time of orientation etc. In order to remove these errors several rock samples of same age are collected and the position of poles is determined after the study of their palaeomagnetism and calculation of average value on the basis of statistical methods.

Based on the above method the positions of poles were determined in Japan, Italy, France etc., on the basis of palaeomagnetic reconstruction of Cenozoic lavas. Blackett and his associates determined the position of poles before 200 million years in British Isles on the basis of palaeomagnetic reconstruction of sandstones. The study revealed considerable changes in the positions of poles in the past. This study, thus, revealed the fact, 'that magnetic poles have changed their positions and there has been considerable wandering in the position of poles.' On the basis of this revelation two inferences may be drawn :

(1) The poles must have changed their positions and the continents and ocean basins might have remained stationary at their places throughout geological time.

(2) Polar wandering has occurred due to continental drift i.e., continents changed their relative positions while magnetic poles remained stationary.

Polar wandering curves are prepared for different continents on the basis of data derived through palaeomagnetic reconstruction. As per rule if there has not been continental drift, then the polar wandering curves of different continents at a certain time period (same time for all the continents) shall be the same, but if the continental drift has occurred then these polar wandering curves would be different for each continent. The magnetic polar wandering curves, when plotted for different continents for same period, differ from each other. This clearly shows that poles have not changed their positions rather continents have changed their positions. Thus, it is concluded that 'the concepts of permanency of continents and ocean basins and polar wandering stand automatically rejected and continental displacement and drift becomes a reality.' It is, thus, validated that if the relative positions of continents have changed, the position of magnetic pole determined on the basis of contemporary rocks of a continent would differ from the position of magnetic pole (of same period) of the other continents. It may be further elaborated. So long as two continents are joined together or are not drifting in relation to one another, the magnetic polar wandering curves for same period would be the same for both the continents. According to A.G. Wegener all the continents were joined together in the form of Pangaea till late Permian period. If this was so, then there should be only one palaeomagnetic pole for all the continents during Palaeozoic era. This inference became true when the palaeomagnetic polar wandering curve was prepared for Palaeozoic Pangaea by joining all the present day continents together so as to conceive the situation in Palaeozoic era.

It is, thus, finally proved that based on polar wandering curves of different periods for different continents on the basis of data derived from palaeomagnetic reconstruction not only the concept of continental drift is validated but the mechanism of disruption of Wegener's Pangaea, separation of different continents and their displacement is also validated.

Polarity Reversal

The study of palaeomagnetism also revealed that magnetization of some rocks was not conformal to the geomagnetic field i.e. the rocks were magnetized in opposite direction of main geomagnetic field. It was further substantiated during the decade 195060 that the occurrence of reversely magnetized rocks was not rare phenomenon rather it was universal phenomenon. The available data of palaeomagnetism reveals the fact that about 50 per cent of the rocks of the crust have got magnetized in opposite direction to the geomagnetic field. There may be two possibilities in this regard

(1) At the time of magnetization of rocks at given time period some rocks might have been magnetized in opposite direction to the geomagnetic field or initially all the rocks were magnetized in the direction of geomagnetic field but at a later date the direction of some rocks might have changed and hence opposite direction of palaeomagnetism of rocks might have become possible. This mechanism of reversal of polarity is called self reversal.

(2) Alternatively, originally the magnetization of reversely magnetized rocks might have taken place in the direction of geomagnetic field but at a later date there might have been reversal in the direction of geomagnetic field itself. This mechanism of reversal of polarity is called geomagnetic field reversal.

The first possibility of reversal of polarity i.e., self reversal of polarity, as referred to above, could not be substantiated on the basis of available field data though Neel suggested a few theoretical possibilities to validate self reversal. Most of the scientists are of the opinion that terrestrial rocks are magnetized always in the direction of geomagnetic field, but there is reversal in the direction of geomagnetic field, i.e., north-south direction of geomagnetic field after certain time becomes south-north. For example, if the geomagnetic field is in normal direction (north-south), all the rocks of all the continents formed at that time are magnetized in normal direction but when the normal direction of geomagnetic field gets reversed (south-north), all the rocks of all the continents at that time (during reversed direction of geomagnetic field) are magnetized again in the direction of geomagnetic field but this time the direction of magnetism of rocks is opposite to the direction of previously formed and magnetized rocks because now the direction of geomagnetic field has got reversed itself. It is generally believed that field reversal occurs at regular interval of time.

Scientists have measured magnetic polarity of rocks upto 4.5 million years which denotes definite and perfect time sequence. The rocks formed at the same time period in all the continents denote same polarity. Polarity events within different geomagnetic polarity epochs have been named after the place where remanent magnetism (palaeomagnetism) was studied first.

Concept of Hary Hess : The concept of sea-floor spreading was first propounded by Prof. Hary Hess of the Princeton university in the year 1960. His concept was based on the research findings of a large number of marine geologists, geochemists, geophysicists etc. Masson of the Scripps Institute of Oceanography obtained significant information about the magnetism of the rocks of sea-floor of the Pacific Ocean with the help of magnetometer. Later on he surveyed a long stretch of the sea-floor of the Pacific Ocean from Mexico to British Columbia along the western coast of North America. When the data of magnetic anomalies obtained during the aforesaid survey were displayed on a chart, there emerged well defined patterns of stripes. Based on these information Hary Hess propounded that the mid-oceanic ridges were situated on the rising thermal convection currents coming up from the mantle. The oceanic crust moves in opposite directions from mid-oceanic ridges and thus there is continuous upwelling of new molten materials (lavas) along the mid-oceanic ridges. These molten lavas cool down and solidify to form new crust along the trailing ends of divergent plates (oceanic crust). Thus, there is continuous creation of new crust along the mid-oceanic ridges. This, according to Hess, proves the fact that sea-floor spreads along the mid-oceanic ridges and the expanding crusts (plates) are destroyed along the oceanic trenches. These facts prove that the continents and ocean basins are in constant motion.

W. G. Vine and Mattheus conducted the magnetic survey of the central part of Carlsberg Ridge in Indian Ocean in 1963 and computed the magnetic profiles on the basis of general magnetism. When they compared the computed magnetic profiles with the profiles of magnetic anomalies plotted on the basis of actual data obtained during the survey, they found sizeable difference between the two profiles. When they plotted the magnetic profiles on the basis of alternate bands of normal and reverse magnetism in separate stripes of 20 km width on either side of the ridge, they found complete parallelism between the computed profiles and observed profiles.

Vine and Mattheus have opined on the basis of the evidences of temporal reversal in the geomagnetic field and the concept of sea-floor spreading as propounded by Deitz and Hess that when molten hot lavas come up with the rising thermal convection currents along the mid-oceanic ridges and get cooled and solidified, these also get magnetized at the same time, in accordance with the then geomagnetic field and thus alternate bands or stripes of magnetic anomalies are formed on either side of the mid-oceanic ridge. In other words, when molten lavas are upwelled along the mid-oceanic ridges, these divide the earlier basaltic layer into two equal halves and these basaltic layers slide horizontally on either side of the mid-oceanic ridges. The findings of Cox, Doell and Dalrympal (1964), Opdyke et. al (1966) and Heritzler (1966) have validated the following facts-

(i) there is reversal in the main geomagnetic field of the earth (known as geocentric dipole magnetic field),

(ii) normal and reverse magnetic amomalies are found in alternate manner on either side of the mid-oceanic ridges,

(iii) there is complete parallelism in the magnetic anomalies on either side of the mid-oceanic ridges and

(iv) there is parallelism in the time sequence of palaeomagnetic epochs and events calculated for 4.5 million years on the basis of magnetism of basaltic rocks or sedimentary rocks.

It may be concluded, on the basis of above discussion, that there is continuous spreading of seafloor. New basaltic crust is continuously formed along the mid-oceanic ridges. The newly formed basaltic layer is divided into two equal halves and is thus displaced away from the mid-oceanic ridge. Alternate stripes of positive and negative magnetic anomalies are found on either side of the mid-oceanic ridges. Such magnetic anomalies (positive and negative) are formed because of temporal reversal in the geomagnetic field. The rocks formed during normal geomagnetic field contain positive magnetic anomalies while the rocks formed during reverse polarity (reversed geomagnetic field) denote negative magnetic anomalies.

The age of magnetic stripes, the rate of seafloor spreading and the time of drifting of different continents are calculated on the basis of above facts. The dating of the magnetic stripes formed upto 4.5 million years before present has been completed on the basis of information obtained from the survey of palaeomagnetism of the sea-floors of different oceans. The rate of sea-floor spreading is calculated on two bases e.g. (i) on the basis of the age of isochrons (isochrons are those lines which join the points of equal dates of the magnetic stripes plotted on the map) and (ii) on the basis of distance between two isochrons. Thus, the rates of spreading (drifting) of different oceans have been determined on the basis of above principles. The Atlantic and Indian Oceans are spreading (expanding) very sluggishly i.e. at the rate of 1.0 to 1.5 cm per year while the Pacific Ocean is expanding at the rate of 6.0 cm per year. It may be pointed out that the rate of seafloor spreading always means the rate of expansion only on one side of the mid-oceanic ridge. For example, if the rate of seafloor spreading is reported to be 1.0 cm per year, the total spreading of the concerned ocean would be 1+1=2 cm per year. The recent studies have shown that the maximum spreading of the Pacific Ocean is 6 to 9 cm per year (total expansion 12 to 18 cm/year) along the eastern Pacific ridge between equator and 30° S latitude, (ii) the southern Atlantic Ocean is spreading along the southern Atlantic ridge at the rate of 2 cm per year (total expansion 4 cm/year) and (iii) the Indian Ocean is expanding at the rate of 1.5 to 3 cm per year (total expansion being 3 to 6 cm/year).

Motion of Plates

All lithospheric plates constantly move with respect to each other with varying rates. Plate motions are currently measured and monitored by using satellites and lasers. It may be mentioned that each plate moves as a single unit having relatively little changes in its middle part. Only the plate margins undergo changes. It is also to be stated that the plate motion is relative with respect to other plate i.e., any change in rate or direction of motion in one plate causes corresponding changes in other plates. If the plates are rigid blocks and move on the surface of the spherical earth, their motion can be explained in terms of Euler's geometrical theorem.

'Euler's geometrical theorem shows that every displacement of a plate from one position to another on the surface of a sphere can be regarded as a simple rotation of the plate about a suitable chosen axis passing through the centre of the sphere' (E.R. Oxburgh, 1979). The rotation axis of plates passes through the centre of the globe. 'All points on the plate travel along small circle paths about the chosen axis (of rotation) in passing from their initial to final positions. It follows that any plate boundary which is conservative (i.e. involves neither plate growth nor destruction) must be parallel to small circle, the axis of which is the axis of rotation for the relative motion' (E.R. Oxburgh, 1979). On the other hand, the margin of the plate, which is not parallel to small circle, becomes either constructive (accreting) or destructive (consuming) plate margin.

'A corollary of Euler's theorem is that the velocity of relative motion across a constructive or destructive boundary is proportional both to angular velocity about the axis of rotation for the motion of the plates, and to the angular distance of the point on the boundary under consideration from the axis of rotation. It implies that velocities vary continuously along all constructive and destructive boundaries, being smallest in 'high rotational latitudes' and greatest in 'low rotational latitudes' (E.R. Oxburgh, 1979).

W.J. Morgan has successfully explained the spreading of equatorial Atlantic and plate movement on the basis of Euler's geometrical theorem. It may be mentioned that mid-Atlantic ridge crest is displaced on either side along numerous transform faults running in almost east-west direction, 'whose importance here is that they represent conservative sectors of plate boundary separating constructive sectors-the spreading parts of the ridge (mid-Atlantic ridge). All active transform faults on the same ridge ought to be segments of co-axial small circles if the plate model is valid' (E.R. Oxburgh, 1979). According to plate model with reference to plate motion based on Euler's geometrical theorem all the great circles, if drawn, must intersect at a single point which would be the pole of rotation. When W.J. Morgan constructed great circles normal to the strike of transform faults of the equatorial Atlantic Ocean, he found that all the great circles except one intersected at one common point P at 57.5°N and 36.5°W. This result thus validated the mechanism of plate motion on the basis of Euler's geometrical theorem. It maybe clarified that the surface area of the earth does not increase due to plate movement rather it remains constant because the addition (accretion) of new basaltic crust at the constructive plate margins along mid-oceanic ridges and consequent sea-floor spreading is suitably compensated by loss of crust due to subduction along the converging (consuming) plate boundaries.

Various Causes

Several mechanisms, sources and possible causes of plate motion (movement) have been suggested by scientists but none of them could be fully substantiated till now due to lack of convincing evidences. A majority of scientists consider thermal convective currents inside the earth as possible driving force for the movement of plates. It may be pointed out that A. Holmes postulated the concept of rising thermal convection currents from within the earth in 1928. The mechanism of thermal convective currents in liquid matter was theoretically studied by Lord Rayleigh. Currently, a host of scientists have accorded acceptance to the mechanism of thermal convective currents on the basis of thermal and pressure conditions of the interior of the earth. I.G. Gass has validated the mechanism of origin and movement of unstable thermal convection currents in the mantle (below earth's crust). According to Gass the viscosity of mantle depends entirely on temperature and pressure. The viscosity of ascending materials caused by upward movement of thermal convective currents due to high temperature decreases and hence the upward flow velocity of the matter increases. The centres of upward movement of hot and liquid matter with ascending currents are generally located below the mid-oceanic ridges. Though the depth of such centres is not correctly known but it is believed that these are located at an average depth of 300-400 km from the earth's surface. The rising convection currents transport hot and liquid matter upward which after reaching the point just below the crust (plates) split and diverge in opposite directions in the form of horizontal flow which is confined to the depth upto 200 km. Thus, the divergence of convection currents (just below the mid-oceanic ridges) with hot and molten matter causes plate movement in opposite directions. On the other hand, two sets of converging thermal convection currents bring two plates together and the plate margins are subducted.

Some scientists are of the view that plate motion is caused due to high gravity force because of creation of additional matter (lava and magma) on either side of the mid-oceanic ridges. This high gravity force causes lateral movement of plates in opposite direction from the ridge crests. According to another view the intrusion of magma in the mid-oceanic ridges from below causes separation of oceanic plates from ridge crests and their displacement in opposite direction.

It may be concluded that this is the weak point (lack of commonly accepted cause of plate motion) in the theory of plate tectonics which is vulnerable to severe criticism. This problem of real cause of plate motion needs scientific solution.

Continental Drift and the Plate Tectonics : On the basis of the evidences of reconstruction of palaeomagnetism and sea-floor spreading it has now been validated that the continents and ocean basins have never been stationary or permanent at their places rather these have always been mobile throughout the geological history of the earth and they are still moving in relation to each other. The scientists have discovered ample evidences to demonstrate the opening and closing of ocean basins. For example, the Mediterranean Sea is the residual of once very vast ocean (Tethys Sea) and the Pacific Ocean is continuously contracting because of gradual subduction of American plate along its ridges. In nutshell it may be opined that continental drift has now become a reality on the basis of plate tectonics.

Mountain Building and the Plate Tectonics : Plate tectonic theory has enabled scientists to explain the problem of origin of folded mountains which was hitherto unresolved till the postulation of this great scientific theory in the decade 1960-70. It may be pointed out that several hypotheses have been propounded to solve this gigantic geological problem (e.g. thermal contraction hypothesis by Jeffreys, continental drift theory by F.B.Taylor and A.G. Wegener, thermal convection current hypothesis by A. Holmes, sliding continent hypothesis by Daly, radioactivity hypothesis by Joly etc) from time to time but none of them could be universally accepted because the exponents of these hypotheses could not present convincing scientific evidences in support of their respective hypotheses. Now, the plate tectonic theory offers convincing explanation for the solution of complex riddle of mountain building.

Vulcanicity and the Plate Tectonics : Based on plate tectonics there is close relationship between plate boundaries and vulcanicity as most of the world's active volcanoes are associated with plate boundaries. About 15 per cent of the world's active volcanoes are found along the constructive plate margins or divergent plate margins (along the mid-oceanic ridges, where two plates move in opposite directions) whereas 80 per cent volcanoes are associated with the destructive/convergent/consuming plate boundaries (where two plates collide). Some volcanoes are also found in intra-plate regions e.g., volcanoes of the Hawaii Island, fault zone of East Africa etc. There are three major belts of volcanoes e.g. (1) mid-Atlantic Ridge zone, (2) circum-Pacific zone and (3) mid-continental zone.

The intensity of volcanic activity is also related to the nature of plate boundaries. Divergent or constructive plate boundaries are associated with quiet volcanic eruption known as fissure eruption. The volcanic lava of constructive plate margins is tholeiite which is in fact a type of basalt having less quantity of potash and is formed due to differential melting. The basaltic lava associated with divergent (destructive) plate boundaries representing circum Pacific belt and mid-continental (Alpine) belt is rich in silica content and is mixed with andesite, dacite and rhyolite. The volcanic lava associated with rift valleys is rich in alkalis. This is also called as alkaline basalt.

Active volcanoes are associated with mid oceanic ridges. Under the influence of rising thermal convection currents oceanic plates (crust) are separated and two plates move in opposite directions from the ridge crests. Because of divergence of two plates the confining pressure of superincumbent load is released and consequently melting-point is lowered which causes partial melting of upper mantle and formation of tholeiite basalt which moves upward through ascending thermal convection currents and appears as fissure flow of basaltic lava. This basaltic tholeiite lava after cooling and solidification forms new oceanic crust. This volcanic mechanism leads to formation of ridges parallel to mid-oceanic ridges.

The newly formed basaltic crust is divided into two equal halves and are placed on either side of the ridge. These parallel basaltic stripes placed on either side of the ridge move away from the mid-oceanic ridge due to sea-floor spreading effected by ascending thermal convection currents and associated upwelling of lava and (basaltic stripes) are accreted at the trailing margins of divergent plates. This is also validated on the basis of parallel but alternate pattern of positive and negative anomalies of palaeomagnetic stripes. Iceland presents an ideal example of this mechanism because it is situated on both the sides of mid-Atlantic ridge i.e., mid-Atlantic ridge (locally called as Reykjanes ridge) passes through the middle of Iceland through which magma upwells from time to time. The eruption of Helgafell volcano in 1973 presents evidence in support of this proposition. There is continuous growth in the surface area of Iceland due to basaltic lava. It is estimated that the island has grown in size by 400 km since the beginning of Tertiary (65 million years B.P.) epoch, which indicates average growth rate of 0.6 cm/yr. The age of lava (basalt) increases away from the ridge as recent lava is found close to the ridge, 2 million year-old lava away from the ridge and 65 million - year old lava at the margin of the island.

The aforesaid inference is also validated on the basis of evidences of volcanic islands situated on the ocean floor. For example, the volcanic islands of Atlantic Ocean arc without doubt associated with the mid-Atlantic ridge. The most active volcanic islands are nearest to the ridge whereas dormant and extinct volcanoes are located at the farthest distance from the ridge. It may be pointed out that volcanic islands are formed near the ridge due to upwelling of magma from below. As the sea floor spreads these volcanic peaks move away from the ridge and magma source. When they move far away from the ridge the supply of magma comes to an end and thus most of these volcanic islands are submerged under sea waves and become sea mounts or guyots. It may be mentioned that not all the volcanic peaks submerge beneath sea waves as a few of them project from 1500 to 3000 m above sea-level. The study of basaltic lava of the volcanic islands of the Atlantic Ocean has revealed the fact that volcanic islands located nearest to the ridge are characterized by recent lava while those located at the farthest distance from the ridge have oldest lava. For example, the oldest lava of Azores islands located on either side of the mid-Atlantic ridge is 4 million years old while the oldest lava of Cape Verde island located near African coast (farthest from the ridge) is 120 million years old.

The island arcs with volcanic peaks and associated oceanic trenches are formed when oceanic plate is subducted below continental belt. Seismic shocks and heat are generated at the depth of 700 km due to friction of continental plate and subducted oceanic plate. Consequently, upper mantle, basaltic crust of ocean floor and overlying sediments get melted and thus magma is formed. It may be pointed out that volcanic peaks of island arcs have been formed of sodium-rich basalt. Such basalt is formed when volcanic eruption occurs in oceanic water. Sodium-rich basalt is covered with andesite of relatively lesser density but rich in silicon in comparison to underlying basalt.

Regarding the origin of andesite-daciterhyolite along the circum-Pacific folded mountain chain two contrasting views have been floated.

(1) Ringwood (1974) has stated that andesitedacite—rhyolite are formed due to partial melting of amphibolite of subducted Benioff zone and melting of quartz eclogite at greater depth in the mantle.

(2) According to Gilluly andesite—dacite—rhyolite are formed due to partial melting of oceanic tholeiite or amphibolite or eclogite and its mixing with sediments of ocean floor such as sandstone, chert and radiolarian ooze.

Apparently, the explanation of volcanoes of Hawai Island does not fit in the framework of plate tectonic theory but the problem may be solved if we look into the entire mechanism involved in the volcanic process in the east Pacific Ocean. The Hawai Island is south-eastern extension of Midway Island-Emperor sea mounts-Kamchatka Island Arcs and is located far away from the East Pacific Ridge but Hawai Island is characterized by active volcanic activities whereas the above mentioned island arcs are dominated by dormant volcanoes and ancient lava (25 to 75 million years old). It is believed that there is active plume (magma source) beneath Hawai Island which ensures continuous supply of molten magma for longer duration of time. There has been upwelling of lava in the Hawai Island for the last 70 million years. Due to plate movements the Pacific Oceanic floor after being separated from East Pacific Ridge continued to move in north-westerly direction at the rate of 9 cm per year with the result volcanic peaks having plume underneath also moved north-westward. Thus, the plume beneath Hawai Island continued to supply lava to the volcanoes of the island. On the other hand, as the other islands moved far away from the centre (plume) of lava supply due to sea-floor spreading, the lava supply dried up and the volcanoes became dormant.

The formation rather emplacement of granites into continental folded mountains and intrusion of batholiths can be explained on the basis of plate tectonic theory. During the collision of plates and formation of folded mountains continental rocks are subducted and reach greater depth where these get melted and form magma There is marked variation in the composition of magma involved in the continental folded mountains and lava of volcanoes of island arcs and andesite lava of subduction zone. Continental rocks are dominated by low density matter e.g., silica and aluminium oxide contents. When melted, the resultant magma is also dominated by such matter (silica and alluminium oxides). Because of low density the resultant magma intrudes in the overlying crust, with the result folded mountains are further uplifted and granitic batholiths having quartz and feldspar minerals are also formed. The Ranchi batholiths of the Chhotanagpur Highlands (Jharkhand) may be associated with Arachaean mountain building.

The origin and characteristics of lava plateaus of the continents can also be explained on the basis of plate tectonic theory. The formation of extensive basaltic lava plateaus of India, Brazil, Columbian plateau of the USA etc. may be related to continental breaking. It is believed that lava plateaus might have been formed due to separation of continents and their movement in relation to other continents e.g. Deccan lava plateau of India due to its separation from Africa-Australia and its northward movement; Columbia lava plateau of the USA due to separation of N. America from Europe and westward movement of the former; Brazilean lava plateau due to separation of S. America from Africa and westward movement of the former etc.

Readers are advised to consult chapter 12 of this book for explanation of origin of volcanoes of circum-Pacific belt, mid-Atlantic belt, mid-continental belt etc., in terms of different types of plate boundaries (e.g. convergent, divergent and conservative plate boundaries).

Earthquakes and the Plate Tectonics : Seismic events can be explained in terms of plate boundaries. From the standpoint of movement and tectonic events and creation and destruction of geo-materials the plate boundaries are divided into (i) constructive plate boundaries, (ii) destructive plate boundaries, and (iii) conservative plate boundaries. Constructive plate boundaries represent the trailing ends of divergent plates which move in opposite directions from the mid-oceanic ridges, destructive plate boundaries are those where two convergent plates collide against each other and the heavier plate boundary is subducted below relatively lighter plate boundary and conservative plate boundaries are those where two plates slip past each other without any collision. Major tectonic events associated with these plate boundaries are ruptures and faults along the constructive plate boundaries, faulting and folding along the destructive plate boundaries and transform faults along the conservative plate boundaries. All sorts of disequilibrium are caused due to different types of plate motions and consequently earthquakes of varying magnitudes are caused.

Normally, moderate earthquakes are caused along the constructive plate boundaries because the rate of rupture of the crust and consequent movement of plates away from the mid-oceanic ridges is rather slow and the rate of upwelling of lavas due to fissure flow is also slow. Consequently, shallow focus earthquakes are caused along the constructive plate boundaries or say along the mid-oceanic ridges. The depth of 'focus' of earthquakes associated with the constructive plate boundaries ranges between 25 km to 35 km but a few earthquakes have also been found to have occurred at the depth of 60 km. It is, thus, obvious that the earthquakes occurring along the mid-Atlantic Ridge, mid-Indian Oceanic Ridge and East Pacific Rise are caused because of movement of plates in opposite directions (divergence) and consequent formation of faults and ruptures and upwelling of magma or fissure flow of basaltic lavas.

Earthquakes of high magnitude and deep focus are caused along the convergent or destructive plate boundaries because of collision of two convergent plates and consequent subduction of one plate boundary along the Benioff zone. Here mountain building, faulting and violent volcanic eruptions (central explosive type of eruptions) cause severe and disastrous earthquakes having the focus at the depth upto 700 km. This process, convergence of plates and related plate collision, explains the maximum occurrence of earthquakes of varying magnitudes along the Firy Ring of the Pacific or the Circum-Pacific Belt (along the western and eastern margins of the Pacific Ocean or say along the western coastal margins of North and South Americas and thus the Rockies-Andes

Mountain Belt and along the eastern coastal margins of Asia and island arcs and festoons parallel to the Asiatic coast). The earthquakes of the mid-continental belt along the Alpine-Himalayan chains are caused due to collision of Eurasian plates and African and Indian plates. The earthquakes of the western marginal areas of North and South Americas are caused because of subduction of American plate beneath the Pacific plate and the resultant tectonic forces whereas the earthquakes of the eastern margins of Asia are originated because of the subduction of Pacific plate under Asiatic plate. Similarly, the subduction of African plate below European plate and the subduction of Indian plate under Asiatic plate causes earthquakes of the mid-continental belt.

Creation of transform faults along the conservative plate boundaries explains the occurrence of severe earthquakes of California (USA). Here, one part of California moves north-eastward while the other part moves south-westward along the fault plane and thus is formed transform fault which causes earthquakes.

5

Formation of Mountains

Mountains are significant relief features of the second order on the earth's surface. A mountain may have several forms viz. (i) mountain ridge, (ii) mountain range, (iii) mountain chain, (iv) mountain system, (v) mountain group and (vi) cordillera. A mountain ridge is a system of long, narrow and high hills. Generally, the slope of one side of a ridge is steep while the other side is of moderate slope but a ridge may also have symmetrical slopes on both the sides. A mountain range is a system of mountains and hills having several ridges, peaks and summits and valleys. In fact, a mountain range stretches in a linear manner. In other words, a mountain range represents a long but narrow strip of mountains and hills. All of the hills of a mountain range are of the same age but there are structural variations in different members of the range. A mountain chain consists of several parallel long and narrow mountains of different periods. Some times, the mountain ranges are separated by flat upland or plateaus. A mountain system consists of different mountain ranges of the same period. Different mountain ranges are separated by valleys. A mountain group consists of several unsystematic patterns of different mountain systems. Cordillera consists of several mountain groups and systems. In fact, cordillera is a community of mountains having different ridges, ranges, mountain chains and mountain systems. The mountainous region of the western part of North America is the best example of a cordillera.

The Classification

1. On the Basis of Height

(i) low mountains; height ranges between 700 to 1,00 m.

(ii) rough mountains; height-1000 m to 1,500 m

(iii) rugged mountains; height-1,500 to 2,000 m

(iv) high mountains; height above 2,000 m

2. On the Basis of Location

(i) Continental mountains

(a) coastal mountains, examples; Applachians, Rockies, Alpine mountain chains, Western and Eastern Ghats of India etc.

(b) inland mountains, examples; Ural mountains (Russia), Vosges and Black Forest block mountains (Europe), Himalayas, Aravallis, Satpura, Maikal, Kaimurs etc. (India), Kunlun, Tienshan, Altai etc. (Asia) etc.

(ii) Oceanic mountains-Most of the oceanic mountains are below water surface (below sea level). Oceanic mountains are located on continental shelves and ocean floors. Some oceanic mountains are also well above the sea level. If the height of the mountains is considered from the oceanic floor and not from sea-level, many of the oceanic mountains will become much higher than the Mount Everest. For example, Mauna Kea volcanic mountain of Hawaii Island is 4200 m high from the sea level but if its height is considered from the sea bottom, its height becomes 9140 m which is higher than the highest mountain, Mount Everest (8848 m AMSL) of the continent. Similarly, the Antilean Mountain system is 3000 m above sea-level but it is also 5400 m below sea-level, and thus its total height from the oceanic floor becomes 8400 m. Most of the oceanic mountains are volcanic mountains.

On the Basis of Mode of Origin :—

(1) Original or tectonic mountains are caused due to tectonic forces e.g., compressive and tensile forces motored by endogenetic forces coming from deep within the earth. These mountains are further divided into 4 types on the basis of orogenetic forces responsible for the origin of a particular type of mountain.

(i) Folded mountains are further divided into 3 sub-types on the basis of their area. These are originated by compressive forces.

(A) young folded mountains

(B) mature folded mountains

(C) old folded mountains

(ii) Block mountains are originated by tensile forces leading to the formation of rift valleys. They are also called as horst mountains.

(iii) Dome mountains are originated by magmatic intrusions and upwarping of the crustal surface. Examples, normal domes, lava domes, batholithic domes, lacolithic domes, salt domes etc.

(iv) Mountains of accumulations are formed due to accumulation of volcanic materials. Thus, these are also called as volcanic mountains. Different types of volcanic cones (e.g. cindercones, composite cones, acid lava cones, basic lava cones etc.) come under this category.

(2) Circum-erosional or relict mountains: examples, Vindhyachal ranges, Aravallis, Satpura, Eastern Ghats, Western Ghats etc. (all from India).

On the Basis of Period of Origin :—

(1) *Pre-Cambrian Mountains :* Examples, Laurentian mountains, Algoman mountains, Kilarnean mountains etc. (North America), mountains of Feno-Scandia, North-West Highlands and Anglesey etc. (Europe).

(2) *Caledonian Mountains :* Mountains formed during Silurian and Devonian periods, examples Taconic mountains of the Applachian system, mountains of Scottland, Ireland and Scandinavia (Europe), Brazilides of South America, Aravallis, Mahadeo, Satpura etc., of India.

(3) *Hercynian Mountains :* Mountains formed during Permian and Permocarboniferous periods, examples: mountains of Iberian peninsula, Ireland, Spanish Messeta, Brittany of France, South Wales, Cornwall, Mendips, Paris basin, Belgian coal-fields, Rhine Mass, Bohemian plateau, Vosges and Black Forest, plateau region of central France, Thuringenwald, Frankenwald, Hartz mountain, Donbas coal-field (all in Europe); Variscan mountains of Asia include Altai, Sayan, Baikal Arcs, Tien Shan, Khingan, mountains of Dzungarian basin, Tarim basin, Nanshan, Alai and Trans Alai mountains of Amur basin, Mongolia and Gobi etc; Australian Variscan mountains include the scattered hills in the Eastern Cordillera, New England of New Southerwales; North American Variscan mountains include Applachians; South American Variscan mountains are Austrian and Saalian folds of San Juan and Mendoza, mountains of Punaarc of Atacama, Gondwanides of Argentina etc.

(4) *Alpine Mountains :* Mountains formed during Tertiary period, examples: Rockies (North America), Andes (South America), Alpine mountain systems of Europe (main Alps, Carpathians, Pyrenees, Balkans, Caucasus, Cantabrians, Apennines, Dinaric Alps etc.), Atlas mountains of northwest Africa; Himalayas and mountains coming out of Pamir Knot of Asia (Taurus, Pauntic, Zagros, Elburz, Kunlun etc.).

Faultblock Mountains

Block mountains, also known as faultblock mountains, are the result of faulting caused by tensile and compressive forces motored by endogenetic forces coming from within the earth. Block mountains represent the upstanding parts of the ground between two faults or on either side of a rift valley or a graben. Essentially, block mountains are formed due to faulting in the ground surface.

Block mountains are generally of two basic types e.g. (i) tilted block mountains having one steep side represented by fault scarp and one gentle side and (ii) lifted block mountains represent real horst and are characterized by flattened summit of tabular shape and very steep side slopes represented by two boundary fault scarps. Block mountains are also called as horst mountains.

Block mountains are found in all the continents e.g. (i) young block mountains around Albert, Warner and Klamath lakes in the Steens Mountain District of Southern Oregon, Wasatch Range in the Utah province etc. in the USA, (ii) Vosges and Black Forest mountains bordering the faulted Rhine Rift valley in Europe, (iii) Salt Range of Pakistan etc. Sierra Navada mountain of California (USA) is considered to be the most extensive block mountain of the world. This mountain extends for a length of 640 km (400 miles) having a width of 80 km (50 miles) and the height of 2,400 to 3,660 m (8,000 to 12,000 feet). There is difference of opinions among the scientists regarding the origin of block mountains. There are two theories for the origin of these mountains viz. (i) fault theory and (ii) erosion theory.

The Fault Theory : Most of the geologists are of the opinion that block mountains are formed due to faulting. The structural patterns of Great Basin Range mountains of Utah province (USA) were closely studied by Clarence King and G.K. Gilbert who named these mountains as faulted blocks (between 1870 and 1875 A.D.). Since then the mountains formed due to large-scale faulting were named block mountains. Later on G.D. Louderback opined that Basin Range mountains were formed due to faulting and tilting in the ground surface. W.M.Davis also advocated for the fault theory of the origin of block mountains. Block mountains are formed in a number of ways.

(i) Block mountains are formed due to upward movement of middle block between two normal faults. The upthrown block is also called as horst. The summital area of such block mountain is of flat surface but the side slopes are very steep.

(ii) Block mountains may be formed when the side blocks of two faults move downward whereas the middle block remains stable at its place. It is apparent that the middle block projects above the surrounding surface because of downward movement of side blocks. Such block mountains are generally formed in high plateaus or broad domes.

(iii) Block mountains may be formed when the middle block between two normal faults moves downward. Thus, the side blocks become horsts and block mountains. Such mountains are associated with the formation of rift valleys.

The Erosion Theory : J.F. Spurr, on the basis of detailed study of Great Basin Range mountains of the USA, opined that these mountains were not formed due to faulting and tilting, rather they were formed due to differential erosion. According to Spurr the mountains, after their origin in Mesozoic era, were subjected to intense erosion. Consequently, differential erosion resulted into the formation of existing denuded Great Basin Range mountains. It may be pointed out that erosion theory of the origin of block mountains is not acceptable to most of the scientists because they believe that denudation may modify mountains but cannot form a mountain. In fact, deformatory process play major role in the origin of block mountains.

Folded Mountains

Folded mountains are formed due to folding of crustal rocks by compressive forces generated by endogenetic forces coming from within the earth. These are the highest and most extensive mountains of the world and are found in ail the continents. The distributional pattern of folded mountains over the globe denotes the fact that they are generally found along the margins of the continents either in north-south direction or east-west direction. Rockies, Andes, Alps, Himalayas, Atlas etc., are the examples of folded mountains. Folded mountains are classified on various bases a follows.

(1) Folded mountains are divided into 2 broad categories on the basis of the nature of folds. (i) Simple folded mountains with open folds. Such mountains are characterized by well developed system of anticlines and synclines wherein folds are arranged in wave-like pattern. These mountains have open and relatively simple fols. (ii) Complex folded mountains represent very complex structure of intensely compressed folds. Such complex structure of folds is called 'nappe'. In fact, complex folded mountains are formed due to the formation of recumbent folds caused by powerful compressive forces.

(2) Folded mountains are classified into (i) young folded mountains (which are least affected by denudational processes) and (ii) mature folded mountains. It may be pointed out that it is difficult to find true young folded mountains because the process of mountain is exceedingly slow process and thus denudational processes start denuding the mountains right from the beginning of their origin. Mature folded mountains are characterized by monoclinal ridges and valleys. This classification is based on the age factor.

(3) On the basis of the period of origin folded mountains are divided into (i) old folded mountains and (ii) new folded mountains. All the old folded mountains were originated before Tertiary period. The folded mountains of Caledonian and Hercynian mountain building periods come under this category. These mountains have been so greatly denuded that they have now become relict folded mountains, for example, Aravallis, Vindhyachal etc. The Alpine folded mountains of Tertiary period are grouped under the category of new folded mountains, for example, Rockies, Andes, Alps, Himalayas etc.

Main Features

(1) Folded mountains are the youngest mountains on the earth's surface.

(2) The lithological characteristics of folded mountains reveal that these have been formed due to folding of sedimentary rocks by strong compressive forces. The fossils found in the rocks of folded mountains denote the fact that the sedimentary rocks of these mountains were formed due to deposition and consolidation of sediments in water bodies mainly in oceanic environment because the argillaceous rocks of folded mountains contain marine fossils.

(3) Sediments are found upto greater depths, thousands of metres (more than 12,000 metres). Based on this fact some scientists have opined that the sediments involved in the formation of sedimentary rocks of folded mountains might have been deposited in deep oceanic areas but the marine fossils found in the rocks belong to such marine organisms which can survive only in shallow water or shallow sea. It means that the sedimentary rocks of folded mountains were deposited in shallow seas.

The sea bottoms were subjected to continuous subsidence due to gradual sedimentation. Thus, the greater thickness of sediments could be possible due to continuous sedimentation and subsidence and consequent consolidation of sediments due to ever increasing superincumbent load.

(4) Folded mountains extend for greater lengths but their widths are far smaller than their lengths, For example, the Himalayas extend from west to east for a length of 2400 km (1500 miles) but their north-south width is only 400 km (250 miles). It means that folded mountains have been formed in long, narrow and shallow seas. Such water bodies have been termed geosynclines and it has been established that 'out of geosynclines have come out the mountains' or 'geosynclines have been cradles of mountains.' According to P.G. Worcester 'all great folded mountains stand on the sites of former geosynclines'.

(5) Folded mountains are generally round in arch shape having one side concave slope and the other side convex slope.

(6) Folded mountains are found along the margins of the continents facing oceans. For example, Rockies and Andes are located along the west ern margins of North and South Americas respectively and face Pacific Ocean. They are located in two directions e.g. north-south (e.g. Rockies and Andes) and west-east directions (e.g. Himalayas). The Alpine mountains are located along the southern margins of Europe facing Mediterranean sea. If we consider former Tethys Sea, then the Himalayas were also located along the margins of the continent.

The Geosynclines

Meaning and Concept : The geological history of the continents and ocean basins denotes the fact that in the beginning our globe was characterized by two important features viz. (i) rigid masses and (ii) geosynclines. Rigid masses representing the ancient nuclei of the present continents, have remained stable for considerably longer periods of time. These rigid masses are supposed to have been surrounded by mobile zones of water characterized by extensive sedimentation. These mobile zones of water have been termed 'geosynclines' which have now been converted by compressive forces into folded mountain ranges.

On an average, a geosyncline means a water depression characterized by sedimentation. It has now been accepted by majority of the geologists and geographers that all the mountains have come out of the geosynclines and the rocks of the mountains originated as sediments were deposited and later on consolidated in sinking seas, now known as geosynclines. If we consider the height and thickness of sediments of the young folded mountains of Tertiary period (e.g. Rockies, Andes, Alps, Himalayas etc.), then it appears that the geosynclines should have been very deep water bodies but the marine fossils found in the sedimentary rocks of these folded mountains belong to the category of marine organisms of shallow seas. It is, thus, obvious that the geosynclines are shallow water bodies characterized by gradual sedimentation and subsidence. Based on above facts geosynclines can now be defined as follows :

'Geosynclines are long but narrow and shallow water depressions characterized by sedimentation and subsidence'.

J.A. Steers (1932) has aptly remarked, 'the geosynclines have been long and relatively narrow depressions which seem to have subsided during the accumulation of sediments in them.'

The following are the general characteristics of geosynclines.

(1) Geosynclines are long, narrow and shallow depressions of water.

(2) These are characterized by gradual sedimentation and subsidence.

(3) The nature and patterns of geosynclines have not remained the same throughout geological history rather these have widely changed. In fact, the location, shape, dimension and extent of geosynclines have considerably changed due to earth movements and geological process.

(4) Geosynclines are mobile zones of water.

(5) Geosynclines are generally bordered by two rigid masses which are called forelands.

The Concept

The concept of geosynclines was given by James Hall and Dana but the concept was elaborated and further developed by Haug. J.A. Steers (1932) has remarked, 'while the theory of geosynclines is due to Haug, the concept of idea belongs to Hall and Dana'. It is desirable to discuss the concept of geosynclines developed by different exponents.

(1) *Concept of Hall and Dana* : Dana studied the folded mountains and postulated that the sediments of the rocks of folded mountains were of marine origin. These rocks are deposited in long, narrow and shallow seas. Dana named such water bodies as geosynclines. He defined, for the first time, geosynclines as long, narrow and shallow and sinking beds of seas.

Hall elaborated the concept of geosynclines as advanced by Dana. He presented ample evidences to show relationship between geosynclines and folded mountains. He opined that the rocks of folded mountains were deposited in shallow seas. According to Hall the beds of geosynclines are subjected to subsidence due to continuous sedimentation but the depth of water in the geosynclines remains the same. Geosynclines are much longer than their widths.

(2) *Concept of E. Haug* : 'If the idea of geosynclines is due to Hall and Dana, the theory of their development is really due to Haug'. He defined geosynclines as long and deep water bodies. According to Haug 'geosynclines are relatively deep water areas and they are much longer than they are wide.' He drew the palaeogeographical maps of the world and depicted long and narrow oceanic tracts to demonstrate the facts that these water tracts were subsequently folded into mountain ranges. He further postulated that the positions of the presentday mountains were previously occupied by oceanic tracts i.e. geosynclines. Geosynclines existed as mobile zones of water between rigid masses. He identified 5 major rigid masses during Mesozoic era e.g. (i) North Atlantic Mass, (ii) Sino-Siberian Mass, (iii) Africa-Brazil Mass, (iv) Australia-India-Madagascar Mass and (v) Pacific Mass. He located 4 geosynclines between these ancient rigid masses e.g. (i) Rockies geosyncline, (ii) Ural geosyncline, (iii) Tethys geosyncline and (iv) Circum-Pacific geosyncline.

According to Haug there is systematic sedimentation in the geosynclines. The littoral margins of the geosynclines are affected by transgressional and regressional phases of the seas. The marginal areas of the geosynclines have shallow water wherein larger sediments are deposited whereas finer sediments are deposited in central parts of the geosynclines. The sediments are squeezed and folded into mountain ranges due to compressive forces coming from the margins of the geosynclines. He has further remarked that it is not always necessary that all the geosynclines may pass through the complete cycle of the processes of sedimentation, subsidence, compression and folding of sediments. Some times, no mountains are formed from the geosynclines inspite of continuous sedimentation for long duration of geological time.

Though the contributions of Haug in this regard are praiseworthy as he developed the concept of geosynclines but his theory suffers from certain serious drawbacks and confusing ideas about them. His palaeogeographical map of Mesozoic era depicted unbelievable larger extent of rigid masses (land areas) in comparison to geosynclines (oceanic areas). Questions arise, as to what happened to such extensive land masses after Mesozoic era ? Where did they disappear ? Haug could not explain these and many more Questions. His geosynclines as very deep oceanic tracts are also not acceptable because the marine fossils found in the folded mountains belong to the group of marine organisms of shallow seas.

(3) *Concept of J.W. Evans* : According to Evans the geosynclines are so varied that it becomes difficult to present their definite form and location. The beds of geosynclines are subjected to gradual subsidence because of sedimentation. The form and shape of geosynclines change with changing environmental conditions. A geosyncline may be narrow or wide. It may be of different shapes. There may be several alternative situations of geosynclines e.g. (i) it maybe between two land masses (example, Tethys geosyncline between Laurasia and Gondwanaland), (ii) it may be in front of a mountain or a plateau (for example, resultant long trench after the origin of the Himalayas, this depression was later on filled with sediments to form Indo-Gangetic Plains), (iii) it may be along the margins of the continents, (iv) it may be in front of a river mouth etc. According to Evans all the geosynclines irrespective of their varying forms, shapes and locations are characterized by twin processes of sedimentation and subsidence. Geosynclines, after long period of sedimentation, are squeezed and folded into mountain ranges.

(4) *Views of Schuchert* : He attempted to classify geosynclines on the basis of their characteristics related to their size, location, evolutionary history etc. He has divided geosynclines into 3 categories.

(i) Monogeosynclines are exceptionally long and narrow but shallow water tracts as conceived by Hall and Dana. The geosynclinal beds are subjected to continuous subsidence due to gradual sedimentation and resultant load. Such geosynclines are situated either within a continent or along its borders. These are called mono because they pass through only one cycle of sedimentation and mountain building. Applachian geosyncline is considered to be the best example of monogeosynclines. In place of the Applachians (USA) there existed a long and narrow Appalachian geosyncline during pre

Cambrian period. The geosyncline was bordered by highland mass known as Applachia in the east. Applachian geosynclines were folded from Ordovician to Permian periods.

(ii) Polygeosynclines were long and wide water bodies. These werc definitely broader than the monogeosynclines. These geosynclines existed for relatively longer period than the monogeosynclines and these have passed through complex evolutionary histories. These are considered to have experienced more than one phase of orogenesis, consequently they 'may have been diversified by the production of one or more parallel geanticlines arising from their floors in the squeezing process'. They originated in positions similar to those of monogeosynclines. Rocky and Ural geosynclines are quoted as the representative examples of polygeosynclines.

(iii) Mesogeosynclines are very long, narrow and mobile ocean basins which are bordered by continents from all sides. They are characterized by great abyssal depth and long and complex geological histories. These geosynclines pass through several geosynclinal phases e.g., phases of sedimentation, subsidence and folding. Mesogeosynclines are similar to the geosynclines conceived by Haug. Tethys geosyncline is the typical example of such type. Mediterranean Sea is the remnant of Tethys geosyncline. This geosyncline was folded into Alpine mountains of Europe and the Himalayas of Asia. The unfolded remaining portion of Tethys geosyncline became Mediterranean sea, an example of median mass of Kober.

(5) *Concept of Arthur Holmes* : Besides describing main characteristics of geosynclins, A. Holmes has also elaborated the causes of the origin of different types of geosynclines. He has also described the detailed processes and mechanisms of sedimentation and subsidence and consequent orogenesis. According to him no doubt sedimentation leads to subsidence but this process cannot account for the greater thickness of sediments in geosynclines rather earth movements can cause subsidence of high magnitude in the geosynclinal beds. He further pointed out that the process of subsidence of the geosynclinal beds was not a sudden process rather it was a gradual process. The deposition of sediments upto the thickness of 12,160 m (40,000 feet) in the Applachian geosyncline could be possible during a long period of 3,000,000,000 years from Cambrian period to early Permian period at the rate of one foot of sedimentation every 7,500 years. Holmes has identified 4 major types of geosynclines and has described the mode of their origin separately as given below —

(i) *Formation of Geosynclines due to Migration of Magma* : According to Holmes the crust of the earth is composed of 3 shells of rocks. Just below the outer thin sedimentary layer lies (i) outer layer of granodiorite (thickness, 10 to 12 km), followed by (ii) an intermediate layer of amphibolite (thickness, 20-25 km), and (iii) a lower layer of eclogite and some peridotite. He has further pointed out that migration of magmas from the intermediate layer to neighbouring areas causes collapse and subsidence of upper or outer layer and thus is formed a geosyncline. It may be summarized that some geosynclines are formed due to displacement of light magmas and consequent subsidence of crustal surface. Present Coral Sea, Tasman Sea, Arafura Sea, Weddell Sea and Ross Sea have been quoted as typical examples of such geosyncline. This concept of Holmes has been severely criticised because the transfer and displacement of magmas cannot cause subsidence to form geosynclines.

(ii) *Formation of Geosynclines due to Metamorphism* : According to Holmes the rocks of the lower layer of the crust, as referred to above, are metamorphosed due to compression caused by converging convective currents. This matamorphism increases the density of rocks, with the result the lower layer of the crust is subjected to subsidence and thus a geosyncline is formed. Caribbean Sea, the western Mediterranean Sea and Banda Sea have been quoted as examples of this category of geosynclines. This concept has been rejected on the ground that compression caused by convergent convective currents would not cause metamorphism rather it would cause melting of rocks due to resultant high temperature.

(iii) *Formation of Geosynclines due to Compression* : Some geosynclines are formed due to compression and resultant subsidence of outer layer of the crust caused by convergent convective currents. Persian Gulf and Indo-Gangetic trough are considered to be typical examples of this group of geosynclines.

(iv) *Formation of Geosynclines due to Thinning of Sialic Layer* : According to Holmes there may be two possibilities if a column of rising con vective currents diverges after reaching the lower layer of the crust in opposite directions. (i) The Sialic layer is stretched apart due to tensile forces exerted by diverging convective currents. This process causes thinning of sialic layer which results in the creation of a geosyncline. The former Tethys geosyncline is considered to have been formed in this manner. (ii) Alternatively, the continental mass may be separated due to enormous tensile force generated by divergent convective currents. Former Ural geosyncline is supposed to have been formed due to this mechanism.

(6) *Views of Others* : Dustar has classified geosynclines into 3 types on the basis of structure of mountain ranges.

(i) Inter-continental geosynclines are always situated between two continental or land masses. Schuchert's mesogeosyncline is similar to this type. Ural Geosyncline is quoted as the representative example.

(ii) Circum-continental geosynclines are generally situated along the margins of the continents. Schuchert's monogeosyncline is the example.

(iii) Circum-oceanic geosynclines are generally found along the marginal areas of the oceans where continental margins meet with the oceanic margins. Stille has named such geosyncline as marginal geosyncline while others have called it special type of geosyncline or unique geosyncline. More extensive geosynclines have been named by Stille as orthogeosynclines. Stille has further classified the geosynclines on the basis of intermittent volcanic activity during their infilling into (i) eugeosynclines and (ii) miogeosynclines. Eugeosynclines have relatively high amount of volcanic products (Greek prefix eu means high status of igneous activity) while miogeosynclines have low volcanic products (mio means low).

Various Stages

The geosynclinal history is divided into three stages viz. (i) lithogenesis (the stage of creation of geosynclines, sedimentation and subsidence of the beds of geosynclines, (ii) orogenesis (the stage of squeezing and folding of geosynclinal sediments into mountain ranges, and (iii) gliptogenesis (the stage of gradual rise of mountains, and their denudation and consequent lowering of their heights). These stages would be elaborated during the discussion of geosynclinal theory of Kober.

Mountain Building Theory : The process of the origin of block mountains, dome mountains, and volcanic mountains (mountains of accumulation) is more or less well understood but the problem of the origin of folded mountains is very much complex and complicated. Different hypotheses and theories have been postulated from time to time by various scientists for the explanation of the origin of folded mountains but none of them could become commonly acceptable to majority of the scientists. Recently, plate tectonic theory has, to larger extent, solved the problem of mountain building at global scale. The hypotheses and theories related to mountain building are divided into two groups, (i) theories based on horizontal forces and (ii) theories based on vertical forces.

(1) The first group includes those theories which postulate the origin of mountains due to horizontal crustal movement and consequent contraction and folding of crustal surface into mountains. This group is further subdivided into two subgroups e.g. (i) the group of contraction theories (i.e. horizontal movements are caused due to contraction of the earth because of cooling) and (ii) the group of drift theories

(i.e. the horizontal movements are caused due to continental displacement and drift). Thermal contraction theory of Jeffreys and geosynclinal Theory of Kober belong to the group of contraction theories whereas Continental Drift theories of F.B. Taylor, and A.G. Wegener, Thermal Convection Current Theory of A. Holmes, Sliding Continents Theory of Daly, Radioactivity Theory of Joly and Plate Tectonic Theory are included in the group of drift theories.

(2) The second group includes those theories which are based on vertical movements coming from within the earth, e.g. Undulation and Oscillation Theory of Harmon.

Geosynclinal Orogen Theory of Kober Objectives : Famous German geologist Kober has presented a detailed and systematic description of the surface features of the earth in his book 'Der Bau der Erde'. His main objective was to establish relationship between ancient rigid masses or tablelands and more mobile zones or geosynclines, which he called 'orogen.' Kober not only attempted to explain the origin of the mountains on the basis of his geosynclinal theory but he also attempted to elaborate the various aspects of mountain building e.g., formation of mountains, their geological history and evolution and development. He considered the old rigid masses as the foundation stones of the present continents. According to him present continents have grown out of rigid masses. He defined the process of mountain building or orogenesis as that process which links rigid masses with geosynclines. In other words, mountains are formed from the geosynclines due to the impacts of rigid masses.

The Orogenetic Force : Kober's geosynclinal theory is based on the forces of contraction produced by the cooling of the earth. He believes in the contraction history of the earth. According to J.A. Steers (1932) 'Kober is definitely a contractionist, contraction providing the motive force for the compressive stress'. In other words, the force of contraction generated due to cooling of the earth causes horizontal movements of the rigid masses or forelands which squeeze, buckle and fold the sediments into mountain ranges.

Base of the Theory : According to Kober there were mobile zones of water in the places of present-day mountains. He called mobile zones of water as geosynclines or orogen (the place of mountain building). These mobile zones of geosynclines were surrounded by rigid masses which were termed by Kober as 'kratogen'. The old rigid masses included Canadian Shield, Baltic Shield or Russian Massif, Siberian Shield, Chinese Massif, Peninsular India, African Shield, Brazilian Mass, Australian and Antarctic rigid masses. According to Kober mid-Pacific geosyncline separated north and south Pacific forelands which were later on foundered to form Pacific Ocean. Eight morphotectonic units can be identified on the basis of the description of the surface features of the earth during Mesozoic era as presented by Kober e.g. (i) Africa together with some parts of Atlantic and Indian Oceans, (ii) Indian Australian land mass, (iii) Eurasia, (iv) North Pacific continent, (v) South Pacific continent, (vi) South America and Antarctica etc.

Kober has identified 6 major periods of mountain building. Three mountain building periods, about which very little is known, are reported to have occurred during pre-Cambrian period. Palaeozoic era saw two major mountain building periods the Caledonian orogenesis was completed by the end of Silurian period and the Variscan orogeny was culminated in Permo-Carboniferous period. The last (6th) orogenic activity known as Alpine orogeny was completed during Tertiary epoch.

Kober has opined that mountains are formed out of geosynclines. According to Kober geosynclines, the places of mountain formation (known as orogen) are long and wide water areas characterized by sedimentation and subsidence. According _to J.A. Steers (1932), 'Kober's views (on geosynclines and orogenesis) are, then, a combination of the old geosynclinal hypothesis of Hall and Dana, which was developed later by Haug, and his own views on orogenesis.'

The Mechanism of the Theory : According to Kober the whole process of mountain building passes through three closely linked stages of lithogenesis, orogenesis and gliptogenesis. The first stage is related to the creation of geosynclines due to the force of contraction caused by cooling of the earth. This preparatory stage of mountain building is called lithogenesis. The geosynclines are long and wide mobile zones of water which are bordered by rigid masses, which have been named by Kober as forelands or kratogen. These upstanding land masses or forelands are subjected to continuous erosion by fluvial processes and eroded materials are deposited in the geosynclines. This process of sediment deposition is called sedimentation. The ever increasing weight of sediments due to gradual sedimentation exerts enormous pressure on the beds of geosynclines, with the result the beds of geosynclines are subjected to gradual subsidence. This process is known as the process of subsidence. These twin processes of sedimentation and resultant subsidence result in the deposition of enormous volume of sediments and attainment of great thickness of sediments in the geosynclines.

The Second Stage is related to mountain building and is called the stage of orogenesis. Both the forelands start to move towards each other be cause of horizontal movements caused by the force of contraction resulting from the cooling of the earth. The compressive forces generated by the movement of forelands together cause contraction, squeezing and ultimately folding of geosynclinal sediments to form mountain ranges. The parallel ranges formed on either side of the geosyncline have been termed by Kober as randketten (marginal ranges).

According to Koberfolding of entire sediments of the geosyncline or part thereof depends upon the intensity of compressive forces. If the compressive forces are normal and of moderate intensity, only the marginal sediments of the geosyncline are folded to form two marginal randketten (marginal ranges) and middle portion of the geosyncline remains unaffected by folding activity (thus remains unfolded). This unfolded middle portion is called zwischengebirge (betwixt-mountains) or median mass. Alternatively, if the compressive forces are acute, the whole of geosynclinal sediments are compressed, squeezed, buckled and ultimately folded and both the forelands are closeted. This process introduces complexity in the mountains because acute compression results in the formation of recumbent folds and nappes.

Kober has attempted to explain the forms and structures of folded mountains on the basis of his typical median mass. 'Really, Kober's typical "orogen" (geosynclines) well explains the origin of mountains'. 'The idea of median mass of Kober fully explains the process of mountain building'. According to Kober the Alpine mountain chains of Europe can well be explained on the basis of median masses. According to him Tethys geosyncline was bordered by European land mass in the north and by African rigid mass in the south. The sediments of Tethys geosyncline were compressed and folded due to movement of European landmass (foreland) and African rigid mass (foreland) together in the form of Alpine mountain system. According to Kober the Alpine mountain chains were formed because of compressive forces coming from two sides (north and south). Betic Cordillera, Pyrenees, Province ranges, Alps- proper, Carpathians, Balkan mountains and Caucasus mountains were formed due to northward movement of African foreland. On the other hand, Atlas mountain (north-west Africa), Apennines, Dinarides, Hellenides and Taurides were formed due to southward movement of European landmass.

Alpine mountains further extend into Asia where mountain ranges follow latitudinal directions e.g., west-east orientation but the latitudinal pattern is broken in north-eastern hill region of India where mountain ranges take southerly trend in the form of Burmese hills. Asiatic Alpine ranges begin from Asia minor and run upto Sunda Island in the East Indies. Kober has also explained the orientation of thrust or compression of Asiatic folded mountains on the basis of his foreland theory. Asiatic folded mountains including the Himalaya were formed due to compression and folding of sediments of Tethys geosyncline caused by the movement of Angaraland and Gondwana Forelands together. Two marginal ranges (randketten) were formed on either side of the geosyncline and unfolded middle portion remained as median mass. According to Kober Asiatic Alpine folded mountains can be grouped into two categories on the basis of orientation of folds i.e. (i) the ranges, which were formed by the northward compression, include Caucasus, Pantic and Taurus (of Turkey), Kunlun, Yannan and Annan ranges, and (ii) the ranges, which were formed by the southward compression, include Zagros and Elburz of Iran, Oman ranges, Himalayas, Burmese ranges etc. Tibetan plateau is a fine example of median mass between Kunlun-Tien-Shan and the Himalayas.

The median mass may be in various forms e.g. (i) in the form of plateau (examples, Tibetan plateau between Kunlun and Himalaya, Iranian plateau between Zagros and

Elburz, Anatolian plateau between Pantic and Taurus, Basin Range between Wasatch ranges and Seirra Navada in the USA) ; (i) in the form of plain (example, Hungarian plain between Carpathians and Dinaric Alps), and (iii) in the form of seas (examples, Mediterranean Sea between African Atlas mountains and European Alpine mountains, Caribbean Sea between the mountain ranges of middle America and West Indies).

Third Stage of mountain building is characterized by gradual rise of mountains and their denudation by fluvial and other processes. Continuous denudation results in gradual lowering of the height of mountains.

The Evaluation of the theory : Though Kober's geosynclinal theory satisfactorily explains a few aspects of mountain building but the theory suffers from certain weaknesses and lacunae.

(1) The force of contraction, as envisaged by Kober, is not sufficient to cause mountain building. In fact, very extensive and gigantic mountains like the Alps, the Himalayas, the Rockies and the Andes cannot be formed by the force of contraction generated by cooling of the earth.

(2) According to Suess only one side of the geosyncline moves whereas the other side remains stable. The moving side has been termed by Suess as backland whereas stable side has been called foreland. According to Suess the Himalayas were formed due to southward movement of Angaraland. The Gondwanaland remained stationary. This observation of Suess gained much favour previously but after the postulation of plate tectonic theory his views have become meaningless and the concept of Kober, that both the forelands move together, has been validated because ample evidences of palaeomagnetism and sea-floor spreading have shown that both Asiatic and Indian plates are moving towards each other.

(3) Kober's theory some how explains the west-east extending mountains but north-south extending mountains (Rockies and Andes) cannot be explained on the basis of this theory. Inspite of a few inherent limitations and weaknesses Kober is given credit for advancing the idea of the formation of mountains from geosynclinal sediments because geosyncline found berth in almost all the subsequent theories even in plate tectonic theory.

Jeffreys Theory of Thermal Contraction

Objectives : Jeffreys, a strong exponent of contraction theory, postulated his 'thermal contraction theory' to explain the origin and evolution of major reliefs of the earth's surface (continents, ocean basins, mountains, island arcs and festoons) but his major objective was to explain the origin and distributional patterns of mountain systems of the globe. Jeffreys was a contractionist. His theory was based on mathematical reasoning. He postulated his contraction theory because he could not find any strong reason in the continental drift theory which advocated horizontal movement of the continents due to tidal force of the sun and the moon and the gravitational force as envisaged by A.G. Wegener.

Orogenetic Force : Jeffreys used the force of contraction resulting partly from cooling of the earth due to loss of heat through radiation from the earth's surface and partly from the decrease of the speed of the earth's rotation. In fact, the forces invoked by Jeffreys are divided into two groups.

(1) Force coming through the cooling of the earth. The earth, after being formed, started cooling due to loss of heat through radiation. This process resulted in the gradual decrease of the size of the earth due to contraction on cooling. The resultant contraction provided adequate force (as believed by Jeffreys) to form various relief features including mountains.

(2) Force coming through decrease in the speed of earth's rotation. About 1600 million years ago the earth completed its one rotation in about 0.84 hour whereas it presently completes one rotation in about 24 hours. The decrease in the rotational speed caused contraction in the equatorial circumference of the earth. It may he concluded that the force of contraction was derived through the contraction of the earth due to (i) cooling of the earth and (ii) due to decrease in the speed of earth's rotation.

The Mechanism : Jeffreys' theory is based essentially on the history of the contraction of the earth. According to Jeffreys the earth began to shrink because of contraction caused by gradual cooling of the earth due to loss of heat through radiation from the very beginning of its origin. He has mathematically calculated the extent of contraction on cooling. A decrease of temperature upto 400°C in the 400 km thick outer shell of the earth would cause shortening of the diameter of the earth by 20 km and the circumference by 130 km due to cooling and resultant contraction. He calculated the maximum shortening of the crust due to contraction to be 200 km and the reduction in surface area upto 5 x 10'6 Cm^2.

According to Jeffreys the earth is composed of several concentric shells (layers). The cooling and resultant contraction take place layer after layer but the cooling is effective upto the depth of only 700 km from the earth's surface. "The region of the earth from the centre to somewhere about 700 kilometres from the surface may have undergone no appreciable change of temperature, and consequently no marked change in volume" (J.A. Steers, 1932). Within the zone of 700 km from the earth's surface every upper layer has cooled earlier and more than the layer immediately below the upper layer. Thus, each upper layer contracted more than the layer just below it. Further, each upper layer continued to cool unless obstructed by the immediate lower layer. The outer layer began to cool first due to loss of heat through radiation. It may be pointed out that there is a limit of cooling beyond which no further cooling is possible. After maximum cooling and resultant contraction of the upper layer lower layer just lying below the upper layer begins to cool and contract, with the result already cooled and contracted upper layer becomes too large to fit in with the still cooling and contracting lower layer. The core of the earth is not affected by cooling because of exceptionally high temperature prevailing there. Thus, the core obstructs the contraction of the layer lying above it. The cooling and contracting layer lying below the already cooled and contracted layer become too big to fit in with the core of the earth. There is such a layer between the upper and lower layer where contraction is such that the intermediate layer can fit in with the lower layer. This layer is called level of no strain.

The layer lying over the level of no strain is too big to fit with the lower layer and hence the upper layer has to collapse on the lower layer so that it canfit with the lower layer. This process (collapse of upper layer on lower layer) results in the decrease in the radius of the earth which causes horizontal compressive stress which leads to buckling and folding of the rocks of upper layer. Thus, the mountains are formed. The lower layer below the level of no strain is too short to fit with the core of the earth and hence the lower layer has to stretch horizontally. This process implies a lateral spreading and thinning out of the materials of the lower layer below the level of no strain. The spreading and thinning of the lower layer introduces a state of stress which causes fractures and fissures resulting into breaking of rocks. This mechanism allows further collapse of the already cooled outer layer and thus already formed mountains are subjected to further rise in height.

Jeffrey's has also explained various aspects of mountain building e.g., period of mountain building, zones of mountain building, direction of mountains, etc.

Period of Mountain Building : According to Jeffreys the process of aforesaid mechanism of mountain building is not always active throughout the geological periods rather is confined to certain periods only. There is continuous accumulation of compressive and tensile forces resulting from contraction of the earth due to cooling and this process continues until the accumulated forces exceed the rock strength. When, this state (when accumulated compressive and tensile forces exceed the rock strength) is reached, folding and faulting are introduced and the process of mountain building sets in and this process continues till the compressive and tensile forces are strong and active. When these forces become weak, mountain building stops and the period of quiescence sets in. Again the process of accumulation of compressive and tensile forces starts and the next process of mountain building begins when these forces again become strong enough to fold the crustal rocks. Thus, two periods of mountain building are separated by a long period of quiescence.

Zones of Mountain Building : According to Jeffreys mountain building depends upon the nature and strength of rocks. The areas having soft and elastic rocks are most affected by the process of mountain building as the rocks are easily folded by compressive forces caused by contraction but the regions having hard and less elastic rocks are affected by tensile forces and thus several faults and fractures are formed because such rocks are easily broken into blocks. It is, thus, apparent that mountain building is localized in certain zones of the globe.

Direction of the Force : According to Jeffreys not all the areas below the earth surface are equally affected by the mechanism of cooling and contraction. The cooling process was more active below the oceanic crust than the continental crust because of dissimilar structure of these two zones. Thus, the rocks below the oceanic crust experienced more cooling and contraction than the rocks below the continental crust. Thus, the force of contraction is directed from the oceanic crust towards the continental crust. This mechanism results in the formation of mountains along the continental margins parallel to, the oceans. Rockies and Andes are the examples of such situation.

Direction of Mountains : According to Jeffreys the compressive force generated by contraction of the earth due to cooling was directed from oceanic areas towards the continental areas almost at right angle and thus the mountain ranges were formed parallel to the oceanic areas. The layout and direction of the Rockies and Andes mountains are very well explained on the basis of this theory because these mountains run north to south along the western margins of North and South America respectively and are parallel to the Pacific Ocean but the west-east extent of the Alps and the Himalayas cannot be explained on the basis of this theory.

The Evaluation : Though Jeffreys has attempted to explain the origin and evolution of surface features of the earth and has presented several evidences in support of his thermal contraction theory but his theory has been severely criticised and attacked on the following grounds.

(1) The force of contraction resulting from the cooling of the earth is not sufficient enough to account for the origin and evolution of major surface reliefs of the globe. A Holmes has remarked that 'the calculated reduction of area (by Jeffreys) is seriously in deficit of the amount to explain mountain building.'

(2) The concept of cooling of the earth in the system of concentric shells (layers) is erroneous and is not acceptable.

(3) The impact of decrease in the speed of rotation of the earth on mountain building is doubtful. J.A. Steers (1932) has aptly remarked, 'It may, in fact, be safely concluded that whatever effects the changing speed of rotation in geological times may have had, it was totally inadequate to influence mountain building in any marked way.'

(4) It is improper to believe that contraction would have been so immense about 200 million years ago so that it might have formed such gigantic mountains of Tertiary period as the Rockies, the Andes, the Alps, the Himalayas etc.

(5) As per thermal contraction theory of Jeffreys the continents and oceans should have been uniformly distributed as the earth was contracted from all sides but presently there is uneven distribution of continents and oceans.

(6) According to this theory the situation of mountains should always be parallel to the oceans. The arrangement of the Rockies and Andes is justified on the basis of this theory but the arrangement of European Alpine mountains and the Himalayas cannot be explained.

(7) If we believe in the competence of the force of contraction to form mountains it cannot produce great ranges of mountains as they are found at present over the globe but it would produce a larger number of small puckers or minor folds.

(8) According to this theory there should not be any definitive distributional pattern of mountains as they may be formed everywhere because all parts of earth's crust experienced contraction but contrary to this mountains are found in certain patterns e.g., along the margins of the continents extending either north-southward or west-eastward.

Daly's Theory of Sliding Continent

Objectives : Daly postulated his theory of sliding continents in his book 'Our Mobile Earth' in the year 1926 to explain the origin and evolution of different relief features of the earth's surface. Though Daly attempted to throw light on major relicfs of the globe but his main objective was to explain the causes and processes of mountain building. He attempted to explain salient aspects of folded mountains e.g., origin, successive upheavals, distributional patterns and orientation and extent.

Orogenetic force : The main force implied by Daly for the origin of the mountains has been the force of gravity. The whole theory of Daly is based on the nature and rate of downward slide of the continents fostered by gravitational force. 'The key to the Daly's views is the idea that there has been downhill sliding movement of continental masses. In other words, the controlling factor has been gravity' (J.A. Steers, 1932). Daly himself proclaimed that his theory based on gravitational force was competent to deal with all the problems of mountain building satisfactorily.

Axioms of the Theory : Daly has assumed certain axioms (self proved facts) in support of his theory. If we look into the history it appears that 'a major part of the theory is based on self proved facts or axioms'. It may be pointed out that Daly did not elaborate his axioms. He admitted himself that his theory can well explain the problems of orogenesis on the force of gravity alone.'

According to Daly a solid crust was formed just after the origin of the earth. He named this solid crust as primitive crust. In early times there existed a series of ancient rigid masses which were generally situated near the poles and around the equator. These rigid masses have been named by Daly as polar and equatorial domes. Thus, there were three belts of rigid masses e.g. (i) north polar domes, (ii) equatorial domes and (iii) south polar domes. These three belts of rigid masses were separated by depressed regions which were called by Daly as mid-latitude furrows and primeval Pacific Ocean. These depressed regions were, in fact, oceanic areas (or say geosynclines) the beds of which were formed of primitive crust which was formed with the origin of the earth.

The crust, according to Daly, composed of granites, was heavier than the rocks of substratum below the crust. The crust was composed of heavier granites while the substratum was formed of lighter glassy basalt. It may be pointed out that this view of Daly is isostatically totally wrong. He further assumed that the water bodies occupied about half of the globe and Tethys geosyncline (northern mid-latitude furrow between north polar dome and equatorial dome) 'was a marked feature throughout much of geological time.' Land masses (polar equatorial domes - rigid masses) projected above the water bodies and the polar and equatorial domes were sloping towards mid-latitude furrows (which were in fact geosynclinal tracts) and the Pacific Ocean.

The Mechanism : Daly has believed in the collapse of the primitive crust but has not elaborated the mechanism of collapse. It may be surmised that the primitive crust would have been probably bad conductor of heat and so the surface temperature would have fallen soon to that of the present time but the loss of heat from the interior into the exterior part continued and hence the interior part contracted away from the outer shell or crust. Consequently, the outer crust would have collapsed on the still contracting interior due to (i) the weight of the oceanic water, (ii) the weight of geosynclinal sediments and (iii) gravitational force of the centre of the earth. It may be pointed out that the impact of gravitational force was more under the oceanic crust than the continental domes because the former was nearer to the earth's centre. It appears (though not described by Daly) that the mid-latitudinal furrows were formed as geosynclines due to collapse of outer crust on the contracting interior of the earth and due to the gravitational force coming from the centre of the earth.

The sediments derived through the erosion of polar and equatorial domes (more precisely continental domes) were deposited by the rivers into the mid-latitudinal furrows and the Pacific Ocean (geosynclines). Continuous sedimentation and weight of the oceanic waters exerted enormous pressure on the beds of oceans (geosynclines) with the result their beds were subjected to continuous subsidence. Thus, downward pressure on the oceanic (geosynclinal) beds due to continuous sedimentation and resultant subsidence of geosynclinal beds caused lateraf pressure on the continental masses, with the result they were transformed into broad continental domes known as polar and equatorial domes. As the oceanic beds were depressed downward due to gravitational force of the earth's centre, and weight of oceanic water and geosynclinal sediments, the size of domes continued to increase. It may be pointed out that gradual increase in the lateral pressure resulting from continuous downward movement of geosynclinal beds was responsible for increase in the size and height of the continental domes.

The sediments of the continental domes began to expand because of increase in the size and height of the domes and consequently sediments of the domes began to lose weight and became lighter in weight. In order to compensate the loss of weight of sediments of the continental domes there began underground flowage of densc materials from below the oceanic (geosynclinal beds) beds towards the continental domes. Because of this process denser materials began to accumulate in the continental domes from below. Because of the repetition of the above processes the continental domes continued to grow in size and height, 'probably not as rapidly in the centre as towards their peripheries'. The increase in the size of domes caused pressure on the crust under the oceanic beds (geosynclinal beds). As the size of domes continued to expand, the resultant pressure on oceanic beds also continued to increase. When the tolerance limit of the oceanic crust to withstand the ever increasing pressure was crossed, the oceanic beds began to rupture and break. Thus, the support of the continental domes was removed due to rupture of the oceanic beds which introduced strong tensional movements due to which larger blocks of continental mass began to slide towards the geosynclines. The geosynclinal sediments were thus squeezed and folded due to compressive force coming from the sliding continental blocks giving birth to folded mountains.

According to Daly the broken continental blocks and parts of oceanic crust founder in the substratum because the density of outer crust is more than the substratum. On the other hand, the geosynclinal sediments do not founder in the substratum rather these float on the substratum because these are less dense than the substratum. Because of this fact geosynclinal sediments are more folded by the compressive forces generated by sliding continental blocks. It is, thus, obvious that greater the amount of slipping of continental blocks, the more geosynclinal sediments are squeezed and more and greater folded mountains are formed. Daly has further pointed out that the foundered continental blocks in the substratum are melted due to high temperature and thus rise in the volume of molten continental blocks causes further rise in the mountains.

The sliding continent theory of Daly also well explains the distributional patterns of folded mountains e.g., north-south and west-east extents. According to Daly folded mountains are formed because of squeezing and folding of geosynclinal sediments by compressive forces caused by sliding of the continental blocks towards the geosynclines. Thus, west-east extending mountains (e.g. Alpine chains and the Himalayas) were formed due to sliding of polar and equatorial domes towards mid-latitude furrow (Tethys geosyncline) and north-south extending mountains (e.g. Rockies and Andes) were formed due to sliding of continental masses towards Pacific Ocean. Similarly, the island arcs and festoons parallel to the Asiatic coast were formed due to sliding of Asiatic mass towards Pacific Ocean.

Evaluation of the Theory : Though the 'sliding continent theory' by Daly is based on well known principle of gravitational force and tries to explain the problem of mountain building in a simple manner yet it does not present coherent account of the problem as the theory does not go into details and there is a wide gap between theoretical and practical aspects of the theory. Major part of the theory is based on self proved facts (axioms). The theory suffers from the following limitations.

(1) The sliding continent theory presents erroneous concepts about the structure of the interior of the earth. His concept, that the outer crust is denser than the substratum, is against the evidences of seismology because it is now proven fact that the density increases with increasing depth in the interior of the earth.

(2) Daly's theory is based on several guesses and surmises. Why did the earth's crust become asymmetrical? Why the continental domes were sloping towards mid-latitude furrows (geosynclines)?. How was the Pacific Ocean formed ? Daly does not offer any convincing explanation to these and many more questions.

(3) This theory presents erroneous views about geosynclines because these are generally considered as long, narrow and relatively shallow depressions of water but Daly's geosynclines were in fact oceans (e.g. mid-latitude furrows and Pacific Ocean). If these are accepted as geosynclines they would have never been filled with sediments and thus no mountains could have been formed.

(4) Daly has also presented confusing ideas and erroneous concepts about the mechanism of mountain building. In fact, this theory does not care for the extension and depth of oceans and amount of sediments deposited in them but expects mountains from every ocean (geosyncline).

(5) The theory provides wrong views about the mechanism and process of gravity. The theory does not throw light on the termination of pulling effects of gravity

and the beginning of the rupture of the beds of the geosynclines. Thus, there is no coherence between different events related to mountain building as envisaged by the sliding continent theory. In fact, the theory presents some piecemeal analysis of mountain building rather than a complete or perfect perspective.

(6) The theory to certain extent believes in such distribution of land and sea (mid latitude furrows and Pacific Ocean and polar and equatorial continental domes) as to suit its own purpose. Wooldridge and Morgan have aptly remarked, 'compiete rejection of the idea may be premature, but it is a fair comment to say that the cause of primary "bulges" which start the slipping has in no sense been satisfactorily indicated.'

Theory of Thermal Convection Current

Objectives : Arthur Holmes postulated his thermal convection current theory in the year 1928-29 to explain the intricate problems of the origin of major relief features of the earth's surface. Holmes' major objectives were not confined to search the mechanism of mountain building based on sound scientific background but were also directed towards finding scientific explanation for the origin of the continents and ocean basins in terms of continental drift as he was opposed to the concept of permanency of the continents and ocean basins as envisaged by the advocates of thermal contraction of the earth. Wooldridge and Morgan have aptly remarked, "The only unifying theory which shows a hopeful sings of reconciling certain of the divergent hypotheses of mountain building and continental drift is that due to Holmes' (S.W. Wooldridge and R.S. Morgan, 1959).

Orogenetic Force : The driving force of mountain building implied by Arthur Holmes is provided by thermal convection currents originating deep within the earth. The main source of the origin of convective currents is excessive heat in the substratum wherein disintegration of radioactive elements generates heat regularly. In fact, the whole theory depends exclusively on the mechanism of convective currents.

Base of the Theory : According to Holmes the earth consists of 3 zones or layers e.g. (i) upper layer of granodiorite (10 to 12 km), (ii) intermediate layer (20 to 25 km) of amphiboliw and (iii) lower layer of eclogite. He has further grouped these three layers into two zones e.g. (i) crust consisting of upper and middle or intermediate layers and crystalline upper part of lower layer and (ii) substratum representing molten part of the lower layer. Crust and substratum are composed of sial and sima respectively. Generally, sial is absent in the oceanic areas.

The origin of convective currents within the earth depends on the presence of radioactive elements in the rocks. The disintegration of radioactive elements generates heat which causes convective currents. According to Holmes there is maximum concentration of radioactive elements in the crust but temperature is not so high because there is gradual loss of heat through conduction and radiation from the upper surface at the rate of 60 calories per square centimeter per year. 'This is approximately equal to the radioactive energy produced by a layer 14 km thick of granite, 16.5 km of granodiorite, 52 km of plateau basalt or gabbro and 60 km of peridotite' (J.A. Steers, 1932). According to Holmes the loss of heat from the earth's surface is compensated by the heat produced by a crustal shell of 60 km thickness. Thus, there is no accumulation of additional heat in the earth's crust inspite of maximum concentration of radioactive elements. On the other hand, though there is very low concentration of radioactive elements in the substratum but the gradual accumulation of heat produced by the radioactive elements causes convective currents. The convective currents depend on two factors e.g. (i) thickness of the crust near the equator and poles and (ii) uneven distribution of radioactive elements in the crust. Ascending convective currents originate under the crust near the equator because of greater thickness of crust whereas descending convection currents are originated under the polar crust because of its shallow depth. The rising convective currents originating from below the continental crust are more powerful than the convective currents originating from below the oceanic crust because of greater concentration of radioactive elements in the continental crust.

The Mechanism : Convective currents, thus, are generated at some places in the substratum. Because of difference of temperature gradient from the equator (greater) towards the poles (low) rising convective currents are formed under the equatorial crust while downward moving (descending) convective currents are generated under the polar crust. The convective currents originating under the continental crust are more powerful than the convective currents originating under the oceanic crust. It may be pointed out that the currents originating under the equatorial crust move towards the poles i.e., towards north and south and thus the crusts are carried away with the convective currents.

There are two situations of rising convective currents when they reach the lower limit of the crustal masses.

(i) The crustal mass, where two rising convective currents diverge in opposite directions, is stretched and thinned due to tensional forces and ultimately the crust is ruptured and broken into two blocks which are carried away by lateral divergent convective currents and the opening between two blocks becomes seas. Thus, divergent convective currents cause continental drift.

(ii) Where two lateral convective currents originating under the continental and oceanic crusts converge. compressive force is generated which causes subsidence in the crustal zones giving birth to geosynclines and closing of sea. It is apparent that divergent connective currents move the crustal blocks away in opposite directions and thus create seas and oceans while convergent convective currents bring crustal blocks together and thus form mountains.

The convective currents are divided into two groups on the basis of their locational aspect e.g. (i) convective currents of rising columns and (ii) convective currents of falling columns. The rising convective currents after reaching the lower limit of the crust diverge in opposite directions. This outward or divergent movement introduces tensional force due to which the crust is stretched, thinned and ultimately broken and the broken crustal blocks are moved apart. The wide open area between two drifting crustal blocks in opposite directions is filled with water and thus an ocean is formed. According to Holmes the equatorial crust was stretched and ruptured due to divergence of rising convective currents which carried the ruptured crustal blocks towards the north and south and Tethys Sea was formed. This phase is called 'Opening of Tethys'. Again two sets of convergent or downward moving (descending) currents brought Laurasia and Gondwanaland together and thus Tethys was compressed and folded into Alpine mountains. This phase is called 'Closing of Tethys'.

'The convective mechanism is not a steady process but a periodic one, which waxes and wanes and then begins again with a different arrangement of centre' (A. Holmes 1952). It means that the convective currents originate at several centres which are not permanent. Geosynclines are formed due to subsidence of crustal blocks mainly continental shelves due to compressive force generated by convergent convective currents moving laterally together under continental and oceanic crusts. In other words, when the continental and oceanic crusts move together and converge under the continental shelves, they descend downward and thus cause immense compression due to which the crust is subjected to subsidence to form geosyncline. The geosynclines which are always located above the convective currents of rising columns under continental and oceanic crusts bring materials in the descending convective currents of falling column. Continuous compression and sedimentation causes gradual subsidence of geosynclines. Holmes has described a cyclic pattern of thermal convective currents which includes the origin of convective currents, formation of geosynclines, sedimentation and orogenesis and further rise in the mountains. According to Holmes the cyclic pattern of convective currents and related mountain building pass through three phases or stages.

First Stage : The first stage is of the longest duration during which convective currents are originated in the substratum. The rising convective currents of two centres converge under the continental shelves and thus form geosynclines due to compression coming from the convergence of two sets of lateral currents. Geosynclines are subjected to continuous sedimentation and subsidence. As the sediments are pressed downward into geosynclines, these go further downward and are intensely heated and metamorphosed. Metamorphism of sediments causes rise in their density which further causes downward movement of the metamorphosed materials. Thus, the falling column of downward moving convective currents is the column of increasing density. Amphibolites are metamorphosed into ecologites. A portion of heat is spent during the process of metamorphism and hence the heat does not accumulate to greater extent. The first stage, characterized by high velocity convective currents, is in fact the preparatory stage of mountain building which is marked by the creation of geosynclines, sedimentation and subsidence of materials partly, caused by compression resulting from convergence of convective currents and partly by increase in the density of materials due to metamorphism.

Second Stage : The second stage is marked by phenomenal increase in the velocity of convective currents but this stage is relatively of short duration. The main cause for the phenomenal increase in the velocity of convective currents is the downward movement of cold materials in the falling column and upward movement (rise) of hot materials in the rising column of convective currents. Increased pressure due to metamorphism of geomaterials in the falling column of descending currents increases the velocity of downward moving convective currents. The high velocity convergent convective currents buckle geosynclinal sediments and thus initiate the process of mountain building. This stage, thus, is called the stage of orogenesis.

Third Stage : The third stage is characterized by waning phase of thermal convective currents due to incoming hot materials in the falling column and upward movement (rise) of colder materials in the rising column. Gradually, the rising column becomes a cold column i.e., cold materials are accumulated at the centre of the origin of rising (upward moving) convective currents due to which these currents cease to operate and the whole mechanism of convective currents comes to an end. The termination of the mechanism of convective currents yields several results e.g. (i) The materials of the falling column start rising because of decrease in the pressure at the top of the falling column due to the end of deposition of materials. This mechanism causes further rise in the mountains. (ii) The depressed and subsided heavier materials in the falling column of descending convective currents start rising due to decrease in the weight and pressure at the top of the falling column. (iii) Eclogite, which was depressed downward, gets melted due to immense heat and thus it expands. This expansion in the volume of molten eclogite causes further rise in the mountains. This stage is known as the stage of gliptogenesis. It is, thus, apparent that the thermal convective currents of Holmes explains all the three stages of mountain building e.g. lithogenesis, orogenesis and gliptogenesis.

Griggs through his experiments has validated the mechanism of convective currents and consequent mountain building.

The Evaluation : Commenting on Holmes thermal convective current theory J.A. Steers (1932) has remarked, 'The theory is interesting, but it depends upon such factors about which little is known'. It may be pointed out that this comment of Steers about 68 years ago is not valid today as there are ample convincing scientific evidences which validate the mechanism of convective currents originating from within the mantle.

Even much appreciated plate tectonic theory is based on thermal convective currents. The theory was criticised, at the time of its postulation in 1928-29, on the following grounds.

(1) Convective current theory, no doubt, is a leading theory in a new direction but whole of the theory depends on such factors about which very little is known. Rising and failing columns are doubtful phenomena and therefore doubtful stage can never be taken for the explanation of natural phenomena.

(2) The whole mechanism of convective currents depends on the heat generated by radioactive elements in the substratum (now mantle) but several scientists have raised doubt about the availability of required amount of heat generated by radioactive elements. If heat, thus, is insufficient, convective currents may not be generated and, therefore, the whole mechanism and working of the theory would not be possible. It may be further pointed out that the rising currents pass on their heat into the crust through conduction. This process also causes loss of heat which may weaken the currents.

(3) The horizontal flow of thermal convective currents under the continental and oceanic crusts is also a doubtful phenomena because of lack of required amount of heat to drive these currents. If horizontal flow of convergent movement of convective currents is not possible, then the falling column would not exist and hence mountain cannot be formed.

(4) The metamorphism of amphibolites into ecologites and resultant downward movement of relatively denser ecologites is also a doubtful phenomenon. Even we accept the metamorphism of amphibolites into ecologites but the resultant increase in density from 3.0 to 3.4 would not be enough to depress and sink ecologites in the falling column. If desired sinking of ecologites is not possible, there would not be proper accommodation of materials brought by the horizontal convergent convective currents into the falling column. If this is so, the whole falling column would be filled with ecologites and the next stage of the mechanism of the convective currents would not work. It is, thus, argued that the theory does not make proper provision for the accommodation of additional materials.

(5) According to this theory convective currents are originated at few centres only under the continental and oceanic crusts but question arises, why are they not

originated at all places ? If this so happens, the horizontal movement of these currents would not be possible. The whole of the continents would be divided into several blocks as the rising convection currents originating from numerous centres would break the crusts and would give birth to volcanic eruptions of various sorts. This observation has been now validated on the basis of plate tectonics as rising convective currents diverge under the mid-oceanic ridges and thus the plate is ruptured and two plates move in opposite directions due to divergent convective currents and fissure flows of lavas occur along the mid-oceanic ridges representing the rupture zone. It may be concluded that the idea of thermal convective currents conceived by A. Holmes about 71 years ago (in 1928-29) proved its worth in 1960s when the scientists were looking forward to search such a force which can explain the movement of plates. Now, the process of mountain building can be very satisfactorily explained on the basis of convective currents though not in the way as conceived by A. Holmes in 1928-29 but on the lines of plate tectonics.

Joly's Theory of Radioactivity

Objectives : Joly postulated his theory based on radioactivity of certain radioactive minerals in the year 1925 in his book, 'Surface History of the Earth' to account for the origin and evolution of surface features of the earth. His theory is also known as 'thermal cycle theory' or 'theory of the surface of the earth'. Though the main objective of Joly's theory was to present a detailed account of the thermal history of the earth and mathematical explanation of the structure of the interior of the earth but he also attempted to explain the problems of mountain building and the continental drift. In fact, 'Joly's views on the earth's surface history are based on such reasonable premises, and are so simple in their conception, that they have met with a great deal of favour' (J.A. Steers, 1932). While commenting on Joly's theory of radioactivity he has remarked, 'It should not be accepted as proved, but retained as an hypothesis which probably contains a certain elements of truth'.

Orogenetic Force : The driving force of mountain building as invoked by Joly is provided by expansion and contraction of the substratum of the earth resulting into transgressional and regressional phases of the seas (geosynclines). The expansion and contraction of the substratum are based on the mechanism of heat generated by radioactive elements of the rocks. It may be pointed out that the theory of A. Holmes and Joly are based on radioactive elements but they sought their help differently e.g. Holmes used radioactive elements to explain the origin of thermal convective currents in the substratum while Joly used them to explain the melting and resolidification of the substratum. He also implied tidal force and friction to explain continental drift.

Base of the Theory : The whole mechanism of Joly's theory is based on the presence of radioactive elements of the rocks of the earth. In order to explain various aspects of the mechanism of radioactive elements Joly has described first the structure of the earth. According to him continents are made of lighter sialic materials the density of which is 2.67 while the oceanic beds are formed of heavier materials of sima having average density of 3.0. Thus, the crust has been assumed to have been composed of sial and substratum of basalt (sima). Besides a few exceptions, sial is not found in oceanic beds.

According to Joly the rocks of the earth contain radioactive elements but their distribution is not uniform in all zones of the earth. Radioactive elements are found in abundance in sialic zone or the continental rocks but the rocks of sima forming the oceanic crusts are less radioactive. Continuous breakdown of certain radioactive elements like uranium, thorium etc., generates heat. It may be pointed out that the actual rate of heat production by radioactive elements is exceedingly small but it becomes sufficient enough to produce appreciable result after long period of accumulation. Though the production of heat is comparatively higher in the continental crust because of more radioactive elements than the oceanic beds but there is no large-scale accumulation of heat in the continental crust due to continuous loss of heat through radiation.

The Mechanism : According to Joly the disintegration of radioactive elements of sialic or continental rocks produces heat but it does not accumulate in the continents or sial because the total loss of heat through radiation from the sialic crust is more than the total heat produced by the radioactive elements. He has further pointed out that temperature increases with increasing depth. After detailed mathematical calculation Joly estimated the amount of temperature at the depth of 30 km to be 1050°C. He estimated the maximum thickness of sial to be 30 km. According to him there is no transfer of heat from sima to overlying sial. He has also estimated the amount of temperature at the outer limit of sima under the continents to be 1050°C. The conditions under the oceans are rather different. Since there is no sial in the oceanic beds, so the heat produced by radioactive elements, though very small, is lost to the oceanic water through conduction but such situation does not exist at greater depths in the substratum (sima) under the oceans. Temperature increases with increasing depth in the substratum (sima) under the oceans because of accumulation of heat produced by radioactive elements. This mechanism causes temperature gradient at great depth in sima (substratum). The temperature becomes equal to the melting point of basalt. There is no transfer of heat from the lower part of sima to the upper part of sima so there is accumulation of heat in the lower layers of sima beneath the oceans. The melting point is 1150°C whereas the temperature at the top of substratum (sima) is 1050°C. If the temperature of the substratum rises to 1150°C it would attain its melting point but the substratum would still remain in solid state unless required amount of latent heat of fusion is provided. Joly has calculated that the required amount of heat to liquefy the substratum would be available in 33,000,000 to 56,000,000 years. If such conditions become possible i.e., if the substratum reaches the molten condition, several changes take place in the earth's structure.

Period of Transgressional Sea : Several interesting events take place when the substratum reaches the molten condition due to accumulation of greater amount of heat produced by the breakdown of radioactive elements.

(1) The expansion of sima due to melting causes increase in the radius of the globe.

(2) Continental masses or sialic masses are raised relative to the centre of the globe.

(3) The density of sima decreases due to melting and hence sialic masses begin to sink in molten sima.

(4) The level of oceanic water rises due to sinking of sialic or continental masses into liquid sima. This mechanism causes extension of oceanic water over the continental margins. This process of expansion of oceanic waters and their encroachment on continental margins is called transgression of sea and the concerned stage is known as the phase of transgressional sea.

(5) Transgression of sea results in sedimentation on the submerged continental margins. Thus, this theory of radioactivity accounts for the origin of geosynclines due to submergence of continental margins during transgressional phase of sea.

(6) The conditions under the oceans are different because there is absence of sial. The increase in the radius and the circumference of the globe due to melting of sima produces tension in the oceanic beds which causes cracks and faults. Molten materials or molten basalts come upward through these cracks and faults. These molten basalts are then solidified and thus oceanic islands are formed. The radioactivity theory, thus, explains the islands of the Pacific and other oceans.

(7) Continental masses easily float over molten sima, consequently they are more affected, by tidal force which causes westward movement of the continents. It is in this way that the radioactivity theory also describes the process of continental drift.

(8) Continental drift changes the position of the continents and the oceans as the former occupy the positions of the latter. This process allows the escape of heat and thus the transgressional phase comes to an end.

Period of Regressional Sea : The phase of regressional sea is characterized by the following events -

(1) The temperature of the substratum decreases because of loss of heat due to continental drift. Thus, the cooling of the substratum results in the resolidification of molten substratum. The cooling of the substratum begins from its upper layer

and continues downward and ultimately the whole of the substratum becomes solid on cooling.

(2) The density of the substratum, which was relatively decreased during its molten stage, again increases to regain its previous value.

(3) The radius and the circumference of the globe, which were increased due to melting of the substratum, are again shortened to their previous position, with the result the continents, which were raised relative to the centre of the globe, are again brought to their previous positions.

(4) Relative increase in the density of the substratum due to resolidification causes contraction of the oceanic beds which results in the withdrawal of oceanic waters from the continental margins. This is called the phase of regressional sea. Because of the withdrawal of oceanic water previously submerged continental margins (during the phase of transgressional sea) rise upward and the deposited sediments are exposed above the water level.

(5) It may be remembered that the oceanic beds were subjected to maximum expansion during the period of transgressional phase due to melting of the substratum. Similarly, the oceanic beds are also subjected to maximum contraction during the period of regressional sea due to resolidification of molten substratum. Thus, contracting beds of two oceans exert lateral compression on the sediments deposited on the continental margins (geosynclines), consequently the sediments deposited during the period of transgressional sea are squeezed, buckled and folded and thus mountains are formed.

Joly has described two parallel processes of mountain building. (i) The sediments deposited in the shallow seas of the continental margins are squeezed and folded due to lateral compression caused by two contracting oceanic beds. (ii) Vertical force is produced during the process of resolidification of the substratum. This vertical force raises the whole mountain system formed during the first process. It is obvious that according to this theory mountains are always formed along the margins of the continents facing oceans. The intensity of lateral pressure and consequent magnitude of folding depend on the amount of contraction of oceanic beds. It may be argued that large oceans would produce more powerful lateral compression and hence greatest mountain would face largest ocean. To some extent this statement is true as the Rockies and the Andes mountains face the Pacific Ocean.

Joly also explains the period of quiescence between two periods of mountain building. The total period of two solid phases of the substratum (solid phase, molten phase and resolidification phase of the substratum) is called one revolution wherein the melting of substratum (sima) takes total time period of 33,000,000 to 56,000,000 years. It may be, thus, inferred that the process of mounting building occurs in cyclic manner wherein the period of mountain building is alternated by the period of quiescence.

The Evaluation : Though the radioactivity theory of Joly based on scientific facts and mathematical calculation was widely appreciated by several scientists but simultaneously it was also severely criticized. A few critics of the theory do not grant theoretical status to the views of Joly rather they take his views as merely descriptive accounts of the earth's interior. In fact, the theory of Joly is a well developed geomorphic story of the earth rather than a theory. J.A. Steers (1932) has remarked that 'the theory is, at first sight, convincing and it certainly does give adequate explanations of many features of the earth's surface'. The following shortcomings have been pointed out by the critics of the theory.

(1) The theory is based on radioactive elements of the rocks of the earth at different depths about which very little is known. Thus, the force of expansion and contraction of the substratum (sima) due to melting and cooling respectively based on radioactive elements is doubtful and perhaps is not enough to form mountains.

(2) Jeffreys did not agree with the 30-km thickness of the continental masses as envisaged by Joly. According to Jeffreys the thickness of the continental crust may not be more than 16 km. If the thickness of the continental crust is accepted to be 16 km then the whole mechanism of Joly's theory would come to a grinding halt as required amount of heat of 1150°C would not be possible at the depth of 16km.

(3) Joly's concept of cyclic nature of mountain building has been disputed by some critics. The theory envisages uniform periods of quiescence between two periods of mountain building but this concept has also been disputed. J.A. Steers has commented that "In short, the very essence of the theory, the approximately equally spaced recurrence of similar conditions, seems to be one of its main drawbacks." He has further remarked that "there seems to be little doubt that mountain building periods have been recurrent to some extent, but it is very doubtful if they have been so regular as Joly's theory would make them" (J.A. Steers, 1932).

(4) This theory envisages two facts about mountain building. (i) 'The greatest mountains must face the greatest oceanic beds.' (ii) Both the margins of the continent must have mountains of the same period and both the margins should be regular. The first fact is validated to some extent but the second fact is not validated.

(5) This theory presents erroneous concept about geosynclines. As per this theory geosynclines are always formed due to submergence of continental margins due to transgression of seas. It means that geosynclines should always be located

around the continents. On the other hand, it has been generally accepted that geosynclines are long, narrow and shallow water bodies which are characterized by continuous sedimentation and subsidence but Joly's geosynclines receive sediments but do not undergo the process of subsidence. Without subsidence the enormous thickness of sediments of the present Alpine mountains cannot be explained.

Theory of Plate Tectonic

Objective : Plate tectonic theory is a comprehensive theory which offers explanations for various relief features and tectonic events viz. mountain building, folding and faulting, continental drift, vulcanicity, seismic events (earthquakes) etc. The theory belongs to a host of scientists of different disciplines. Plate tectonic theory is, in fact, the outcome of combined efforts of many scientists of different countries working together and separately. The theory came into light in the 1960s. It envisages the formation of mountains due to collision of plate boundaries.

Orogenetic Force : The orogenetic force to form mountains is provided by the compressive forces caused by the collision of two convergent plates along the destructive plate boundaries. Thermal convective currents originating in the mantle have been accepted as the competent force for the movement of plates. The plates move in different directions relative to each other under the impact of thermal convective currents. Plate movements take place in accordance with the Euler's geometrical theorem which envisages the movement of plates in the form of simple rotation along a pole of rotation.

Base of the Theory : The rigid lithospheric slabs or rigid and solid land masses having a thickness of about 100 km composed of earth's crust and some portion of upper mantle are technically called 'plates'. The term 'plate' was first used by Canadian geologist J.T. Wilson in 1965. The whole mechanism of the evolution, nature and motion and resultant reactions of plates is called 'plate tectonics'. Plate tectonic theory, a great scientific achievement of the decade of 1960s, is based on two major scientific evidences e.g. (i) evidences of palaeomagnetism and (ii) evidences of sea-floor spreading. Six major and 20 minor plates have been identified so far (e.g. Eurasian plate, Indian-Australian plate, American plate, Pacific plate, African plate and Antarctic plate).

McKenzie and Parker discussed in detail the mechanism of plate motions on the basis of Euler's geometrical theorem in 1967. Hary Hess (1960) elaborated the mechanism of plate movement on the basis of the evidences of sea-floor spreading. W.J. Morgan and Le Pichon elaborated the various aspects of plate tectonics in 1968.

Three types of plate boundaries (see fig. 6.7 in chapter 6 on the origin of continents and ocean basins) have been identified e.g. (i) destructive plate boundaries, (ii) constructive plate boundaries and (iii) conservative plate boundaries.

(1) Constructive plate boundaries also called as 'divergent plate boundary' or 'accreting plate boundary' represent zones of divergence along the mid-oceanic ridges and are characterized by continuous addition (accretion) of materials as there is constant upwelling of molten materials (basaltic lavas) from below the mid-oceanic ridges. These basaltic lavas are cooled and solidified and are added to the trailing margins of the divergent plates and thus new oceanic crust is continuously formed. In fact, oceanic plates split apart along the mid-oceanic ridges and move in opposite directions and thus transform faults are formed.

(2) Destructive plate boundaries also known as 'consuming plate boundaries' or 'convergent plate boundaries' are those where two plates collide against each other and the leading edge of one plate having relatively lighter material overrides the other plate and the overridden plate boundary of relatively denser material is subducted or thrust into the upper mantle and thus a part of the crust is lost in the mantle. This mechanism results in constant loss of crustal materials.

(3) Conservative plate boundaries also known as 'shear plate boundaries' are those where two plates slip past each other without any collision along the transform fault and thus crust is neither created nor destroyed.

The Mechanism : According to plate tectonic theory mountains are formed due to collision of two convergent plates. Mountains are always formed along the destructive plate boundaries. It is obvious that the process of mountain building is associated with destructive plate boundaries of two convergent plates. The plate tectonic theory envisages the formation of mountains due to compression of sediments caused by the collision of two convergent plate boundaries. Two plates moving together under the impact of thermal convective currents collide against each other and the plate boundary having relatively denser materials is subducted under the other plate boundary of relatively lighter materials. This subduction zone is also called Benioff zone. The subduction of plate boundary causes lateral compressive force which ultimately squeezes and folds the sediments and materials of the margins of the plates and thus mountains are formed. The subducted part of the plate after reaching a depth of 100 km or more in the mantle is liquefied and thus expands in volume because of conversion of the portion of plate into magma. This expansion of molten materials causes further rise in the mountains.

The convergence and consequent collision of plate boundaries occurs in three situations viz. (i) collision of two oceanic plates, (ii) collision of two continental plates and (iii) collision of oceanic-continental plates.

(1) *Convergence (Collision) of Two Oceanic Plates* : The collision of two oceanic plates and subduction of the boundary of the plate of relatively denser materials results in the formation of the fold mountain ranges of island arcs and festoons, for

example, island arcs and festoons formed by Japanese islands, Phillippines etc. around the western margin of the Pacific Ocean off the east coast of Asia. The fold mountain ranges of island arcs and festoons 'form where a section of the ocean floor is subducted in the ocean basin away from a continent i.e., where ocean floor crust is on either side of the convergent plate boundary' (M.J. Bradshaw et al. 1978).

The best example of the formation of mountains due to collision of two oceanic plates is the situation of Japanese island arc. Mountains of Japan range in height from 3000 m to 4000 m AMSL. It may be pointed out that all the mountains of Japan are of volcanic origin. Though Japanese mountains exhibit a number of characteristic features of folded mountains but they can no longer by regarded as fold mountains like the Alps and the Himalayas. Honshu Island represents the most characteristic example of the situation of the convergence of two oceanic plates.

Honshu is bordered by Japan Trench in the east and Japan Sea in the west. The western part of the island is more frequented by volcanic activities than the eastern part. The island is characterized by two belts of metamorphic rocks on either side. It is believed that the Japan Trench was formed due to subduction of Pacific Oceanic plate under the oceanic crust to the east of Japan. According to plate tectonic theory the subducted portion of plate after reaching a depth of 100 km or more starts melting due to high temperature prevailing in the upper mantle. The magma, thus formed, ascends and appears as volcanic eruption about 200 km away from the oceanic trench. Since Japan is very close to the Japan Trench and hence western part of Japan is more frequented by volcanic activities. This process is still continuing as the Pacific plate is being continuously subducted under the oceanic crust along the Japan Trench. The eruptions of volcano in the month of June, 1991 in Japan after a dormant period of about 200 years and the eruption of Mt. Pinatuboon June 9,1991 in Manila, Phillippines, validate the authenticity of this theory of plate tectonics. The volcanic eruptions caused by subduction of oceanic plates under the oceanic crust off the Japanese coast resulted into continuous accumulation of volcanic rocks and consequent increase in the height of island arc and thus the formation of volcanic mountains could be possible.

(2) *Convergence (Collision) of Continental and Oceanic Plates* : The collision of continental and oceanic convergent plates results in the formation of cordillera type of folded mountains e.g. the western cordillera of North America (including the Rockies). When one continental and the other oceanic plates collide due to their convergence along subduction or Benioffzone, the oceanic plate boundary being heavier due to comparatively denser materials is subducted below the continental plate

boundary. The sediments deposited on the continental margins are squeezed and folded due to compressive forces caused by the subduction of oceanic plate. The Rockies and the Andes mountains were formed due to subduction of the Pacific ocean plate under the American continental plate.

(3) *Convergence (Collision) of two Continental Plates* : When two convergent plates composed of continental crusts collide against each other, the continental plate having relatively denser materials is subducted under the other continental plate having comparatively lighter materials than the former. The resultant lateral compression squeezes and folds the sediments deposited on either side of the continental plate margins and the sediments of the geosynclines lying between two convergent continental plates and thus forms gigantic folded mountains e.g,. the Alps and the Himalayas.

The origin of the Alpine mountains of Europe and Asia are well explained on the basis of this mechanism (collision of two convergent continental plate boundaries) of plate tectonics. There existed a long Tethys geosyncline between Eurasian plate in the north and Africa-Indian plate in the south during Mesozoic era. The geosynclinal sediments of Tethys sea were squeezed and folded into Alpine-Himalayan mountain chains due to lateral compressive forces caused by the convergence and collision of Eurasia and African- Indian continental plates during Cenozoic era. It maybe pointed out that the formation of Alpine-Himalayan mountain chains could be possible due to continued collision of continental plates and consequent orogenesis along several subduction zones for long period of time.

About 70-65 million years ago there was an extensive geosyncline, known as Tethys geosyncline, in the place of the Himalayas. Tethys geosyncline was bordered by Asiatic plate in the north and Indian plate in the south. Tethys geosyncline began to contract in size due to movement of Indian and Asiatic plates together. About 60-30 million years ago the Indian plate came very close to Asiatic plate. The Indian plate began to actively subduct under the Asiatic plate. The convergence and collision of Asiatic and Indian plates and consequent subduction of Indian plate under the former caused lateral compression due to which the sediments of Tethys geosyncline were squeezed and folded into three parallel chains of the Himalayas about 30-20 million years ago. It has been estimated that the crust has been shortened by 500 km between Asiatic and Indian plates due to convergence of two plates and subduction of Indian plate.

Alpine mountains of Europe were formed due to convergence and collision of European and African plates. Since the collision of these two continental plates was very complex and hence the structure of the European Alpine mountains is also very complex. The African plate is still moving northward and is being subducted under European plate to the south of Aegean arc. Similarly, Indian plate is also being continuously subducted under Asiatic plate.

The Evaluation : The overwhelming majority of the scientists all over the world is of the view that plate tectonic theory has almost solved the problem of the origin of continents and ocean basins and of mountain building. In fact, the continental drift has now become a reality on the basis of evidences of palaeomagnetism and sea-floor spreading. Plate tectonic theory also satisfactorily explains the cyclic pattern of mountain building.

It may be pointed out that 4 major periods of mountain building have been identified e.g. (i) pre Cambrian orogeny, (ii) Caledonian orogeny, (iii) Hercynian orogeny and (iv) Tertiary orogeny. All the earlier theories of mountain building, as discussed in the preceding pages, suffer from a common defect that they, some how, do attempt to explain the origin of the folded mountains of Tertiary period but they do not throw any light on the mountains older than Tertiary period. It may be mentioned that plates are always in motion due to which some times all the land masses unite together to form Pangaea and again break up and move away relative to each other and new distributional pattern of continents and ocean basins is evolved.

The past history of the earth upto 200 million years has been reconstructed on the basis of evidences of palaeomagnetism. About 200 million years before present all the continents were united together in the form of Pangaea II (a super continent). It is believed that before the situation of Pangaea II, the continents were separated from each other. These continents might have moved relative to each other in such a way that they might have been united together to form a super continent. It is believed that the continents moved together due to plate motions and were united together in the form of Pangaea I during Pre Cambrian period, about 700 million years ago. About 600-500 million years ago Pangaea I was disrupted. About 460 million years ago the Atlantic Ocean began to close down due to convergence of American and Eurasian plates and the Caledonian mountains were formed. About 300 million years ago the Atlantic Ocean was completely closed and the orogenesis of the Applachian mountains was completed during Permian period. At the same time Hercynian mountains of Europe were formed About 200 million years ago all the continents were again united to form Pangaea II. About 150 million years ago Pangaea was again disrupted and the Atlantic ocean was reopened. The Alpine mountains were formed due to plate movements during Tertiary period.

The only point of argument and question is related to the competent force responsible for the movement of plates and drifting of continents. Most of the scientists still rely on the thermal convective currents originating from the mantle as the probable adequate force to move the plates (continents) in different directions relative to each other.

6

Valley Development

Meaning and Concept

River Valleys, Graded River and Profile of Equilibrium : Valleys are considered as 'negative landforms' of the earth's surface and are classified on the basis of processes which are involved in their formation e.g. (i) river valleys or fluvially originated valleys or normal valleys or simply valleys (formed by running water- rivers), (ii) glacial valleys (formed by mountains glaciers), (iii) tectonic valleys such as fault and rift valleys (formed by earth movements) etc. It may be mentioned that whenever only valley word is used, this always means river valley, which is generally divided into two type viz. (i) longitudinal valley and (ii) transverse valley. The entire span from the mouth to the source of a particular river is called longitudinal valley whereas transverse valley refers to cross section of the longitudinal valley at any point. The cross-section of the river valley is called transverse profile while the longitudinal section of the valley is called longitudinal profile wherein the study of channel slope (gradient) becomes more important. On the other hand, cross-section or transverse profile involves the study of width and depth of the valley at a given point of the longitudinal profile. Rivers are continuously engaged in the development of their valleys through three major activities viz. (i) valley deepening, (ii) valley widening, and (iii) valley lengthening.

Development Forms

It may be mentioned that the present forms of the valleys of the rivers (e.g. Ganga, Yamuna, Narmada, Godawari, Krishna, Cauvery etc.) have not developed suddenly but have evolved slowly and gradually through several stages of valley development. The water received through rainfall on the earth's surface becomes runoff which may be of two types viz. sheet flow and laminar (linear) flow. The concentrated linear flow forms rills which form narrow furrows which are transformed into gullies and streamlets and rivers. Thus, it is apparent that initially the valleys are very narrow but are changed, at a later date, into well developed wide and deep valleys due to continued erosion. It is also evident that sufficient runoff and ground slope are prerequisite conditions for the development of river valleys. It is important to note that some times rivers and valleys are used as synonym e.g. young, mature and old rivers are equated with young, mature and old valleys respectively but it is not always necessary that the valley of a young river will also be young because in a region of polycyclic relief (where several cycles of erosion have taken place) young valleys are found within old valleys of the rivers. Such situation of topographic discordance (valley in valley topography) in any region arises due to rejuvenation. The development of river valley is related to three processes (mechanisms) of erosion viz. (i) valley floor is deepened through vertical erosion or down-cutting (also known as valley incision) - the entire mechanism is called valley deepening, (ii) valley is widened (increase in valley width) through lateral erosion-the mechanism is called valley widening, and ; (iii) the longitudinal distance is increased through headward and mouthward (generally seaward) erosion-the mechanism is called valley lengthening.

Valley Deepening : Valley deepening involves erosion of valley floor through the mechanism of vertical erosion or down-cutting by the process of pot-hole drilling. The nature and magnitude of vertical erosion of valley floor depends on the nature, size and calibre (angularity) of erosional tools (e.g. 'boulders, cobbles, pebbles etc.) of various kinds, lithological characteristics, channel gradient, flow velocity, transporting capacity of the river, kinetic energy etc. The erosional tools of fairly large size and high calibre, when caught into water eddies, drill the valley floor through the mechanism of pot-hole drilling resulting into the formation of numerous pot holes (cylindrical holes or depressions) of various sizes in the valley floor. On an average the valley deepening includes the mechanisms of hydraulic action, corrosion of valley floor, pot-hold drilling, solution (corrosion) etc. Valley deepening becomes most active when channel gradient is steep, discharge and flow velocity is very high, transporting capacity of the river is maximum, rocks are relatively less resistant and there is gradual rise in the landmass.

Rivers resort to maximum downcutting of their valley floors in the youthful stage of normal cycle of erosion. Valleys become very narrow and deep with almost vertical side walls due to continuous active downcutting of valley floors at exceedingly fast rate. The valley side slopes become convex in plan. Thus, the resultant juvenile valleys are V-shaped and are called gorges and canyons. The valley floors are studded with numerous pot holes which are the result of pot hole drilling. It may be pointed out that though valley deepening dominates in the juvenile stage but valley widening also takes place through weathering and slumping.

Valley Widening : Though valley widening also occurs in the youthful stage of normal cycle of erosion but it becomes most active during mature stage. Valley widening occurs in two ways viz. (i) widening of the upper part of the valley due to erosion of valley walls through lateral erosion and (ii) widening of valley floors. These two mechanisms of valley widening operate together. The valley widening takes place through the following processes-(i) Undercutting (cliffing) of the lower parts of the valley sides through abrasion (corrosion) and hydraulic action causes collapse and slumping of upper portions of the valley walls. The repetition of this mechanism causes gradual retreat of valley walls and consequent widening of the valleys. Though this mechanism of valley widening is active in all stages of valley development but it is most active during mature and old stages. (ii) Sheet-wash of the valley walls through corrosive (solution) mechanism of water causes slow but gradual erosion of valley walls. (iii) Development of rills and gullies along the banks of the rivers (mostly alluvial rivers) results in the erosion of riparian tracts and active widening of river valleys. (iv) Meandering is the most effective mechanism of valley widening in the late mature and old stages of fluvial cycle of erosion etc.

Transverse valley profile (cross-section of the valley) may be of two forms viz. symmetrical, when both the valley sides are almost of uniform slope, and asymmetrical, when one valley side is of steep slope and the other one is of gentle slope. The asymmetrical shape of the cross profile of the valley generally results from the following factors -

(i) Structural controls-deep valleys with steep valley sides develop in the areas having resistant rocks while broad valleys evolve in the localities of soft and less resistant rocks (valley sides are of gentle slope). If resistant and soft rocks are found in alternate manner (horizontally and not vertically) then narrow and deep valleys and open and wide valleys are formed together.

(ii) If there is alternate arrangement of resistant and soft rock beds vertically across the valley, resistant rocks are less eroded than soft rocks and hence stepped cross-profile of the valley, characterized by structural benches, is developed. These are also called valley benches. It may be pointed out that such situation (stepped valley) is local.

(iii) Faulting of crustal rocks causes asymmetry in the transverse section of river valleys.

(iv) Uniclinal (homoclinal) shifting of valley walls in the region of uniclinal structure causes asymmetry in the cross profile of the valley.

(v) Meandering process of the rivers produces asymmetrical valleys wherein one side of the valley is of very steep slope, known as cliff slope, while the opposite side is of very gentle slope, known as slip-off slope.

Valley Lengthening : The increase in the length of the rivers, valley lengthening, is accomplished in the following manner -

(i) *Headward Erosion* : Slumping of the headwalls of the source point of the rivers causes gradual upslope retreat of the valley heads resulting in slow but gradual upslope increase in the valley length. It may be pointed out that not all the streams are always actively engaged in headward erosion.

(ii) *River Capture* : The length of the captor stream increases due to capturing of upstream sections of captured streams.

(iii) *Meandering* : Mechanism of the streams becomes most active during monadnock stage (old stage) of river cycle. The length of the rivers increases enormously because of development of numerous big meanders in the longitudinal course of the rivers.

(iv) *Regressional* : Phase of sea and consequent emergence of previously submerged (under sea water) land is eroded down by the streams, which drain into the sea, and thus the streams extend their courses and mouths upto new coast-line.

(v) *Gradual Seaward Growth* of delta and simultaneous extension of river-mouth towards the sea through mouthward erosion causes valley lengthening.

The Classification

River valleys have been classified variously on different bases by geomorphologists from time to time, as elaborated below—

(1) *On the Basis of Stages of Geomorphic Cycle* : It may be mentioned that the evolution and development of river valleys is a slow and gradual process. The valley development passes through three successive stages of cycle of erosion viz. youth, mature and old stages wherein the form and shape of transverse and longitudinal sections of the valleys undergo the processes of valley transformation. On this basis river valleys are divided into 3 types viz. (i) young, (ii) mature, and (iii) old valleys. Young valleys are narrow and very deep and are bordered by very steep wall-like valley walls. These are of 'V' shape and valley sides are of convex slope. The valleys of gorges and canyons are typical examples. These are the result of accelerated rate of vertical erosion (valley deepening or valley incision). Mature valleys, characterized by rectilinear slope of valley sides, are the result of lateral erosion and are produced during mature stage of the geomorphic cycle. Old valleys, characterized by flat shape and concave but very gentle valley side slope, are the result of monadnock stage of cycle of erosion.

(2) *Genetic Classification* : J.W. Powell first described (1875) consequent streams which followed regional slopes. Later on W.M. Davis classified river valleys into

consequent, subsequent, obsequent, resequent, insequent valleys etc. Consequent valleys represent consequent streams which flow according to regional slope and are master streams. These valleys are also called strike valleys because they develop according to regional slope. Such valleys develop on uplifted landmass mainly folded mountains, volcanic cones, coastal plains etc. The valleys developed along the dip slope are called subsequent or dip valleys. Valleys of obsequent streams which flow in opposite direction to consequent streams are called obsequent valleys. These valleys also develop due to river capture. Resequent valleys develop at much later date and are well adjusted to regional slope. Insequent valleys develop across geological structure and relief barriers (e.g. ridges, mountain ranges etc.)

(3) *On the Basis of Structural Control* : River valleys are largely controlled by geological structures e.g. folded, faulted, uniclinal, domal structures etc. On the basis of geological structures river valleys are classified into (i) homoclinal valleys, developed on uniclinal structure, are characterized by asymmetrical valleys sides (one side is very steep while the other side is gentle) ; (ii) anticlinal valleys, developed along the anticlinal axis of folds ; (iii) synclinal valleys, developed along the synclinal axis ; (iv) rift valleys, developed in the depression caused by two faults; (v) fault line valleys, developed along faultlines ; (vi) joint valleys (valleys of small streams which develop along joints of rocks) etc. Valleys of the Damodar, the Son, the Narmada, the Tapi rivers etc. are examples of rift valleys.

(4) *On the Basis of Structural Trends* :.Valleys of such streams which flow across the geological structure are called insequent valleys which are of two main types viz. antecedent valleys and superimposed valleys.

(5) *On the Basis of Base* : Level Changes - On an average, the development of valleys is controlled by changes in base-level of erosion. Thus, valleys are divided into (i) drowned valleys (due to rise in sea level and transgression of sea) and (ii) rejuvenated valleys (due to fall in sea level and consequent accelerated rate of valley deepening).

Tabular Presentation of Valley Types

1. On the basis of stages of geomorphic cycle	(i)	young valleys
	(ii)	mature valleys
	(iii)	old valleys
2. Genetic classification	(i)	consequent valleys
	(ii)	subsequent valleys
	(iii)	obsequent valleys
	(iv)	resequent valleys
	(v)	insequent valleys

3. On the basis of structural control	(i)	uniclinal valleys
	(ii)	anticlinal valleys
	(iii)	synclinal valleys
	(iv)	rift valleys
	(v)	fault-line valleys
	(vi)	joint valleys
4. On the basis of	(i)	antecedent valleys structural trend
	(ii)	superimposed valleys
5. On the basis of	(i)	drowned valleys base-level changes
	(ii)	rejuvenated valleys

The Profile

Longitudinal Profile and Graded Curve : The longitudinal course of a river from its source to mouth is called longitudinal or simply long profile or valley thalweg. In fact, long profile of a river represents channel gradient of the river from its source to the mouth. Each river tries to develop such a longitudinal course (profile) that it may be able to transport the bed-load downstream. The longitudinal channel course is generally smooth curve which rises upstream. A river is supposed to ultimately remove topographic irregularities by penultimate stage (monadnock stage) and to develop smooth curve from source to mouth. The maximum limit of vertical erosion (valley deepening) is determined by grand base level which represents sea-level. Thus, rivers always try to erode their valleys down to base level of erosion near sea coast and the valley floor becomes concave in such a way that it rises upstream or headward. Thus, the rivers always try to develop smooth concave curve of their channels. The concavity of long profile of a river results from the fact that there is minimum erosion in the upper and lower courses of the river while there is maximum erosion in the middle segment. Lack of required amount of erosion tools (load) and water in the source segment and gentle gradient and very low flow velocity in the lower segment of a river are responsible for minimum erosion whereas sufficient load, channel gradient and flow velocity in the middle course of a river cause maximum erosion of valley floor. This is why a river develops a smooth concave curve of longitudinal course. When a river develops such a course that channel gradient is such that resultant flow velocity is able to transport entire load, the resultant longitudinal curve of valley thalweg is called graded curve and the river after attaining graded curve is called graded river and the long profile of the river becomes profile of equilibrium as there is balance between transporting capacity of the river and total load to be transported i.e. balance between available energy and work to be done.

Concept of Grade : The usage of the term 'grade' in fluvial system does not simply means gradient or slope but means continuous curve of descent of a stream floor downstream which has such a gradient (slope) throughout longitudinal course of the stream that it can transport all the loads downstream. The river having attained such condition is called graded river and its curve as a graded curve. It is generally believed that G.K. Gilbert was first to use the word 'grade' in 1876 followed by W.J. McGee in 1891, W.M. Davis in 1894, H. Gannett in 1896 and W.D. Johnson in 1901, but 'Davis seems mainly, and perhaps wholly, responsible for that deductive elaboration of the concept which has promoted so voluminous and so recurrent a discussion' (G. H. Dury, 1966).

According to Davis 'it is evidently desirable to employ the term 'grade' for the balanced condition of a mature or old river ... the balance between erosion and deposition ... is brought about by the changes in the capacity of a river to do work, and in the quantity of work that a river has to do. The changes continue until the two quantities... reach equality, and then the river may be said to be graded' (Davis, 1902).

G.H. Dury (1966) has summarized the concept of grade as conceived by W.M. Davis as follows :

(a) 'Grade is the balanced condition of a mature or old river : the balance is that between capacity to do work, and quantity of work to be done. It is expressed by an equivalence of erosion and deposition'.

(b) 'The slope of the graded river is the slope (profile) of equilibrium: this slope permits the most effective transport of load'.

(c) 'Once the graded condition is attained, slope can be altered only by changes in the volume/load relationship : such change, operating slowly, is expectable in the normal cycle.'

(d) 'Load increases in quantity and coarseness during youth, in quantity but probably not in coarseness during maturity, and after full maturity decreases both in quantity and in coarseness'.

(e) 'Grade is first established in downstream reaches, where valley deepening reduces downstream gradient, until eventually the excess capacity of the river to do work is wholly offset.'

According to J.E. Kesseli (1941) 'a graded stream be taken as one without waterfalls or rapids. That is to say, he defines grade in terms of transporting power, load, velocity, or tendency to cut or fill' (G.H. Duty, 1966).

According to J.H. Mackin (1948) 'a graded stream is one in which, over a period of years, slope is delicately adjusted to provide, with available discharge and with prevailing channel characteristics, just the velocity required for the transportation of load supplied from the drainage basin. The graded stream is a stream in equilibrium, its diagnostic characteristic is that any change in any of the controlling factors will cause a displacement of the equilibrium in a direction that will tend to absorb the effects of the change'. According to him transporting power of the stream depends on velocity which depends on slope. It is inferred from Mackin's writing, that his 'graded stream neither cuts (erodes)nor fills (deposits)'.

Graded River Controlling Factors : Several factors like channel gradient, volume of water, discharge, flow velocity, quantity and nature of sediments (load), base level of erosion, lithology etc., are considered significant variables which are responsible for the grading of a river and attainment of graded curve and profile of equilibrium.

L.B. Leopold and T. Maddock (1953) have classified variables which control hydraulic geometry of a stream into 3 broad categories viz. (i) independent variables, (ii) semi-dependent variables, and (iii) dependent variables.. Independent variables include stream discharge, sediment load and ultimate base level of erosion. It is argued that stream has little control over these factors, rather it must adjust to them' (A.L. Bloom, 1979). Semi-dependent variables comprise channel width, channel depth, bed roughness, grain size of the sediment load, velocity and channel behaviour i.e., tendency of a river for meandering or braiding its course. 'These are Semi-dependent in as much as they are partly determined by three independent variables (as elaborated above), but they are also capable of mutual self-regulation in a river' (A.L. Bloom, 1979). 'The shape of the cross section (width/depth ratio) determines the distribution of velocity and of shear; for a given width and discharge, total load depends on the ratio of velocity to depth; and velocity/depth ratio is provided in part by adjustment of bed roughness, which is itself a function of grain-size and of suspended-sediment concentration' (G.H. Dury, 1966). Changes in bed roughness, which itself is determined by grain-size, introduce changes in velocity/depth ratio. According to Leopold and Maddock differences in the concavity of longitudinal profile of a stream are directly related to 'differences in the relationship of velocity and depth to discharge'. Meandering of the streams depends on nature and velocity of moving water of the channel, size and shape of channel, erodibility of riparian tracts, bedload and suspended load, discharge etc. On the other hand, meandering influences (increases) channel length, which in turn decreases channel slope (gradient), which determines velocity (low channel gradient ® low velocity ; steep channel gradient ®high velocity) and velocity in turn determines transporting capacity of the stream (high velocity ® high transporting capacity ; low velocity ® low transporting capacity). Thus, meandering property of a river is both dependent and independent variable. Dependent variable includes a single variable viz. downstream slope of the water surface.

River Channel Grading

It is evident from the above discussion that grading of a river denotes perfect balance between transporting capacity of the river and total load to be transported by it. In other words, at each and every point of the longitudinal course of the river transporting capacity is such that total sediment load can be transported downstream or velocity of the river is such that erosional and depositional works are balanced. The river having attained such condition throughout its course is called graded river. It may be mentioned that the graded condition is not suddenly attained rather it is attained slowly and gradually. It may be further pointed out that it is not only the slope (channel gradient) factor which is most important for grading of a river but a critical adjustment among volume of water, discharge, velocity, slope and sediment load is also prerequisite condition.

Let us take the case of variation in sediment load and related adjustment of river channel. If the supply of sediment load decreases in any part of the river, more energy becomes available for valley incision because the river is under-loaded. Thus, resultant valley incision causes decrease in channel gradient. This condition continuous till the eroded materials are transported downstream. Such condition, when erosion exceeds deposition, is called stage of degradation wherein channel gradient becomes less than general slope denoting inequilibrium condition. Contrary to this, if sediment load increases and exceeds the transporting capacity of the river, 'there occurs deposition of extra sediments, which the river is unable to transport downstream.

This condition is called stage of aggradation (when deposition exceeds erosion) wherein slope increases and this condition continues (i.e slope continues to increase) till the slope becomes such that resultant velocity provides required transporting capacity so that all the sediment load is transported downstream. Thus, if balance (adjustment) between transporting capacity of the river (available energy) and total load to be transported downstream (work to be done) is attained, the river is said to be graded.

Alternatively, if sediment load and channel gradient remain constant, any increase in the volume of water will increase discharge and velocity which will increase transporting capacity and hence the rate of erosion will be accelerated leading to the stage of degradation and consequent inequilibrium condition. Conversely, if volume of water decreases, the discharge and velocity will also decrease which would result in decreased transporting capacity of the river and resultant deposition of sediments leading to the development of the stage of aggradation and inequilibrium condition.

If sediment load and volume of water remain constant, increase in channel gradient will lead to increase in flow velocity and erosion and hence there would be the stage of degradation. On the other hand, decrease in channel gradient would result in decrease in velocity and deposition of sediment load leading to the development of the stage of aggradation.

Thus, it may be concluded that the course of a river is said to be graded course or graded curve when there is balance between velocity (and hence transporting capacity) and sediment load to be transported downstream. The longitudinal profile of such a graded river is called graded profile and profile of equilibrium indicating a condition of balance between erosion and deposition throughout longitudinal course of river i.e., from its source to mouth.

It is necessary to point out some misconceptions about graded curve, graded river and profile of equilibrium.

(1) 'Grade is a condition, not an altitude or a certain slope angle ...... A gradded river is in a steady state only with regard to short-term changes. Over a time scale of millions of years, typical of time intervals in which landscapes evolve, the potential energy of an undisturbed river system gradually approaches zero, and the rate of change of the system also decreases. The river remains at grade, but the characteristics of the graded condition change with time' (A. Bloom, 1979).

(2) Graded river does not mean steep or low gradient. A river may be graded at higher gradient while the other river may be graded at lower gradient.

(3) A graded river, in fact, is seldom loaded to the capacity. Thus, a graded river, in reality, is not such river which neither cuts (erodes) nor fills (deposits). A graded river does not mean that there is balance between erosion and deposition in each and every part of the longitudinal course of a river because there is every likelihood that erosion may be dominant in one segment of the river while deposition may be most active in the other part of the same river.

(4) It has been repeatedly discussed above that in order to attain graded stage of a river there must be adjustment (balance) between transporting capacity and sediment load to be transported. But question arises, adjustment between what ? Whether in power (energy) or quantity ? Generally, it is believed that there should be adjustment between transporting power of a river and sediment load to be carried by the river downstream. It is inferred from this corollary that if there is decrease in sediment load, that part of river energy which was previously expended in transporting the load, is now spared and is expended in erosional work resulting in increased rate of erosion. It may be conceived on this basis that if there is total absence of sediment load in a particular river, it would have maximum corrosive power (capacity) but this inference is erroneous, because it is not only the flow velocity and sediment load of a river which alone control erosion and deposition. Thus, the energy spared from transportation of sediment load may not be a factor of erosion because erosion becomes negligible in the absence of load. 'Not only the total mass of the load but the size or calibre of load

are responsible for corrosive capacity (power) of the stream'. Inspite of steep channel gradient and hence high velocity and less quantity of sediment load there is minimum erosion in the upper segment (course) of a river. Similarly, high load but gentle channel gradient and hence low velocity causes minimum erosion in the lower reach (segment or course) of a river. Contrary to these two conditions, there is maximum erosion in the middle reach of the river because of availability of required sediment load, flow velocity and channel gradient. This is why the erosion level (valley floor) of the river from its source to mouth is not straight but is smooth curve.

(5) The gradual decrease of slope of graded curve of a river downstream may be theoretically sound but in practice it seldom occurs as some parts of the river may be graded while other parts may be ungraded.

(6) Grading of the river begins near the mouths of the rivers and proceeds upstream. The overall grading of the river is controlled by grand base level of erosion which is determined by sea-level. Besides, there are some local and temporary base levels which represent lakes, confluences of tributary streams, resistant rock beds etc. The maximum vertical erosion (valley incision) near the mouths of the rivers is determined by grand base level i.e., sea-level. General grading of the river from its mouth to source is attained in stages having a time span of millions of years. When sea-level becomes permanent base level for the entire span of a river and all the local and temporary base levels are eliminated from a river course, then the river attains its general grading.

(7) Generally, graded curve of a river is considered regular from source to mouth but a few geomorphologists do not subscribe to this view point. For example, J.H. Mackin is of the view that there is variation in slope gradient of a graded river. The longitudinal course of a river does not consist of single regular graded curve but consists of several segments but each segment has such slope that the resultant velocity is able to transport all sediment load downstream. Thus, according to Mackin a graded profile is, in fact, transportation slope. Such profile is neither influenced by corrosive power nor by the resistance of bedrock. A graded curve can never be a mathematical curve.

(8) A graded stream does not mean, as understood by some geomorphologists, that it has attained its lowest gradient or minimum slope over which it flows. It is also believed by many that after the attainment of profile of equilibrium, vertical erosion or valley incision by the stream stops and the profile of equilibrium denotes limit of vertical erosion but such connotations are erroneous. The slope of graded river changes with time because there is variation in the quantity and calibre of sediment load, volume of water, discharge etc., but it is also true that

the changes are so gradual and slow that a river, once graded, remains graded except some temporary disturbances.

Curve : Regraded and Disturbed : As already stated above a graded curve and profile of equilibrium is the result of adjustment between volume of water, transporting capacity (velocity) and sediment load. In fact, the graded profile of a river is in a delicate balance. A slight change in any of the independent, Semi-dependent and dependent variables (as elaborated earlier) which control and determine the grading of a river may disturb the equilibrium condition of the river and the river again tries to regrade its course according to new conditions. Decrease or increase in sediment load influences deposition and erosion. A river deposits sediments when there s increase in sediment load and decrease in the volume of water. Conversely, a river starts eroding its valley when there is decrease in sediment load or increase in the volume of water. Thus, a river always tries to develop profile of equilibrium by making adjustment (balance) between erosion and deposition. This balance is so delicate that a slight change in any part of a river or river system causes disturbance and the graded curve (balance or equilibrium) is disturbed with the result the river has to readjust with new conditions. If a river succeeds in attaining rebalance the curve of such a river is called regraded curve and its profile is called regraded profile of equilibrium.

The disturbance in the graded profile of a river may be caused due to a variety of factors but increase or decrease in channel gradient is the most significant causative factor. Changes in channel gradient are generally effected by rejuvenation and deposition of sediments.

Rejuvenation Effects

Rejuvenation simply means sudden and phenomenal increase in the erosive power of the streams and consequent accelerated rate of downcutting (valley incision) caused by steepening of channel gradient either due to negative change (fall) in sea level or upliftment of land mass in a river course. Rejuvenation may be effected either at a the mouth of the river or in the middle course or in the headwaters of the river.

1. *Rejuvenation at the Mouth of a River* : Rejuvenation at the mouth of a river occurs when there is fall (negative change) in sea level after the river has developed graded curve and profile of equilibrium. Consequently, the graded profile of the river is disturbed because channel gradient is steepened at the mouth of the river due to lowering of sea level which results in the acceleration of flow velocity. Thus, the erosive power of the river increases and hence the river starts fresh vigorous valley deepening at its mouth according to new base level of erosion determined by new sea level (B′). Now the river tries to develop its new curve according to

new sea level (B′) which is lower in height than the old sea level (B). The rejuvenated river starts regrading its course from its mouth and the mechanism of regrading proceeds upstream. Wherever the new curve intersects the old curve, break in long profile is formed. Such breaks are called nick points or head of rejuvenation. These nicks recede upstream as the mechanism of regrading of the river course proceeds upstreams. Ultimately, all such nick points (water falls) are eliminated and the entire span of longitudinal profile of the river is regraded and profile of equilibrium is reestablished.

2. *Rejuvenation in the Middle Course of a River* : The rejuvenation in the middle course of a graded river down the confluence of a tributary occurs when the supply of sediment load by the concerned tributary to the receiving master stream decreases significantly and thus the master stream down the confluence becomes underloaded and available extra energy, which was previously spent in transporting the sediment load, is spared and is now expended in erosional work, with the result erosive power of the stream is increased, which causes rejuvenation and thus the previously graded profile gets disturbed. The master river now degrades and deepens its valley downstream from the confluence, with the result the gradient of upstream section is steepened, which causes headward erosion. Headward erosion continues until the gradient of upstream section of the river is adjusted to the previous gradient and the entire span of the long profile is regraded.

3. *Rejuvenation in the Headwaters of the River* : Significant decrease in the supply of sediment load in the uppermost section, say headwaters, of the stream results in increased erosion as the stream becomes underloaded and hence the headwater section is rejuvenated. Since the supply of sediments has decreased in the headwaters of the river, this condition (decrease in sediment load) prevails throughout the entire span of longitudinal course of the river. Consequently, the entire span of the river valley is deepened so that sediment load produced by increased erosion equals the transporting capacity of the river. Whenever the river becomes able to attain this condition, the disturbed graded curve is regraded and the river reattains its profile of equilibrium.

Deposition Effects

Deposition of extra sediments results in increase of the level of valley floor and hence there is decrease in channel gradient, with the result the graded curve of the river gets disturbed. Sedimentation at the mouth of the river which drains into the sea, results in the formation of delta provided that other conditions are favourable (i.e. sea waves are not very active). Gradually delta grows seaward and thus the length of longitudinal course of the river also increases. Lengthening of river course lessens channel gradient, which results in marked decrease in velocity and transporting capacity of the river, with the result the river has to deposit sediments in all parts of its course in order to regrade its course at higher elevation. As the delta grows in size, deposition by the river also increases in all parts of its course.

If volume of water, sediment load, and slope remain constant, the graded condition of a river becomes more or less permanent but if any juvenile tributary stream meets the master stream and brings sufficient sediment load, the master stream becomes unable to transport additional sediment load downstream, with the result additional sediments are deposited down the confluence resulting in the increase in channel gradient and consequent disturbance in the graded curve of the river. Sedimentation in the downstream section from the confluence with the juvenile tributary stream results in decrease in channel gradient in upstream section. Consequently, sedimentation in the upstream section from the confluence continues until the entire span of the longitudinal course attains such slope which allows transportation of all sediment load. Thus, the river is regraded but at higher elevation.

It is, thus, apparent that graded curve of the river is disturbed by a host of causative factors but the river tries to remove such obstacles which disturb its graded profile and ultimately regrades its profile. The regraded curve may be at higher elevation or lower elevation, depending on local conditions.

7

Process of Eruption

Vulcanicity and Landforms

Meaning and Concept : The terms volcanoes, mechanism of volcanoes and vulcanicity are more or less synonym to common man but these have different connotations in geology and geography. 'A volcano is a vent, or opening, usually circular or nearly circular in form, through which heated materials consisting of gases, water, liquid lava and fragments of rocks are ejected from the highly heated interior to the surface of the earth' (P.G. Worcester, 1948). 'A volcano is essentially a fissure or vent, communicating with the interior, from which flows of lava, fountains of incandescent spray or explosive bursts of gases and volcanic ashes are erupted at the surface.' On the other hand, 'the term vulcanicity covers all those processes in which molten rock material or magma rises into the crust or is poured out on its surface, there to solidify as a crystalline or semicrystalline rock' (S.W. Wooldridge and R.S. Morgan, 1959). Some scientists have also used the term of vulcanism as synonym to the term of vulcanicity. For example, P.G. Worcester (1948) has maintained that 'vulcanism includes all phenomena connected with the movement of heated material from the interior to or towards the surface of the earth.'

It is apparent from the above definitions of volcano and vulcanicity (vulcanism) that the later (vulcanicity) is a broader mechanism which is related to both the environments, endogenetic and exogenetic. In other words, vulcanicity includes all those processes and mechanisms which are related to the origin of magmas, gases and vapour, their ascent and appearance on the earth's surface in various forms. It is evident that the vulcanicity has two components which operate below the crustal surface and above the crust. The endogenetic mechanism of vulcanicity includes the creation of hot and liquid magmas and gases in the mantle and the crust, their expansion and upward ascent, their intrusion, cooling and solidification in various forms below crustal surface (e.g. batholiths, lacoliths, sills, dykes, lopoliths, phacoliths etc.) while the exogenous mechanism includes the process of appearance of lava, volcanic dusts and ashes, fragmental material, mud, smoke etc., in different forms e.g., fissure flow or lava flood (fissure or quiet type of volcanic eruption), violent explosion (central type of volcanic eruption), hot springs, geysers, fumaroles, solfatara, mud volcanoes etc. It may be, thus, concluded that the vulcanicity is a broader mechanism which includes several events and processes which work below the crust as well as above the crust whereas volcano is a part of vulcanicity (vulcanism).

Volcanic Components

Volcanoes of explosive type or central eruption type are associated with the accumulated volcanic materials in the form of cones which are called as volcanic cones or simply volcanic mountains.

There is a vent or opening, of circular or nearly circular shape, almost in the centre of the summital part of the cone. This vent is called as volcanic vent or volcanic mouth which is connected with the interior part of the earth by a narrow pipe, which is called as volcanic pipe. Volcanic materials of various sorts are ejected through this pipe and the vent situated at the top of the pipe. The enlarged form of the volcanic vent is known as volcanic crater and caldera. Volcanic materials include lavas, volcanic dusts and ashes, fragmental materials etc.

There is a wide range of variations in the mode of volcanic eruptions and their periodicity. Thus, valcanoes are classified on the basis of (i) the mode of eruption and (ii) the period of eruption and the nature of their activities.

(1) Classification on the Basis of the Mode of Eruptions

(i) Central eruption type or explosive eruption type

(a) Hawaiin type

(b) Strombolian type

(c) Vulcanian type

(d) Peleean type

(e) Visuvius type

(ii) Fissure eruption type or quiet eruption type

(a) Lava flood or lava flow

(b) Mud flow

(c) Pumaroles

(2) Classification on the Basis of Periodicity of Eruptions

(a) Active volcanoes

(b) Dormant volcanoes

(c) Extinct volcanoes

Classification Basis of Eruptional Nature : Volcanic eruptions occur mostly in two ways viz. (i) violent and explosive type of eruption of lavas, volcanic dusts, volcanic ashes and fragmental materials through a narrow pipe and small opening under the impact of violent gases and (ii) quiet type or fissure eruption along a long fracture or fissure or fault due to weak gases and huge volume of lavas. Thus, on the basis of the nature and intensity of eruptions volcanoes are divided into two types e.g. (1) central eruption type or explosive eruption type and (2) fissure eruption type or quiet eruption type.

Volcanoes of Central Eruption Type : Central eruption type or explosive eruption type of volcanoes occurs through a central pipe and small opening by breaking and blowing off crustal surface due to violent and explosive gases accumulated deep within the earth. The eruption is so rapid and violent that huge quantity of volcanic materials consisting of lavas, volcanic dusts and ashes, fragmental materials etc., are ejected upto thousands of metres in the sky. These materials after falling down accumulate around the volcanic vent and form volcanic cones of various sorts. Such volcanoes are very destructive and are disastrous natural hazards. Explosive volcanoes are further divided into 5 sub-types on the basis of difference in the intensity of eruption, variations in the ejected volcanic material and the period of the action of volcanic events as given below.

Hawaiin Type of Volcanoes — Such volcanoes erupt quietly due to less viscous lavas and non-violent nature of gases. Rounded blisters of hot and glowing mass/boll of lavas (blebs of molten lava) when caught by a strong wind glide in the air like red and glowing hairs. The Hawaiin people consider these long glassy threads of red molten lava as Pele's hair (Pele is the Hawaiin goddess of fire). Such volcanoes have been named as Hawaiin type because of the fact that such eruptions are of very common occurrence on Hawaii island. The eruption of Kilavea volcano of the southern Hawaii island in 1959-60 continued for seven days (from November 14 to 20, 1959) when about 30 million cubic metres of lavas poured out. The intermittent eruptions continued upto December 21, 1959, when the volcano became dormant. It again erupted on January 13, 1960 and about 100 million cubic metres of lavas were poured out of one kilometre long fissure.

Strombolian Type of Volcanoes—Such volcanoes, named after Stromboli volcano of Lipari island in the Mediterranean Sea, erupt with moderate intensity. Besides lava, other volcanic materials like pumice, scoria, bombs etc., are also ejected upto greater height in the sky. These materials again fall down in the volcanic craters. The eruptions are almost rhythmic or nearly continuous in nature but some times they are interrupted by long intervals.

Vulcanian Type of Volcanoes—These are named after volcano of Lipari island in the Mediterranean Sea. Such volcanoes erupt with great force and intensity. The lavas are so viscous and pasty that these are quickly solidified and hardened between two eruptions and thus they crust over (plug) the volcanic vents. These lava crusts obstruct the escape of violent gases during next eruption. Consequently, the violent gases break and shatter the lava crusts into angular fragments and appear in the sky as ash-laden volcanic clouds of dark and often black colour assuming a convoluted or cauliflower shape.

Peleean Type of Volcanoes—These are named after the Pelee volcano of Martinique Island in the Caribbean Sea. These are the most violent and most explosive type of volcanoes. The ejected lavas are most viscous and pasty. Obstructive domes of lava are formed above the conduits of the volcanoes. Thus, every successive eruption has to blow off these lava domes. Consequently, each successive eruption occurs with greater force and intensity making roaring noise. The most disastrous volcanic eruption of Mount Pelee on May 8, 1902 destroyed the whole of the town of St. Pierre killing all the 28,000 inhabitants leaving behind only two survivors to mourn the sad demise of their brethren. Such type of disastrous violent eruptions are named as nuee ardente meaning thereby 'glowing cloud' of hot gases, lavas etc., coming out of a volcanic eruption. The nuee ardente spread laterally out of the mountain (Mount Pelee) with great speed which caused disastrous avalanches on the hillslopes which plunged down the slope at a speed of about 100 kilometres per hour. The annihilating explosive eruption of Krakatoa volcano in 1883 in Krakatoa Island located in Sunda Strait between Java and Sumatra is another example of violent volcanic eruption of this type.

Visuvious Type of Volcanoes—These are more or less similar to Vulcanian and Strombolian types of volcanoes, the difference lies only in the intensity of expulsion of lavas and gases. There is extremely violent expulsion of magma due to enormous volume of explosive gases. Volcanic materials are thrown up to greater height in the sky. The ejected enormous volume of gases and ashes forms thick clouds of 'cauliflower form.' The most destructive type of eruption is called as Plinian type because of the fact that such type of eruption was first observed by Plini in 79 A.D.

Fissure Eruption Type of Volcanoes—Such volcanoes occur along a long fracture, fault and fissure and there is slow upwelling of magma from below and the resultant lavas spread over the ground surface. The speed of lava movement depends on the nature of magma, volume of magma, slope of ground surface and temperature conditions. The Laki fissure eruption of 1783 in Iceland was so quick and enormous that huge volume of lavas measuring about 15 cubic kilometres was poured out from a 28km long fissure. The lava flow was so enormous that it travelled a distance of 350 kilometres.

Periodicity Based Classification : Volcanoes are divided into 3 types on the basis of period of eruption and interval period between two eruptions of a volcano e.g. (i) active volcanoes, (ii) dormant volcanoes and (iii) extinct volcanoes.

(i) Active volcanoes are those which constantly eject volcanic lavas, gases, ashes and fragmental materials. It is estimated that there are about more than 500 volcanoes in the world. Etna and Stromboli of the Mediterranean Sea are the most significant examples of this category. Stromboli Volcano is known as Light House of the Mediterranean because of continuous emission of burning and luminous incandescent gases. Most of the active volcanoes are found along the mid-oceanic ridges representing divergent plate margins (constructive plate margins) and convergent plate margins (destructive plate margins represented by the eastern and western margins of the Pacific Ocean). The latest eruption took place from Pinatubo volcano in June 1991 in Phillipines.

(ii) Dormant volcanoes are those which become quiet after their eruptions for some time and there are no indications for future eruption but suddenly they erupt very violently and cause enormous damage, to human health and wealth. Visuvious volcano is the best example of dormant volcano which erupted first in 79 A.D., then it kept quiet upto 1631 A.D. when it suddenly exploded with great force. The subsequent eruptions occurred in 1803, 1872, 1906, 1927, 1228 and 1929.

(iii) Extinct volcanoes are considered extinct when there are no indications of future eruption. The crater is filled up with water and lakes are formed. It may be pointed out that no volcano can be declared permanently dead as no one knows, what is happening below the ground surface.

Discharged Volcanic Materials : Volcanic materials discharged during eruptions include gases and vapour, lavas, fragmental materials and ashes.

Vapour and Gases : Steam and vapour constitute 60 to 90 per cent of the total gases discharged during a volcanic eruption. Steam and vapour include (i) phreatic vapour and (ii) magmatic vapour whereas volcanic gases include carbon dioxide, nitrogen oxides, sulphur dioxide, hydrogen, carbon monoxide etc. Besides, certain compounds are also ejected with the volcanic gases e.g. sulphurated hydrogen, hydrochloric acid, volatile chlorides of iron, potassium and other metallic matter.

Magma and Lava : Generally, molten rock materials are called magmas below the earth's surface while they are called lavas when they come at the earth's face. Lavas and magmas are divided on the basis of silica percentage into two groups e.g. (i) acidic magma (higher percentage of silica and (ii) basic lava (low percentage of silica). Lavas and magmas are also classified on the basis of light and dark coloured minerals into (i) felsic lava and (ii) mafic lava. Basaltic or mafic lava is characterized by maximum fluidity.

Basaltic lava spreads on the ground surface with maximum flow speed (from a few kilometres to 100 kilometres per hour, average flow speed being 45 to 65 km per hour) due to high fluidity and low viscosity. Basaltic lava is the hottest lava (1,000° to 1,200°C). Lava flow is divided into two types on the basis of Hawaiin language e.g. (i) pahoehoe and (ii) as lava flow or block lava flow. Pahoehoe lava has high fluidity and spreads like thin sheets. This is also known as ropy lava. On the other hand, as lava is more viscous. Pahoehoe lava, when solidified in the form of sacks or pillows, is called pillow lava.

Fragmental or Pyroclastic Materials : Pyroclastic materials thrown during explosive type of eruption are grouped into three categories. (i) Essential materials include consolidated forms of live lavas. These are also known as tephra which means ash. Essential material are unconsolidated and their size is upto 2 mm. (ii) Accessory materials are formed of dead lavas, (iii) Accidental materials include fragmental materials of crustal rocks. On the basis of size pyroclastic materials are grouped into (i) volcanic dust (finest particles), (ii) volcanic ash (2 mm in size), (iii) lapilli (of the size of peas) and (iv) volcanic bombs (6 cm or more in size), which are of different shapes viz. ellipsoidal, discoidal, cuboidal, and irregularly rounded. The dimension of average volcanic bombs ranges from the size of a base ball or basket ball to giant size. Some times the volcanic bombs weigh 100 tonnes in weight and are thrown upto a distance of 10 km.

Global Volcanic Distribution : Like earthquakes, the spatial distribution of volcanoes over the globe is well marked and well understood because volcanoes are found in a well defined belt or zone. Thus, the distributional pattern of volcanoes is zonal in character. If we look at the world distribution of volcanoes it appears that the volcanoes are associated with the weaker zones of the earth's crust and these are closely associated with seismic events say earthquakes. The weaker zones of the earth are represented by folded mountains (western cordillera of North America, Andes, mountains of east Asia and East Indies) with the exceptions of the Alps and the Himalayas, and fault zones. Volcanoes are also associated with the meeting zones of the continents and oceans. Occurrences of more volcanic eruptions along coastal margins and during wet season denote the fact that there is close relationship between water and volcanic eruptions. Similarly, volcanic eruptions are closely associated with the activities of mountain building and fracturing.

Based on plate tectonics, there is close relationship between plate margins and vulcanicity as most of the world's active volcanoes are associated with the plate boundaries. About 15 per cent of the world's active volcanoes are found along the constructive plate margins or divergent plate margins (along the mid-oceanic ridges where two plates move in opposite directions) whereas 80 per cent volcanoes are associated with the destructive or convergent plate boundaries (where two plates collide). Besides, some volcanoes are also found in intraplate regions e.g. volcanoes of the Hawaii Island, fault zones of East Africa etc.

Like earthquakes, there are also three major belts or zones of volcanoes in the world viz. (i) circum-Pacific belt, (ii) mid-continental belt and (iii) mid-oceanic ridge belt.

Circum-Pacific Belt : The circum-Pacific belt, also known as the 'volcanic zones of the convergent oceanic plate margins', includes the volcanoes of the eastern and western coastal areas of the Pacific Ocean (or the western coastal margins of North and South Americas and the eastern coastal margins of Asia), of island arcs and festoons off the east coast of Asia and of the volcanic islands scattered over the Pacific Ocean. This volcanic belt is also called as the Fire Girdle of the Pacific or the Fire Ring of the Pacific. This belt begins from Erebus Mountain of Antarctica and runs northward through Andes and Rockies mountains of South and North Americas to reach Alaska from where this belt turns towards eastern Asiatic coast to include the volcanoes of island arcs and festoons (e.g. Sakhalin, Kamchatka, Japan, Phillippines etc.). The belt ultimately merges with the mid-continental belt in the East Indies. Most of high volcanic cones and volcanic mountains are found in this belt. Most of the volcanoes are found in chains e.g. the volcanoes of the Aleutian Island, Hawaii Island, Japan etc. About 22 volcanic mountains are found in group in Ecuador wherein the height of 15 volcanic mountains is more than 4560 m AMSL. Cotopaxi is the highest volcanic mountain of the world (height being 19,613 feet). The other significant volcanoes are Fuziyama (Japan), Shasta, Rainierand Hood (western cordlliera of North America), a valley of ten thousand smokes (Alaska), Mt. St. Helens (Washington, USA), Kilavèa (Hawaiiland), Mt. Taal, Pinatubo and Mayon of Phillippines etc.

Here volcanic eruptions are primarily caused due to collision of American and Pacific plates and due to subduction of Pacific Plate below Asiatic plate.

Mid-Continental Belt : This belt is also known as 'the volcanic zones of convergent continental plate mergins'. This belt includes the volcanoes of Alpine mountain chains and the Mediterranean Sea and the volcanoes of fault zone of eastern Africa. Here, the volcanic eruptions are caused due to convergence and collision of Eurasian plates and African and Indian plates. The famous volcanoes of the Mediterranean Sea such as Stromboli, Visuvious, Etna etc., and the volcanoes of Aegean Sea are included in this belt. It may be pointed out that this belt does not have the continuity of volcanic eruptions as several gaps (volcanic - free zones) are found along the Alps and the Himalayas because of compact and thick crust formed due to intense folding activity. The important volcanoes of the fault zone of eastern Africa are Kilimanjaro, Meru, Elgon, Birunga, Rungwe etc.

Mid-Atlantic Belt : This belt includes the volcanoes mainly along the mid-Atlantic ridge which represents the splitting zone of plates. In other words, two plates diverge in opposite directions from the mid-oceanic ridge. Thus, volcanoes mainly of fissure eruption type occur along the constructive or divergent plate margins (boundaries). The most active volcanic area is Iceland which is located on the mid-Atlantic ridge. This belt begins from Hekla volcanic mountain of Iceland where several fissure eruption type of volcanoes are found. It may be pointed out that since Iceland is located on the mid-Atlantic ridge representing the splitting zone of American plate moving westward and Eurasian plate moving eastward, and hence here is constant upwelling of magmas along the mid-oceanic ridge and wherever the crust becomes thin and weak, fissure flow of lava occurs because of fracture created due to divergence of plates. The Laki fissure eruption of 1783 A.D. was so quick and enormous that huge volume of lavas measuring about 15 cubic kilometres was poured out from 28-km long fissure. Recently, Hekla and Helgafell volcanoes erupted in the year 1974 and 1973 respectively. Other more active volcanic areas are Lesser Antilles, Southern Antilles, Azores, St. Helena etc. The dreadful and disastrous eruption of Mount Pelee occurred on May 8, 1902 in the town of St. Pierre on the Martinique Island of West Indies in the Caribbean Sea. All the 28,000 inhabitants, except two persons, were killed by the killer volcanic eruption.

Intra-Plate Volcanoes : Besides the aforesaid well defined three zones of volcanoes, scattered volcanoes are also found in the inner parts of the continents. Such distributional patterns of volcanoes are called as intraplate volcanoes, the mechanism of their eruption is not yet precisely known. Vulcanicity also becomes active in the inner parts of continental plates. Massive fissure eruption occurred in the northwestern parts of North America during Miocene period when 1,00,000 cubic kilometres of basaltic lavas were spread over an area of 1,30,000 km^2 to form Columbian plateau. Similarly, great fissure flows of lavas covered more than 5,00,000 km^2 areas of Peninsular India. Parana of Barazil and Paraguay were formed due to spread of lavas over an area of 7,50,000 km^2.

Causes of Eruptions

As stated earlier the volcanic eruptions are associated with weaker zones of the earth surface represented by mountain building at the destructive or convergent plate margins and fracture zones represented by constructive or divergent plate boundaries at the splitting zones of mid-oceanic ridges and the zones of transform faults represented by conservative plate boundaries. The mechanism of vulcanicity (vulcanism) and volcanic eruptions is closely associated with several interconnected processes such as (i) gradual increase of temperature with increasing depth at the rate of 1°C per 32 m due to heat generated from the disintegration of radioactive elements deep within the earth, (ii) origin of magma because of lowering of melting point caused by reduction in the pressure of overlying superincumbent load due to fracture caused by splitting of plates and their movement in opposite direction, (iii) origin of gases and vapour due to heating of water which reaches underground through percolation of rainwater and melt-water (water derived through the melting of ice and snow), (iv) the ascent of magma forced by enormous volume of gases and vapour and (v) finally the occurrence of volcanic eruptions of either violent explosive central type or quiet fissure type depending upon the intensity of gases and vapour and the nature of crustal surface. Theory of plate tectonics now very well explains the mechanism of vulcanism and volcanic eruptions. In fact, volcanic eruptions are very closely associated with plate boundaries. It may be pointed out that the types of plate movements and plate boundaries also determine the nature and intensity of volcanic eruptions. Most of the active fissure volcanoes are found along the mid-oceanic ridges which represent splitting zones of divergent plate boundaries. Two plates move in opposite directions from the mid-oceanic ridges due to thermal convective currents which are originated in the mantle below the crust (plates). This splitting and lateral spreading of plates creates fractures and faults (transform faults) which cause pressure release and lowering of melting point and thus materials of upper mantle lying below the mid-oceanic ridges are melted and move upward as magmas under the impact of enormous volume of accumulated gases and vapour. This rise of magmas along the mid-oceanic ridges (constructive or divergent plate boundaries) causes fissure eruptions of volcanoes and there is constant upwelling of lavas. These lavas are cooled and solidified and are added to the trailing ends of divergent plate boundaries and thus there is constant creation of new basaltic crust. The volcanic eruptions of Iceland and the islands located along the mid-Atlantic ridge are caused because of sea-floor spreading and divergence of plates. It is obvious that divergent or constructive plate boundaries are always associated with quiet type of fissure flows of lavas because the pressure release of superincumbent load due to divergence of plates and formation of fissures and faults is a slow and gradual process.

It is apparent from the above discussion that the mid-oceanic ridges, representing splitting zones, are associated with active volcanoes wherein the supply of lava comes from the upper mantle just below the ridge because of differential melting of the rocks into tholeiitic basalts. Since there is constant supply of basaltic lavas from below the mid-oceanic ridges and hence the volcanoes are active near the ridges but the supply of lavas decreases with increasing distance from the mid-oceanic ridges and therefore the volcanoes become inactive, dormant and extinct depending on their distances from the source of lava supply, e.g. mid-oceanic ridges. This fact has been validated on the basis of the study of the basaltic floor of the Atlantic Ocean and the lavas of several Islands. It has been found that the islands nearer to the mid-Atlantic Ridge have younger lavas whereas the islands away from the ridge have older lavas. For example, the lavas of Azores islands situated on either side of the mid-Atlantic Ridge are 4-million year old whereas the lavas of Cape Verde Island, located far away from the said ridge, are 120 million year old.

Destructive or convergent plate boundaries are associated with explosive type of volcanic eruptions. When two convergent plates collide along Benioff zone (subduction zone), comparatively heavier plate margin (boundary) is subducted beneath comparatively lighter plate boundary. The subducted plate margin, after reaching a depth of 100 km or more in the upper mantle, is melted and thus magma is formed. This magma is forced to ascend by the enormous volume of accumulated explosive gases and thus magma appears as violent volcanic eruption on the earth's surface. Such type of volcanic eruption is very common along the destructive or convergent plate boundaries which represent the volcanoes of the circum-Pacific belt and the mid-continental belt. The volcanoes of the island arcs and festoons (off the east coast of Asia) are caused due to subduction of oceanic crust (plate) say Pacific plate below the continental plate, say Asiatic plate near Japan Trench.

Harms and Effects

Volcanic eruptions cause heavy damage to human lives and, property through advancing hot lavas and fallout of volcanic materials; destruction to human structures such as buildings, factories, roads, rails, airports, dams and reservoirs through hot lavas and fires caused by hot lavas; floods in the rivers and climatic changes. A few of the severe damages wrought by volcanic eruptions may be summarized as given below

(1) Huge volumes of hot and liquid lavas moving at considerably fast speed (recorded speed is 48 km per hour) bury human structures, kill people and animals, destroy agricultural farms and pastures, plug rivers and lakes, burn and destroy forest etc. The great eruption of Mt. Loa on Hawaii poured out such a huge volume of lavas that these covered a distance of 53 km down the slope. Enormous Laki lava flow of 1783 A.D. travelled a distance of 350 km engulfing two churches, 15 agricultural farms and killing 24 per cent of the total population of Iceland. The cases of Mt. Pelee eruption of 1902 in Martinique Island (in Caribbean Sea) (total death 28,000) and St. Helens eruption of 1980 (Washington, USA) are representative examples of damages done by lava movement. The thick covers of green and dense forests on the flanks of Mt. St. Helens were completely destroyed due to severe forest fires kindled by hot lavas.

(2) Fallout of immense quantity of volcanic materials including fragmental materials (pyroclastic materials), dusts and ashes, smokes etc. covers large ground surface and thus destroys crops, vegetation and buildings, disrupts and diverts natural drainage systems, creates health hazards due to poisonous gases emitted during the eruption, and causes killer acid rains.

(3) All types of volcanic eruptions, if not predicted well in advance, causes tremendous losses to precious human lives. Sudden eruption of violent and explosive type

through central pipe does not give any time to human beings to evacuate themselves and thus to save themselves from the clutches of death looming large over them. Sudden eruption of Mt. Pelee on the Island of Martinique, West Indies in the Caribbean Sea, on May 8, 1902 destroyed the whole of St. Pierre town and killed all the 28,000 inhabitants leaving behind only two survivors to mourn the sad demise of their brethren. The heavy rainfall, associated with volcanic eruptions, mixing with falling volcanic dusts and gases causes enormous mud flow or 'lahar' on the steep slopes of volcanic cones which causes sudden deaths of human beings. For example, great mud flow created on the steep slopes of Kelut volcano in Japan in the year 1919 killed 5,500 people.

(4) Earthquakes caused before and after the volcanic eruptions generate destructive tsunamis seismic waves which create most destructive and disastrous sea waves causing innumerable deaths of human beings in the affected coastal areas. Only the example of Krakatoa in 1883 would be sufficient enough to demonstrate the disastrous impact of tsunamis which generated enormous sea waves of 30 to 40 m height which killed 36,000 people in the coastal areas of Java and Sumatra.

(5) Volcanic eruptions also change the radiation balance of the earth and the atmosphere and thus help in causing climatic changes. Greater concentration of volcanic dusts and ashes in the sky reduces the amount of insolation reaching the earth's surface as they scatter and reflect some amount of incoming shortwave solar radiation. Dust veils, on the other hand, do not hinder in the loss of heat of the earth's surface through outgoing long wave terrestrial radiation. The ejection of nearly 20 cubic kilometres of fragmental materials, dusts and ashes upto the height of 23 km in the sky during the violent eruption of Krakatoa volcano on August 27, 1883, formed a thick dust veil in the stratosphere which caused a global decrease of solar radiation received at the earth's surface by 10 to 20 per cent.

(6) A group of scientists believes that volcanic eruptions and fallout of dusts and ashes cause mass extinction of a few species of animals. Based on this hypothesis the mass extinction of dinosaurs about 60 million years ago has been related to increased world-wide volcanic activity. Acid rains accompanied by volcanic eruptions cause large scale destruction of plants and animals.

Topography Produced by Vulcanicity : Numerous types of landforms are created due to cooling and solidification of magmas below the earth's surface and lavas at the earth's surface and due to accumulation of fragmental materials, dusts and ashes with lavas such as different types of volcanic cones. The cones and craters are not always permanent landforms because they are changed and modified during every successive eruption. Explosive type of volcanic eruptions helps in the formation of several types of volcanic cones whereas fissure flows result in the formation of lava plateaus and lava plains due to accumulation of thick layers of basaltic lavas over extensive areas. The topographic features produced by the entire process of vulcanicity are grouped into two broad categories viz. (i) extrusive topography and (ii) intrusive topography.

(1) Extrusive Volcanic Topography
- (i) From explosive type of eruptions
  - (a) Elevated forms, e.g. volcanic cones
  - (b) Depressed forms, e.g. craters and calderas
- (ii) From fissure eruptions
  - (a) Lava plateaus and domes
  - (b) Lava plains

(2) Intrusive Volcanic Topography
- (i) intrusive lava domes,
- (ii) batholiths,
- (iii) lacoliths,
- (iv) phacoliths,
- (v) lopoliths,
- (vi) sills,
- (vii) dikes,
- (viii) volcanic plugs and stocks etc.

Various Elevated Forms

Volcanic Cones

(i) Cinder or ash cones are usually of low height and are formed of volcanic dusts and ashes and pyroclastic matter (fragmental materials). The formation of cinder cones is initiated due to accumulation of finer particles around volcanic vent in the form of tiny mound, say 'ant mount' which varies in height from a few centimetres to a few metres in the beginning. The size of the cone gradually increases due to continuous accumulation of volcanic materials minus lavas. Some times, the rate of growth of the cone is so high that it gains height of 100 m or more within a week. The slopes of cinder cones range between 30° and 45°. Larger particles are arranged near the craters and rest at the angle between 40° and 45° and the finer particles are deposited at the outer margins of the cones. Since such cones are formed of unconsolidated larger particles and are seldom compacted by lavas and hence they are permeable to water.

Such cones are on an average less susceptible to erosion and hence they maintain their original forms for hundreds of years provided that they are not destroyed by ensuing violent explosion. The volcanic cones of Mt. Jorullo of Mexico, Mt. Izalco of San Salvador, Mt. Camiguin of Luzon Island of Phillippines etc., are typical examples of cindercones.

(ii) Composite cones are the highest of all volcanic cones. These are formed due to accumulation of different layers of various volcanic materials and hence these are also called as strato-cones. In fact, these cones are formed due to deposition of alternate layers of lava and fragmental (phyroclastic) materials wherein lava acts as cementing materials for the compaction of fragmental materials. The cone becomes comparatively resistant to erosion if it is coated by thick layer of lava. On the other hand, if the outer layer is composed of fragmental materials, the composite cone is subjected to severe erosion. Most of the highest symmetrical and extensive volcanic cones of the world come under this category e.g. Mt. Shasta, Mt. Ranier, Mt. Hood (USA), Mt. Mayon of Phillippines, Mt. Fuziyama of Japan, Mt. Cotopaxi of Ecuador etc.

(iii) Parasite cones- Several branches of pipes come out from the main central pipe of the volcano when the volcanic cones are enormously enlarged. Lavas and other volcanic materials come out from these minor pipes and these materials are deposited around newly formed vents located on the outer surface of the main cone and thus several smaller cones are formed on major cone. These cones are called parasite cones because the supply of lava for these cones comes from the main pipe. These cones are also known as adventive or lateral cones. Shastina cone is a parasite cone of Mt. Shasta of the USA.

(iv) Basic lava cone is formed of light and less viscous lava with less quantity of silica. In fact, when the lava coming out of fissuse flow is deficient in silica and is characterized by high degree of fluidity, it cools and solidifies after spreading over larger area. Thus, a long cone with significantly low height is formed. Such cones are also called as shield cones because of their shapes resembling a shield. Since these cones are composed of basaltic lavas, they are also called as basic lava cones. These are also known as Hawana type of cones.

(v) Acid lava cones are formed where the lavas coming out of volcanic eruptions are highly viscous and rich in silica content. In fact, such viscous lavas have very low mobility and hence they are immediately cooled and solidified after their appearance on the earth's surface. Thus, high cones of steep slopes are formed. Such cones are very often known as Strombolian type of cones.

(vi) Lava domes are in fact similar to shield cones in one way or the other. Lava domes differ from shield cones as regards their size. Actually, lava domes are larger and more extensive in size than the shield cones. These are formed due to accumulation of solidified lavas around the volcanic vents. Based on the mode of origin and the place of formation lava domes are divided into 3 categories e.g. (A) plug dome (formed of lavas due to filling of volcanic vents), (B) endogenous dome (formed of silica rich viscous lavas) and (c) exogenous dome (formed of silica-deficient lava with high degree of fluidity).

(vii) Lava plugs are formed due to plugging of volcanic pipes and vents when volcanoes become extinct. These vertical columns of solidified lavas appear on the earth's surface when the volcanic cones are eroded away. The lava-filled volcanic piple is called as volcanic neck. Generally, volcanic necks are cylindrical shaped and measure 50 to 60 m in height (above the ground surface) and 300 to 600 m in diameter. Sometimes diatreme term is used to indicate volcanic neck or pipe filled with breccia. 'Shiprock' which towers 515 metres (1700 feet) over the surrounding, flatlying sedimentary rocks of New Mexico, is an excellent example of a diatreme exposed by the erosion of its enclosing sedimentary rocks' (F. Press and R. Siever 1974).

Depressed Forms

(i) *Craters* : The depression formed at the mouth of a volcanic vent is called a crater or a volcanic mouth, which is usually funnel shaped. The slope of the crater depends upon the volcanic cone in which crater is formed. Normally, a crater formed in a cinder cone slopes at the angle between 25° and 30°. The size of a crater increases with increase and expansion of its cone. A crater may be differentiated from a caldera on the basis of size and mode of formation. An average crater measures 300 m in diameter and 300 m in depth but there is wide range of variations in craters from the standpoint of their size e.g., craters range from small craterlets having a diameter of a few hundred metres to large craters having the diameter of a few kilometres. The crater of extinct Aniakchak volcano of Alaska has a diameter of 9.6 km (6 miles) and the side walls are 364 m to 912 m (1200 to 3000 feet) high. If the Crater Lake of the state of Oregon (USA) is accepted as a crater, it becomes one of the most extensive craters of the world, though many scientists consider it as an example of a caldera. When a crater is filled with water, it becomes a crater lake.

When the crater of volcano becomes very extensive and if there are few eruptions of very small intensity after long time, several smaller cones are formed within the extensive older crater and thus several small-sized craters are formed at the mouth of each volcanic vent inside the extensive crater. Such craters or craterlets are called 'nested craters' or 'craters within the crater' or 'grouped craters'. Such craters are formed only when the next eruption is smaller in intensity than the previous one. The craters formed at the mouth of volcanic vents of parasite cones developed over an extensive volcanic cone is called adventive crater. Three smaller craters are found within the extensive crater of Mt. Taal of Phillippines. Similarly, three and two craters are found within the craters of Visuvius and Etna volcanoes.

(ii) *Calderas* : Generally, enlarged form of a crater is called caldera. There are two parallel concepts for the origin of calderas. According to the first group of scientists a caldera is an enlarged form of a crater and it is surrounded by steep walls from all sides. The caldera is formed due to subsidence of a crater. This concept has been propounded by the U.S. Geological Survey. It is believed according to this concept that Aso crater of Japan and Crater Lake of the USA are the result of subsidence. The second group of scientists has opined that the calderas are formed due to violent and explosive eruptions of volcanoes.

Daly, the leading advocate of 'eruption hypothesis' of the origin of calderas, believes that the topographic features formed by subsidence are 'volcanic sinks.' According to the advocates of this hypothesis if calderas are formed due to subsidence there should not be any deposit of pyroclastic materials and volcanic ashes related to a particular volcanic cone near the caldera but evidences have revealed that the remains of volcanic materials related to a particular cone are found not only near the concerned caldera but are also found several kilometres away from the caldera. For example, volcanic materials have been found at the distance of 128 km from the caldera of Crater Lake (USA). The significant calderas of the world are (figure in the brackets denote dimension in kilometres) Lake Toba of Sumatra (50 km x 50 km) in Sumatra, Aira (25 km x 24 km) in Japan, Lake Kutchaio (26 km x 20 km) in Japan, Tarso Yega (20 km x 14 km) in Shara (Africa), Aso San (23 km x 14 km) in Japan, Alban (11 km x 10 km) in Italy, Crater Lake (10 km x 10 km) in USA, Krakatoa (7 km x 6 km) in Indonesia, Kilauea (5 km x 3 km) in Hawaii etc. Smaller calderas housed in a big caldera are called nested calderas or grouped calderas.

Intrusive Topography : When gases and vapour are not very much strong during volcanic activity, the ascending magmas do not erupt as lavas rather these are intruded in viods below the crustal surface and after cooling and solidification assume several interesting forms like batholiths, lacoliths, phacoliths, lopoliths, sills and dykes. These intrusive volcanic forms are seen only when the superincumbent loads of overlying country rocks are removed through prolonged erosion.

Geysers : Geyser, in fact, is a special type of hot spring which spouts hot water and vapour from time to time. The word geyser has been derived from an Icelandic word 'geyser' which means gusher or spouter. This word was used to indicate the spouting water of a hot spring of Iceland known as Great Geyser or Gesih Geyser, representing a minor form of the broader process of vulcanicity, has been variously defined by the scientists. For example, Arthur Holems has defined geyser in the following manner: "Geysers are hot springs from which a column of hot water and steam is explosively discharged at intervals, spouting in some cases to heights of hundreds of feet." According to P.G. Worcester "Geysers are intermittent hot springs that from time to time spout steam and hot water from their craters."

The difference between hot springs and geyser lies in the fact that there is continuous spouting of hot water from the former while there is intermittent (with interval) spouting of water from the alter. A geyser spouts water from a small and narrow vent which is connected by a circuitous pipe with the underground aquifers. This pipe is called as geyser pipe or geyser tube. The length of geyser tube ranges between 30 to 100 m at different places. The temperature of water coming out of a geyser ranges between 75° to 90°C.

Geysers are classified into two types viz. (i) pool type of geyser and (ii) nozzle type of geyser. When a geyser spouts water through an open and relatively large pool, it is called pool type of geyser. Such geysers spout larger volume of water and vapour through long geyser tubes. No deposits are possible around the geyser pools. Nozzle type of geysers spout water and vapour through a very small and constricted vent. Emitted materials are deposited around the geyser vents and thus geyser cones are formed.

Some scientists do not agree to accept hot springs and geysers as two separate forms of vulcanicity rather they believe that both are the same, the difference is only of periodicity of spouting of water. Thus, they have grouped geysers into two categories viz. (1) non-continuous geysers or geysers with intermittent spouting and (2) continuously active geysers. The intermittent geysers are further divided into (i) geysers of equal intervals between two successive period of spouting (wherein interval period between two successive active periods of spouting is certain and fixed, such geysers are, thus, considered to be reliable as regards the periods of interval and spouting, example, Old Faithful Geyser of the Yellow Stone National Park, USA), (ii) variable geysers (wherein the interval period between two successive periods of spouting is not certain), (iii) long-period geysers (wherein the active period of spouting is longest of all the geysers, ranging between a few minutes to one hour, example, Grand Geyser of Iceland spouts water for 30 minutes in continuation before the next interval period starts) and (iv) feeble geyser (wherein the active period of water spouting is very short). Continuously active geysers are, in fact, hot springs which spout water without any interval. The Excelsior Geyser of the Yellow Stone National Park of the USA is the example of this category.

There is no certain observable distributional pattern of geysers over the globe as they are found in almost all the continents and in almost all the climatic zones. The geysers of the USA, Iceland and New Zealand are most widely studied geysers. Geysers are found in groups in the Yellow Stone National Park (USA). About one hundred geysers have been named and another hundred geysers are known to the scientists. There are four major basins of geysers viz. (i) Norris Basin, (ii) Upper Lake Basin, (iii) Lower Lake Basin and (iv) Heart Lake Basin. The major geyser of New Zealand is located in the western region of the northern Island which is also dominated by volcanic activities. The geysers and hot springs are spread over an area of 1786 km^2 (5000 square miles) in Iceland. The most significant geyser of Iceland is Grand Geyser.

The Fumaroles

Fumarole means such a vent through which there is emission of gases and water vapour. It appears from a distant place that there is emission of enormous volume of smokes from a particular centre. Thus, smoke or gas emitting vents are called fumaroles. In fact, fumaroles are directly linked with volcanic activities. Emission of gases and vapour begins after the emission of volcanic materials is terminated in an active volcano. Some times the emission of gases and vapour is continuous but in majority of the cases emission occurs after intervals. It is believed that gases and vapour are generated due to cooling and contraction of magma after the termination of the eruption of a volcano. These gases and vapour appear at the earth's surface through a narrow and constricted pipe (tube). It may be pointed out that fumaroles are the last signs of the activeness of a volcano.

Numerous fumaroles are found in groups near Katmai volcano of Alaska (USA). Here fumaroles are found in groups in extensive valley zone, which is called a valley of ten thousand smokes' which means fumaroles appear from 10,000 vents the diameter of which is around 3 metres. Here fumaroles are found along a linear fracture. Elsewhere, fumaroles are found along the volcanic craters. The temperature of vapour emitted from fumaroles is around 645°C. It may be mentioned that vapour constitutes 98.4 to 98.99 per cent of the total gases emitted from fumaroles. Other gases include carbon dioxide, hydrochloric acid, hydrogen sulphide, nitrogen, some oxygen and ammonia. Some minerals are also emitted with gases and vapour from fumaroles. Sulphur is the most important mineral. Fumaroles dominated by sulphur are called solfatara or sulphur fumaroles.

8

Isostatic Adjustment

Theory of Isostasy

Different relief features of varying magnitudes e.g. mountains, plateaus, plains, lakes, seas and oceans, faults and rift valleys etc. standing on the earth's surface are probably balanced by certain definite principle, otherwise these would have not been maintained in their present form. Whenever this balance is disturbed, there start violent earth movements and tectonic events. Thus, 'isostasy simply means a mechanical stability between the upstanding parts and lowlying basins on a rotating earth'.

The word isostasy, derived from a German word 'isostasios' (meaning thereby 'in equipoise'), was first proposed by American geologist Dutton in 1859 to express his view to indicate 'the state of balance which he thought must exist between large upstanding areas of the earth's surface, mountain ranges and plateaus, and contiguous lowlands, etc.' (S.W. Wooldridge and R.S. Morgan, 1959). According to Dutton the upstanding parts of the earth (mountains, plateaus, plains and ocean basins) must be compensated by lighter rock material from beneath so that the crustal reliefs should remain in mechanical stability. According J.A. Steers (1961), 'this doctrine states that wherever equilibrium exists on the earth's surface, equal mass must underlie equal surface areas.'

Concept Discovery

Though the concept of isostasy came in the mind of geologists all of sudden but its concept grew out of gradual thinking in terms of gravitational attraction of giant mountainous masses. Pierre Bouguer during his expedition of the Andes in 1735 found that the towering volcanic peak of Chimborazo was not attracting the plumb line as it should have done. He thus maintained that the gravitational attraction of the Andes 'is much smaller than that to be expected from the mass represented by these mountains'. Similar discrepancies were noted during the geodetic survey of the Indo-Gangetic plain for the determination of latitudes under the supervision of Sir George Everest, the then Surveyor General of India, in 1859. The difference of latitude of Kalianpur and Kaliana (603 km due northward) was determined by both direct triangulation method and astronomical method. Kaliana was only 96 km away from the Himalayas. The difference between two results amounted to 5.23 seconds as given below :

Result obtained through triangulation = 5023′ 42.294".

Result obtained through astronomical method = 5° 23′ 37.058"

Difference = 5.236"

This discrepancy between two methods was attributed to the attraction of the Himalayas due to which the plumbbob used in the astronomical determination of latitude was deflected.

This interpretation, thus, brought the fact before the scientists that the enormous mass of the Himalaya was responsible, through its attractional force, for the difference in the results of two methods. Later on the matter was referred to Archdeacon Pratt for further investigation and clarification. He attempted to estimate the amount of attraction of the Himalayas on the basic assumption that all the mountains had the average density of 2.75. Thus, Pratt based on minimum estimate of the mass of the Himalayas calculated the gravitational effects on the plumbbob at two places (Kaliana and Kalianpur) and to his dismay he discovered that the difference was surprisingly more than actually worked out during the survey.

Gravitational deflection at Kaliana = 27.853"

Gravitational deflection at Kalianpur = 11.968"

difference = 15.885"

Thus, the difference of 15.885" was in fact more than 3 times the observed deflection of 5.236" during the survey. Pratt's calculation of the difference of the gravitational deflections brought another fact before the scientists that the Himalaya was not exerting the attraction according to its enormous mass. This interpretation gave birth to another problem-What reason is behind low attractional force of the Himalayas ? The following explanations were offered for this question.

(1) The Himalayas are hollow and are composed of bubbles and not the rocks. Due to this fact the weight and density of the Himalayas would be low and thus their gravitational force would also be low. This was the reason for the difference in

the results of two locations as referred to above. This explanation cannot be accepted because such a high mountain, if composed of bubbles, cannot stand on the earth's surface.

(2) If the mountains are not hollow, the visible mountain mass must be compensated by deficiency of mass from below. In other words, the density of the rocks of the mountains 'must be relatively low down to considerable depth.' Thus, the total weight would be low and consequently the attractional force would also be low.

(3) The rocks of the Himalayas are of low density in themselves and thus their attraction is also low.

(4) It was suggested 'that there is such a level below the surface of the earth below which there is no change in the density of the rocks', density varies only above this level. Thus, all columns have equal mass along this level. It was therefore suggested on this basis that 'bigger the column, lesser the density, and smaller the column, greater the density.'

Thus, the debate on the discrepancies of the gravitational deflections of the plumb-line and numerous explanations for these discrepancies resulted into the postulation of the concept of isostasy by different scientists, the views of a few of them are presented below.

Sir George Airy's Concept : According to Airy the inner part of the mountains cannot be hollow, rather the excess weight of the mountains is compensated (balanced) by lighter materials below. According to him the crust of relatively lighter material is floating in the substratum of denser material. In other words. 'sial' is floating in 'sima'. Thus, the Himalayas are floating in denser glassy magma. According to Airy 'the great mass of the Himalayas was not only a surface phenomenon : the lighter rocks of which they are composed do not merely rest on a level surface of denser material beneath, but, as a boat in water, sink into the denser material' (J.A. Steers, 1961). In other words, the Himalayas are floating in the denser magma with their maximum portion sunk in the magma in the same way as a boat floats in water with its maximum part sunk in the water. This concept in fact involves the principle of floatation. For example, an iceberg floats in water in such a way that for every one part to be above water-level, nine parts of the iceberg remain below water level. If we assume the average density of the crust and the substratum to be 2.67 and 3.0 respectively, for every one part of the crust to remain above the substratum, nine parts of the crust must be in the substratum. In other words, the law of floatation demands that 'the ratio of freeboard to draught is 1 to 9.' It may be pointed out that Airy did not mention the example of the floatation of iceberg. He simply maintained that the crustal parts (landmasses) were floating, like a boat, in the magma of the substratum.

If we apply the law of floatation, as stated above, in the case of the concept of Airy, then we have to assume that for the 8848 m height of the Himalaya there must be a root, 9 times more in length than the height of the Himalaya, in the substratum. Thus, for 8848-m part of the Himalaya above, there must be downward projection of lighter material beneath the mountain reaching a depth of 79,632 m (roughly 80,000 m).

Joly applied the principle of floatation for the crust of the earth taking the freeboard to draught ratio as 1 to 8. According to him 'for every emergent part of the crust above the upper level of the substratum there are eight parts submerged' (J.A. Steers, 1961). If we apply Joly's view of flotation to the concept of Airy, there would be downward projection of the Himalaya upto a depth of 70,784 m (8848 m x 8) in the substratum.

Thus, according to Airy the Himalayas were exerting their real attractional force because there existed a long root of lighter material in the substratum which compensated the material above. Based on above observation Airy postulated that 'if the land column above the substratum is larger, its greater part would be submerged in the substratum and if the land column is lower, its smaller part would be submerged in the substratum.' According to Airy the density of different columns of the land (e.g. mountains, plateaus, plains etc.) remains the same. In other words, density does not change with depth, that is, 'uniform density with varying thickness.'

This means that the continents are made of rocks having uniform density but their thickness or length varies from place to place. In order to prove this concept Airy took several pieces of iron of varying lengths and put them in a basin full of mercury. These pieces of iron sunk upto varying depths depending on their lengths. The same pattern may be demonstrated by taking wooden pieces of varying lengths. If put into the basin of water these would sink in the water according to their lengths.

Though the concept of Sir George Airy commands great respect among the scientific community but it also suffers from certain defects and errors. If we accept the Airy's views of isostasy, then every upstanding part must have a root below in accordance with its height. Thus, the Himalaya would have a root equivalent to 79,632 m (if we accept the freeboard to draught ratio as 1 to 9) or 70,784 m (if the freeboard to draught ratio is taken as 1 to 8). It would be wrong to assume that the Himalaya would have a downward projection of root of lighter material beneath the mountain reaching such a great depth of 79,632 m or 70,784 m because such a long root, even if accepted, would melt due to very high temperature prevailing there, as temperature increases with increasing depth at the rate of 1°C per 32 m.

"Quite recently, however, the fundamental concept of Airy, the continental masses floating as lighter (sial) blocks in a heavier (sima) substratum, has been rejuvenated, largely through the influence of Heiskanen's work, so that it is now probably true to say that most geologists favour Airy's explanation" (J.A. Steers, 1961, p. 75).

Archdeacon Pratt's Concept : While studying the difference of gravitational deflection of 5.236 seconds during the geodetic survey of Kaliana and Kalianpur Archdeacon Pratt calculated the gravitational force of the Himalaya after taking the average density of the Himalaya as 2.75 and came to know that the difference should have been 15.885 seconds. He, then, studied the rocks (and their densities) of the Himalaya and neighbouring plains and found that the density of each higher part is less than a lower part. In other words, the density of mountains is less than the density of plateaus, that of plateau is less than the density of plain and the density of plain is less than the density of oceanic floor and so on. This means that there is inverse relationship between the height of the reliefs and density.

According to Pratt there is a level of compensation above which there is variation in the density of different columns of land but there is no change in density below this level. Density does not change within one column but it changes from one column to other columns above the level of compensation. Thus, the central theme of the concept of Pratt on isostasy may be expressed as 'uniform depth with varying density.' According to Pratt equal surface area must underlie equal mass along the line of compensation.

Bowie has opined that though Pratt does not believe in the law of floatation, as stated by Sir George Airy but if we look, minutely, into the concept of Pratt we certainly find the glimpse of law of floatation indirectly. Similarly, though Pratt does not believe directly in the concept of 'root formation' but very close perusal of his concept on isostasy, does indicate the glimpse of such idea (root formation) indirectly. While making a comparative analysis of the views of Airy and Pratt on isostasy Bowie has observed that 'the fundamental difference between Airy's and Pratt's views is that the former postulated a uniform density with varying thickness, and the latter a uniform depth with varying density.'

Hayford and Bowie's Concept : Hayford and Bowie have propounded their concepts of isostasy almost similar to the concept of Pratt. According to them there is a plane where there is complete compensation of the crustal parts. Densities vary with elevations of columns of crustal parts above this plane of compensation. The density of the mountains is less than the ocean floor. In other words, the crust is composed of lighter material under the mountains than under the floor, of the oceans. There is such a zone below the plane of compensation where density is uniform in lateral direction. Thus, according to Hayford and Bowie there is inverse relationship between the height of columns of the crust and their respective densities (as assumed by Archdeacon Pratt) above the line of compensation. The plane of compensation (level of compensation) is supposedly located at the depth of about 100 km. The columns having the rocks of lesser density stand higher than the columns having the rock of higher density.

Bowie made a comparative study of the views of Airy and Pratt on isostasy and concluded that there was a great deal of similarity in their views. In fact, 'both the views appeared to him similar but not the same'. Bowie could observe a glimpse of the concept of root formation and law of floatation of Airy, though indirectly, in the views of Pratt. The concept of Hayford and Bowie, that the crustal parts (various reliefs) are in the form of vertical columns, is not tenable because the crustal features are found in the form of horizontal layers.

Joly's Concept : Joly, while presenting his views on isostasy in 1925, contradicted the concept of Hayford and Bowie. He disapproved the view of Hayford and Bowie about the existence of level of compensation at the depth of about 100 km on the ground that the temperature at this depth would be so high that it would cause complete liquefaction and thus level of compensation would not be possible. He further refuted the concept of Hayford and Bowie that 'density varies above the level of compensation but remains uniform below the level of compensation' on the ground that such condition would not be possible in practice because such condition would be easily disturbed by the geological events and thus the level of compensation would be disturbed. According to Joly there exists a layer of 10-mile (16 km) thickness below a shell of uniform density. The density varies in this zone of 10-mile thickness. It, thus, appears that Joly assumed the level of compensation as not a linear phenomenon but a zonal phenomenon. In other words, he did not believe in a 'line (level) of compensation' rather he believed in a 'zone of compensation' (of 10-mile thickness). Thus, we also find a glimpse of the law of floatation (it may be remembered that Joly did not mention this, we only infer the idea of floatation from Joly's concept) in Joly's concept which is closer to the Airy's concept rather than the concept of Hayford and Bowie.

'This is in close agreement with floatation idea; the areas of low density in the 10-mile layer correspond with downward projections of the light continental crust, while those of high density represent the intervening areas filled with material of the heavier understratum' (S.W. Wooldridge and R.S. Morgan, 1959).

Holmes Concept : The views of Arthur Holmes on isostasy, to a greater extent, are compatible with the views of Airy. Following Airy Holmes has also assumed that upstanding crustal parts are made of lighter materials and in order to balance them major portions of these higher columns are submerged in greater depth of lighter materials (of very low density). According to Holmes the higher columns are standing because of the fact that there is lighter material below them for greater depth whereas there is lighter material below the smaller columns upto lesser depth.

Global Isostatic Adjustment

It may be pointed out that there is no complete isostatic adjustment over the globe because the earth is so unresting and thus geological forces (endogenetic forces) coming from within the earth very often disturb such isostatic adjustment. Moreover, recently a few scientists have even questioned the concept of isostasy. Even there is disagreement among the scientists about local or regional nature of isostasy. It appears from the result of various expeditions, experiments and observations that if the isostatic adjustment does not occur at local level, it does exist at extensive regional level. It is necessary that there must be balance at local level, it maybe and it may not be. The endogenetic forces and resultant tectonic events cause disturbances in the ideal condition of isostasy but nature always tends towards the isostatic adjustment.

For example, a newly formed mountain due to tectonic activities is subjected to severe denudation. Consequently, there is continuous lowering of the height of the mountain. On the other hand, eroded sediments are deposited in the oceanic areas, with the result there is continuous increase of weight of sediments on the sea-floor. Due to this mechanism the mountainous area gradually becomes lighter and the oceanic floor becomes heavier, and thus the state of balance or isostasy between these two areas gets disturbed but the balance has to be maintained. It may be stated that the superincumbent pressure and weight over the mountain decreases because of continuous removal of material through denudational processes. This mechanism leads to gradual rise in the mountain. On the other hand, continuous sedimentation on the sea-floor causes gradual subsidence of the sea-floor. Thus, in, order to maintain isostatic balance between these two features there must be slow flowage of relatively heavier materials of substratum (from beneath the seafloor) towards the lighter materials of the rising column of the mountain at or below the level of compensation. Thus, the process of redistribution of materials ultimately restores the disturbed isostatic condition to complete isostatic balance. Commenting on the validity of the above mechanism of the isostatic adjustment, Wooldridge and Morgan (1959) have remarked, "that some such mechanism operates is indeed very likely ; geologists have irrefutable evidence that sediments can depress the floor of a loaded sea to a limited extent, and some species of sub-crustal flow has been invoked on many other grounds. But clearly we are not justified in regarding the crust as composed of columns, moving up and down independently ; such a conception flouts the facts of observation, and even it did not, it would, on the geological side, create many more problems than it solved' (S. W .Wooldridge and R.S. Morgan, 1959, p. 26).

Some times the endogenetic forces act so suddenly and violently that the state of isostatic balance is thrown out of gear all of sudden and hence the isostatic adjustment through the process of flowage of materials from the substratum is not maintained.

Similarly, some times climatic changes occur at such an extensive global scale that there is accumulation of thick ice sheets on the land surface and thus increased burden causes isostatic disturbance. For example, extensive parts of North America and Eurasia were subsided under the enormous weight of accumulation of thick ice sheets during Pleistocene glaciation but the landmasses began to rise suddenly because of release of pressure of superincumbent thick load of ice sheets due to deglciation and consequent melting of ice sheets about 25,000 years ago and thus the isostatic balance was disturbed. According to an estimate major parts of Scandinavia and Finland have risen by 900 feet. The land masses are still rising at the rate of one foot per 28 years under the process of isostatic recovery. The isostatic adjustment in these areas could not be achieved till now.

The Rocks

The materials of the crust or lithosphere are generally called as rocks. The word lithosphere, in fact, means 'rocksphere' as the literal meaning of 'lithos' is rock. The smallest component of the crust or the lithosphere is element. As regards the whole earth eight most abundant elements (iron, oxygen, silicon, magnesium, nickel, sulphur, calcium and ammonium, constitute 99 per cent of the total mass of the earth whereas only four elements (iron, oxygen, silicon and magnesium) account for 90 per cent of total mass of the earth. On the other hand, the eight most abundant elements which constitute 99 per cent of total mass of the crust are oxygen; silicon, aluminium, iron, magnesium, calcium, potassium and sodium.

Table Important Elements of the Whole Earth and the Crust

Whole Earth		*Earth's Crust*	
Elements	***Percentage***	***Elements***	***Percentage***
1. Iron	35	1. Oxygen	46
2. Oxygen	30	2. Silicon	28
3. Silicon	15	3. Aluminium	8
4. Magnesium	13	4. Iron	6
5. Nickel	2.4	5. Magnesium	4
6. Sulphur	1.9	6. Calcium	2.4
7. Calcium	1.1	7. Potassium	2.3
8. Aluminium	1.1	8. Sodium	2.1
Others, less than	1.0	Others, less than	1.0

More than one element of the earth's crust are organized to form compounds which are known as minerals and minerals are organized to form rocks. The important mineral groups are silicates, carbonates, sulphides, metal oxide etc.

(1) The silicate minerals are very important rock making minerals. The most outstanding rockforming silicate mineral groups are quartz, feldspar, and ferromagnesium. Quartz is composed of two elements viz. silicon and oxygen and is generally a hard and resistant mineral. The most abundant and most important rock forming silicate mineral is feldspar which is also very important economically because it is used in ceramics and glass industry. Feldspar is very weak mineral and is easily broken down and decomposed due to chemical weathering and is changed into clays as hydrated alumino silicates. When silicon and oxygen combine with iron and magnesium, ferro-magnesium minerals are formed. Ferro-magnesium minerals are easily weathered and eroded away and are easily altered and removed. The rocks having abundant ferro-magnesium minerals provide weak structure for the construction of buildings, roads, dams, reservoirs, tunnels etc.

(2) Carbonate group of minerals is very much succeptible to chemical weathering and erosion in humid areas. Calcite is the most important mineral of this group. Limestones and marbles having abundant calcite are corroded by the surface and groundwater and extensive caves are formed below the ground surface. Such areas provide very weak structures for construction sites e.g., construction of buildings, roads, dams, reservoirs, air-strips, tunnels etc.

(3) Sulphide minerals include pyrites, iron sulphides etc. When these minerals come in contact with water or air, these form ferric hydroxides and sulfuric acids which cause serious environmental problems.

(4) Metallic elements like iron, aluminium etc., after reacting with atmospheric oxygen form metal oxides which are commercially very important.

Rocks, thus, representing the geomaterials of the earth's crust, are composed of two or more minerals. Rocks play very important role in deter mining the characteristic features of several types of erosional landforms because the nature and magnitude of erosion largely depends on the structure and composition of rocks. The fundamental dictum of famous American geomorphologist W.M. Davis that 'the landscape is a function of structure, process and time (stages)' lays more emphasis on the dominant role of rocks in the evolution of landforms. According to A. K. Lobeck 'a rock should be conceived as a product of its environment. When the environment is changed, the rock changes'. Rocks are also very helpful in dating the age of the earth as 'rocks are the books of earth history and fossils are the pages'. S. W. Wooldridge and R.S. Morgan (1959) have aptly remarked, 'Rocks whether igneous or sedimentary, constitute on the one hand the manuscripts of the past earth history, on the other, the basis for contemporary scenery.'

Rocks Classification : The crustal rocks are classified on several grounds e.g., mode of formation, physical and chemical properties, locations etc.

Classification on the Basis of Mode of Formation : The rocks are divided into three broad categories on the basis of their mode (method) of formation.

(i) Igneous rocks, formed due to cooling, solidification and crystalization of molten earth materials known as magma (below the earth's surface) and lava (on the earth's surface), e.g. basalt, granites etc.

(ii) Sedimentary rocks, formed through the lithification and compression and cementation of the sediments deposited in a particular place mainly aquatic areas, e.g., sandstones, limestones, conglomerates etc.

(iii) Metamorphic rocks, formed due to change either in the form or composition of either igneous or sedimentary rocks provided that there is no disintegration of preexisting rocks, e.g. slate, quartzite, marble etc.

Igneous Rocks

The word igneous has been derived from a Latin word 'ignis', meaning there by fire. It does not mean that the origin of igneous rocks is associated with fire in any way. In fact, the igneous rocks are formed due to cooling, solidification and crystalization of hot and molten materials known as magmas and lavas. Since the magmas and lavas are so hot that they look like red pieces of fire but this is not the case. Igneous rocks are also called as primary rocks because these were originated first of all the rocks during the formation of upper crust of the earth on cooling, solidification and crystallization of hot and liquid magmas after the origin of the earth. Thus, all the subsequent rocks were formed, whether directly or indirectly, from the igneous rocks in one way or the other. This is why igneous rocks are also called as parent rocks. It is believed that the igneous rocks were formed during each period of the geological history of the earth and these are still being formed.

Characteristics of Igneous Rocks :—

(1) In all, igneous rocks are roughly hard rocks and water percolates with great difficulty along the joints. Some times the rocks become so soft, due to their exposure to environmental conditions for longer duration, that they can be easily dug out by a spade (e.g. basalt).

(2) Igneous rocks are granular or crystalline rocks but there are much variations in the size, form and texture of grains because these properties largely depend upon

the rate and place of cooling and solidification of magmas or lavas. For example, when the lavas are quickly cooled down and solidified at the surface of the earth, there is no sufficient time for the development of grains/crystals. Consequently, either there are no crystals in the resultant basaltic rocks or if there are some crystals at all, they are so minute that they cannot be seen without the help of a microscope. Contrary to this, if magmas are cooled and solidified at a very slow rate inside the earth, there is sufficient time for the full development of grains, and thus the resultant igneous rocks are characterized by coarse grains.

(3) Igneous rocks do not have strata like sedimentary rocks. When lava flows in a region occur in several phases, layers after layers of lavas are deposited and solidified one upon another and thus there is some sort of confusion about the layers or strata but actually these are no strata rather these are layers of lavas. Such examples may be seen anywhere in the Western Ghats where several lava flows during Cretaceous period resulted into the formation of thick basaltic cover having numerous layers of lavas of varying compositions. One can see such lava layers near Khandala or along the deeply entrenched valleys of the Koyna river, the Krishna river, the Saraswati river etc., in and around Mahabaleshwar plateau.

(4) Since water does not penetrate the rocks easily and hence igneous rocks are less affected by chemical weathering but basalts are very easily weathered and eroded away when they come in constant touch with water. Coarse grained igneous rocks are affected by mechanical or physical weathering and thus the rocks are easily disintegrated and decomposed.

(5) Igneous rocks do not contain fossils because (i) when the ancient igneous rocks were formed due to cooling and solidification of molten rock materials at the time of the origin of the earth, there was no life on newly born earth and (ii) since the igneous rocks are formed due to cooling and solidification of very hot and molten materials and hence any remains of plants or animals (fossils) are destroyed because of very high temperature.

(6) The number of joints increases upward in any igneous rock. The joints are formed due to (i) cooling and contraction, (ii) expansion and contraction during mechanical weathering, (iii) decrease in superincumbent load due to removal of materials through denudational processes and (iv) earth movement caused by isostatic disturbances. Whenever these joints are plugged by minerals, the rocks become quite hard and resistant to weathering and erosion.

(7) Igneous rocks are mostly associated with the volcanic activities and thus they are also called as volcanic rocks. Igneous rocks are generally found in the volcanic zones.

Classification of Igneous Rocks : There are vast variations in the igneous rocks in terms of chemical and mineralogical characteristics, texture of grains, forms and size of grains, mode of origin etc. Thus, the igneous rocks are classified on several grounds in a variety of ways as follows—

(1) The most traditional method of the classification of the igneous rocks is based on the amount of silica (SiO2). Thus, the igneous rocks are divided into two broad categories e.g. (i) acidic igneous rocks having more silica, e.g. granites, and (ii) basic igneous rocks having lower amount of silica, e.g. gabbro. It may be pointed out that silica content is not a measure of acidity.

(2) On the basis of the chemistry and mineralogical composition (light and dark minerals) the igneous rocks are classified into two dominant groups e.g. (i) felsic igneous rocks composed of the dominant minerals of the light group such as quartz and feldspar having rich content of silica. The word 'felsic' has been derived from fell(s), feldspar plusic, meaning thereby the dominance of feldspar mineral, (ii) maficigneous rocks composed of the dominant mineral of dark group such as pyroxenes, amphiboles and olivines, all of which have rich contents of magnesium and iron. The word 'mafic' has been derived from magnesium and (ferrous) for iron and is meaning thereby the dominance of magnesium and ferrous (iron) and (iii) ultramafic igneous rocks are characterised by the abundance of pyroxenes and olivine minerals, examples, periodotite (rich in pyroxene and olivine), dunite (rich in olivine) etc.

(3) The igneous rocks are also classified on the basis of texture of grains into 5 major groups—

   (i) Pegmatitic igneous rocks (very coarse grained igneous rocks) include very large crystals several metres across. Examples, granites.

   (ii) Phaneritic igneous rocks (coarse grained igneous rocks). The word phaneritic has been derived from Greek word 'phanero', meaning thereby visible.

   (iii) Aphanitic igneous rocks (fine grained igneous rocks). The word aphanitec has been derived from the Greek word, 'aphan', meaning thereby invisible, that is the grains of the aphanites are so minute that they cannot be seen by bare eyes.

   (iv) Glassy igneous rocks (without grains of any size).

   (v) Porphyritic igneous rocks (mix-grained igneous rocks).

(4) The igneous rocks are more commonly classified on the basis of the mode of occurrence into two major groups.

(i) Intrusive Igneous Rocks (a) Plutonic igneous rocks (b) Hypabyssal igneous rocks

(ii) Extrusive Igneous Rocks (a) Explosive type (b) Quiet type

Intrusive Igneous Rocks : When the rising magmas during a volcanic activity do not reach the earth's surface rather they are cooled and solidified below the surface of the earth, the resultant igneous rocks are called intrusive igneous rocks. These rocks are further subdivided into two major groups of plutonic intrusive igneous rocks and hypabyssal intrusive igneous rocks on the basis of the depth of the place of cooling of magmas from the earth's surface. When the magmas are cooled and solidified very deep within the earth, the resultant rocks become plutonic but when the magmas are cooled just below the earth's surface, the rocks are called as hypabyssal igneous rocks.

(i) Plutonic igneous rocks are formed due to cooling of magmas very deep inside the earth. Since the rate of cooling of magmas is exceedingly slow because of high temperature prevailing there and hence there is sufficient time for the full development of large grains. Thus, the plutonic igneous rocks are very coarse-grained (pegmatites) rocks. Granite is best representative example of this category.

(ii) Hypabyssal igneous rocks are formed due to cooling and solidification of rising magma during volcanic activity in the cracks, pores, crev ices, and hollow places just beneath the earth's surface, the resultant rocks are called as hypabyssal igneous rocks. The magmas are solidified in different forms depending upon the hollow places such as batholiths, laccoliths, phacoliths, lopoliths, sills, dikes etc. It should be remembered that these should not be taken as the types of igneous rocks because these are different shapes of solidified magmas.

(A) Batholiths are long, irregular and undulating forms of solidified intruded magmas. They are usually dome-shaped and their side walls are very steep, almost vertical. The upper portion of batholiths are seen when the superincumbent cover is removed due to continued denudation but their bases are never seen because they are buried deep within the earth. When exposed to the surface they are subjected to intense weathering and erosion and hence their surfaces become highly irregular and corrugated. Numerous batholithic domes were intruded in the Dharwarian sedimentaries in many parts of the peninsular India during pre-Cambrian period. Many of such batholithic domes have now been exposed well above the surface in many parts of the Chhotanagpur plateau of India mainly Ranchi plateau where such batholithic domes are called as Ranchi Batholiths. Murha pahar near Pithauriya village, to the north-west of Ranchi city, is a typical example of exposed Ranchi batholithic domes.

(B) Lacoliths-The word laccolith has been derived from German word, 'laccos' meaning thereby 'lithos' or rocks. Lacoliths are formed due to injection (intrusion) of magmas along the bedding planes of horizontally bedded sedimentary rocks. Lacoliths are of mushroom shape having convex summital form. The ascending gases during a volcanic eruption force the upper strata of the flat layered sedimentary rocks to arch up in the form of a convex arch or a dome. Consequently, the gap between the arched up or domed upper strata and the horizontal lower strata is injected with magma and other volcanic materials.

(C) Phacoliths are formed due to injection of magma along the anticlines and synclines in the regions of folded mountains.

(D) Lopoliths-The word lopolith has been derived from German word 'lopas' meaning thereby a shallow basin or bowl shape body. When magma is injected and solidified in a concave shallow basin whose central part is sagged downward, the resultant form of solidified magma is called a lopolith. The rocks of lopoliths are generally coarse-grained because of slow process of cooling of magmas.

(E) Sills-The word 'sills' has been derived from an Anglo-Saxon word 'syl' meaning thereby a ledge. The sills are usually parallel to the bedding planes of sedimentary rocks. In fact, sills are formed due to injection and solidification of magmas between the bedding planes of sedimentary rocks. Thick beds of magmas are called sills whereas thin beds of magma are termed as 'sheets'. The thickness of sills ranges between a few centimetres to several metres. When sills are tilted together with the sedimentary beds due to earth movements and are exposed to exogenous denudational processes, they form significant landforms like cuesta, hogbacks and ridges.

(F) Dykes represent wall-like formation of solidified magmas. These are mostly perpendicular to the beds of sedimentary rocks. The thick ness of dykes ranges from a few centimetres to several hundred metres but the length extends from a few metres to several kilometres. A well defined dyke is observable across the palaeochannel and valley of the Narmada river near Dhunwadhar Falls (Bheraghat) near Jabalpur city. The relative resistance of dykes in comparison to the surrounding country-rocks gives birth to a few interesting landforms e.g. (i) If the rocks of dykes are weaker and less resistant than the country rocks, the upper portion of dykes is more eroded than the countryrocks, with the result a depression is formed, which, when filled up with water, is called a 'dyke lake; (ii) If the rocks of dykes are more resistant than the country, rocks upstanding ridges and hills are formed because of more erosion of the country-rocks, and (iii) If the rocks of dykes and country-racks are of uniform resistance, both are uniformly dissected and hence no significant landform in developed but the height is gradually reduced.

Extrusive Igneous Rocks : The igneous rocks formed due to cooling and solidification of hot and molten lavas at the earth's surface are called extrusive igneous rocks. Gener ally, extrusive igneous rocks are formed during fissure eruption of volcanoes resulting into flood basalts. These rocks are also called as volcanic.

Rocks : Extrusive igneous rocks are generally fine grained or glassy basalts because laves after coming over the earth's surface are quickly cooled and solidified due to comparatively extremely low temperature of the atmosphere and thus there is no enough time for the development of grains or crystals. Basalt is the most significant representative example of extrusive igneous rocks. Gabbro and obsidian are the other important examples of this group. Extrusive igneous rocks are further divided into two major subcategories on the basis of the nature of the appearance of leaves on the earth's surface e.g. (i) explosive type and (ii) quiet type.

(i) *Explosive Type :* The igneous rocks formed due to mixture of volcanic materials ejected during explosive type of violent volcanic eruptions are called explosive type of extrusive igneous rocks. Volcanic materials include 'bombs' (big fragments of rocks), 'lapilli' (fragments of the size of a peas) and volcanic dusts and ashes. Fine volcanic materials, when deposited in aquatic condition, are called 'tuffs'. The mixture of larger and smaller particles after deposition is called 'breccia' or 'agglomerate.' These are more susceptible to erosion because these are not well consolidated.

(ii) *Quiet Type :* The appearance of lavas through minor cracks and openings on the earth's surface is called 'lava flow'. These lavas after being cooled and solidified form basaltic igneous rocks. Flood basalts resulting from several episodes of lava flow during fissure flows of volcanic eruption form extensive 'lava-plateau' and 'lava-plains' wherein several layers of basalts are deposited one upon another.

The thickness of lavas of the Columbia plateau of the states of Washington and Oregon (USA), spread over an area of about 6,45,000 Km^2 (2,50,000 square miles), measures about 1,216 m (4,000 feet). The extensive lava flows during Cretaceous period covered an area of about 7,74,000 Km^2 (3,00,000 square miles) of Peninsular India. Several beds of basaltic lavas are clearly observable all along the exposed sections of the Western Ghats mainly near Khandala (between Bombay and Pune) and over Mahabaleshwar plateau.

Classification on Basis of Chemical Composition : Though the chemical composition of igneous rocks varies significantly from one group to another group but each type of igneous rock contains some amount of silica. Thus, on the basis of silica content, igneous rocks are divided into the following four types.

(i) Acid igneous rocks are those which carry silica content between 65 to 85 per cent. The average density varies from 2.75 to 2.8. Quartz and white and pink feldspar are the dominant minerals. Acid igneous rocks generally lack in iron and magnesium. On an average acid igneous rocks are hard and relatively resistant to erosion. Granite is the most significant example of this group of rocks. These rocks are light in weight and are used as building materials because of their less erosivity.

(ii) Basic igneous rocks contain silica content between 45 to 60 per cent. Their average density ranges from 2.8 to 3.0. Such igneous rocks are dominated by ferro-magnesium minerals. There is very low amount of feldspar. The rock is heavy in weight and dark in colour because of the dominance of iron content. Basic igneous rocks are easily eroded away when these come in regular contact with water. These rocks are fine-grained igneous rocks. Basalt, gabbro, dolerite etc., are the typical examples of this group.

(iii) Intermediate igneous rocks are those in which silica content is less than the amount present in the acid igneous rocks but more than the basic igneous rocks. The average density ranges between 2.75 and 2.8. Diorite and andesite are the representative examples of this group of rocks.

(iv) Ultra-basic igneous rocks carry silica content less than 45 per cent but their average density varies from 2.8 to 3.4. Peridotite is the typical example of this group of rocks.

Classification on Basis of The Texture of Grains : The texture of the crystals (grains) of igneous rocks depends on 3 basic factors viz. (i) source region of the origin of magmas and lavas and places of their cooling and solidification ; (ii) rate of cooling and solidification of magmas and lavas and (iii) quantity of water and gases (vapour) with hot and molten magmas and lavas. If magmas and lavas are cooled slowly and gradually the grains are well developed but if they are cooled and solidified at a very faster rate, grains are not well developed. The rate of cooling of magmas and lavas also depends upon several factors viz. (i) When magmas are cooled deep within the earth, the rate of cooling is exceedingly slow because of very high temperature prevailing there and hence very large and coarse grains are formed. (ii) If lavas are cooled at the surface of the earth, the rate of cooling is very fast because of very low temperature (in comparison to the temperature of lavas) of outside environment and hence either grains are not formed at all, or if they are formed, they are so minute that they cannot be seen without the aid of a microscope. (iii) If magmas and lavas are associated with larger proportion of water vapour and gases, the rate of their cooling and consequent solidification is slowed down and hence larger grains are formed.

On the basis of the size of grains (texture) igneous rocks are generally divided into (i) coarse grained igneous rocks (plutonic igneous rocks come under this category, granite is the example), (ii) fine-grained igneous rocks (extrusive igneous rocks fall under this group, basalt is the example) and (iii) medium-grained igneous rocks (hypabyssal rocks are generally medium grained igneous rocks).

Alternatively, igneous rocks are divided into six sub-types on the basis of textural characteristics of the rocks e.g. (1) pegmatitic igneous rocks (or very coarse-grained igneous rocks; examples: plutonic igneous rocks e.g. pegmatitic granites, pegmatitic diorite, pegmatitic synite etc.) ; (2) phaneritic igneous rocks (or coarse-grained igneous rocks ; plutonic igneous rocks ; examples, granites, diorties etc.) ; (3) aphanitic igneous rocks (or fine-grained igneous rocks ; grins are so minute that they cannot be seen without the help of a microscope ; examples: basalt, felsite and the rocks of sills and dykes) ; (4) glassy igneous rocks (or grainless igneous rocks; usually there is general absence of grains; examples: pitch stones, obsidians, pumice, perlite etc.) ; (5) porphyritic igneous rocks (or mix-grained igneous rocks and (6) fragmental igneous rocks (consisting of bombs, lapilli, breccia, volcanic dusts and ashes, tuffs etc.).

It is desirable to discuss in brief the major characteristics of granites and basalts which represent the intrusive and extrusive igneous rocks respectively.

The Granites

Granites are the most significant example of the plutonic intrusive igneous rocks which are formed deep within the earth. Since the rate of cooling and solidification of magmas inside the earth is very slow because of very high temperature prevailing underground and hence granites become coarse-grained due to full development of large-sized grains. Granites are composed essentially of the minerals of quartz, feldspar, and mica but the most abundant mineral is feldspar, mainly orthoclase. Some times, the minerals are uniformly distributed and all of them are almost of the same size. Besides, albite, biotite, muscovite and hornblende are also found in granite rocks.

The granite family includes numerous types of rocks. These granitic rocks are differentiated on the basis of their texture and mineral composition, for example, hornblende granite (when hornblende mineral is most dominant), rhyolite granite, pumice granite, absidian granite, pitch-stone granite etc. From the standpoint of chemical composition granites are acidic rocks wherein silica content ranges between 65 to 85 per cent. Granites are generally light in weight as their density varies from 2.75 to 2.8.

Table : Mineral Composition of Granites

Minerals	Feldspar	Quartz	Mica	Hornblende	Iron	Others
Percentage	52.3	31.3	11.5	2.4	2.0	0.55

There is also wide range of colour variation in different types of granites. The colour variation is caused mainly because of the number of different minerals present in the rocks and the size of grains. Generally, granites are of light colour but if orthoclase mineral is present in abundance, the granites become pink to yellow or slightly reddish in colour. If dark coloured hornblende or biotite is a dominant mineral, the granites become of dark black or dark grey colour.

Granites are generally resistant to erosion but when the rocks are well jointed, they are easily weathered and a very peculiar landform, 'tor', is formed.

The Basalts

Basalt is a very fine-grained, dark-coloured extrusive igneous rock which is formed due of cooling and solidification of molten lavas at the surface of the earth. Some times, the cooling of lavas takes place so rapidly that no time is available for the crystallization of basalt and hence no grains are formed, with the result the rock becomes glassy basalt. Basalts having grains, though very small rather minute, are called aphanitic basalts. Chemically, basalts contain 45 to 65 per cent of silica content. Though the rock is heavy in weight but is more susceptible to chemical weathering and fluvial erosion. The dark colour of basalts is because of the abundance of iron. Feldslpar is the most dominant mineral (46.2 per cent). Besides, augite (36.9 per cent), olivine (7.6 per cent), mineral iron (9.5 per cent) etc. (others - 2.4 per cent) are other constituent minerals of basalts. Some times, polygonal cracks are developed in basalts due to contraction on cooling of lavas. Columnar jointings in basalts give birth to peculiar landforms characterized by uneven terrain surfaces.

Sedimentary Rocks

Sedimentary rocks, as the word implies, are formed due to aggregation and compaction of sediments. The word 'sedimentary' has been derived from Latin word 'sedimentum' which means 'settling down'. Sedimentary rocks are also called as stratified or layered rocks because these rocks have different layers or strata of different types of sediments. Some times, layers are absent in some sedimentary rocks, for example loess. The sediments and debris derived through the disintegration and decomposition of the rocks by the agents of weathering and erosion are gradually deposited in water bodies. Thus, layers after layers of sediments and debris are regularly deposited. Continuous sedimentation increases the weight and pressure and thus different layers are consolidated and compacted to form sedimentary rocks.

According to P.G. Worcester (1948) 'sedimentary rocks, as sediment implies, are composed largely of fragments of older rocks and minerals, that have been more or less thoroughly consolidated and arranged in layers and strata.'

Characteristics **:** Though most of sedimentary rocks are deposited due to continuous deposition of sediments in water bodies (lakes, ponds, basins, rivers and seas) but some times these are also formed at the land surface, e.g. loess, rocks of sand dunes, alluvial fans and cones. The following are the main characteristics of sedimentary rocks.

(1) Sedimentary rocks are formed of sediments derived from the older rocks, plant and animal remains and thus these rocks contain fossils of plants and animals. The age of the formation of a given sedimentary rock may be determined on the basis of the analysis of the fossils to be found in that rock.

(2) Sedimentary rocks are found over the largest surface areas of the globe. It is believed that about 75 per cent of the surface area of the globe is covered by sedimentary rocks whereas igneous and metamorphic rocks cover the remaining 25 per cent area. Inspite of their largest coverage the sedimentary rocks constitute only 5 per cent of the composition of the crust whereas 95 per cent of the crust is composed of igneous and metamorphic rocks. Thus, it is obvious that 'the sedimentary rocks are important for extent, not for depth in the earth's crust.'

(3) The deposition of sediments of various types and sizes to form sedimentary rocks take place in certain sequence and system. The size of sediments decreases from the littoral margins to the centre of the water bodies or sedimentation basins. Different sediments are consolidated and compacted by different types of cementing elements e.g., silica, iron compounds, calcite, clay etc.

(4) Sedimentary rocks contain several layers or strata but these are seldom crystalline rocks.

(5) Like igneous rocks sedimentary rocks are not found in massive forms such as batholiths, lacoliths, dykes etc.

(6) Layers of sedimentary rocks are seldom found in original horizontal manner. Sedimentary layers are generally deformed due to lateral compressive and tensile forces. The beds are folded and found in anticlinal and synclinal forms. Tensile and compressive forces also create faults due to dislocation of beds.

(7) Sedimentary rocks may be well consolidated, poorly consolidated and even unconsolidated. The composition of the rocks depends upon the nature of cementing elements and rock forming minerals.

(8) Sedimentary rocks are characterized by different sizes of joints. These are generally perpendicular to the bedding planes.

(9) The connecting plane between two consecutive beds or layers of sedimentary rocks is called 'bedding plane'. The uniformity of two beds along a bedding plane is called conformity (i.e. when beds are similar in all respect). When two consecutive beds are not uniform or conformal, the structure is called unconformity. In fact, 'an unconformity is a break in a stratigraphic sequence resulting from a change in conditions that caused deposition to cease for a considerable time' (J.D. Collison and D.B. Thompson, 1982). There are several types of unconformity e.g. (i) non-conformity (where sedimentary rocks succeed igneous or matamorphic rocks), (ii) angular unconformity (where horizontal sedimentary beds are deposited over previously folded or tilted strata), (iii) disconformity (where two conformable beds are separated by mere changes of sediment type), (iv) paraconformity (where two sets of conformable beds are separated by same types of sediments) etc.

(10) Sedimentation units in the sedimentary rocks having a thickness of greater than one centimetre are called beds. The upper and lower surfaces of a bed are called bedding planes or bounding planes. Some times, the lower surface of a bed is called sole while the upper surface is known as upper bedding surface. There are further sedimentary units within a bed. The units having a thickness of more than one centimetre are called as layers or strata whereas the units below one centimetre thickness are known as laminae. Thus several strata and laminae make up a bed. When the beds are deposited at an angle to the depositional surface, they are called cross beds and the general phenomena of inclined layers are called cross lamination or cross bedding.

(11) Soft muds and alluvia deposited by the rivers during flood period develop cracks when baked in the sun. These cracks are generally of polygonal shapes. Such cracks are called as mud cracks or sun cracks.

(12) Most of the sedimentary rocks are permeable and porous but a few of them are also nonporous and impermeable. The porosity of the rocks depends upon the ratio between the voids and the volume of a given rock mass.

The Classification

1. On The Basis of The Nature of Sediments

(1) Mechanically formed or clastic rocks

(i) Sandstones

(ii) Conglomerates

(iii) Clay rock

(iv) Shale

(v) Loess

(2) *Chemically formed sedimentary rocks*

(i) Gypsum

(ii) Salt rock

(3) *Organically formed sedimentary rocks*

(i) Limestones

(ii) Dolomites

(iii) Coals

(iv) Peats

2. On The Basis of Transporting Agents

(1) *Argillaceous or aqueous rocks*

(i) Marine rocks

(ii) Lacustrine rocks

(iii) Riverine rocks

(2) *Aeolian sedimentary rocks*

(i) Loess

(3) Glacial sedimentary rocks

(i) Till

(ii) Moraines

Mechanically Formed Sedimentary Rocks : Previously formed rocks are subjected to mechanical or physical disintegration and thus the rocks are broken into fragments of different sizes. These are called fragmental rock materials or clastic materials which become source materials for the formation of clastic sedimentary rocks. These materials are obtained, transported and deposited at suitable places by different exogenous processes (geological agents) like running water (rivers), wind, glaciers, and sea waves. These materials are further broken down into finer particles due to their mutual collision during their transportation. These materials after being deposited and consolidated in different water bodies (sedimentation basins, lakes, seas, rivers etc.) form sedimentary rocks known as clastic sedimentary rocks. Sandstones, conglomerates, silt, shale, clay etc., are important members of this group.

(1) *Sandstones :* Sandstones are formed mostly due to deposition, cementation and consolidation of sand grains. Sand grains are divided into five categories on the basis of their size.

Table : Classification of Sands by Grain size

Sand Types	*Grain Size (mm)*
(i) Very coarse sand	1.0 to 2.0
(ii) Coarse sand	0.5 to 1.0
(iii) Medium sand	0.25 to 0.5
(iv) Fine sand	0.125 to 0.25
(v) Very fine sand	0.0625 to 0.125

When sand grains are deposited in water bodies and are aggregated and consolidated by cementing elements (e.g. silica, calcium, iron oxide, clay etc.), sandstones are formed. The colour of sandstones varies according to the nature and amount of cementing elements and minerals. Sandstones become red or gray when cemented by iron oxides but these become white or gray when calcium carbonate dominates. Sandstones become hard and resistant to erosion when cemented by silica. On an average sandstones are porous rocks and water easily percolates through them. On the basis of textural and mineralogical characteristics sandstones are classified into (i) quartz arenites (arenite from Latin word arena, meaning thereby sand) composed entirely of quartz grains, (ii) arkose sandstones (feldspar being the dominant mineral), (iii) lithic arenites (composed of finegrained rock fragments, mostly derived from shales, slates, schists and volcanic rocks) and (iv) graywacke sandstones (composed of quartz, feldspar and rock fragments surrounded by a fine-grained clay matrix).

Conglomerates : Conglomerates are formed due to cementation and consolidation of pebbles of various sizes together with sands. 'The term conglomerate is applied to cemented fragmental rocks containing rounded fragments such as pebbles and boulders; if the fragments are angular or sub angular, the rock is called breccia' (A. Homes and D.L. Holmes, 1978). Polished and rounded fragments are called pebbles having a diameter upto 4 mm while those fragments which have the diameter upto 256 mm are called boulders. The rock fragments after being cemented by clay form gravels. Though gravels are found in layers but there is general absence of uniformity. When the rounded fragmental materials are cemented by quartz, the resultant rocks become conglomerates. If conglomerates are formed due to their cementation by silica, they become very hard rocks and resistant to erosion.

Clay Rock and Shale : Clay rocks are formed due to deposition and cementation of fine sediments. The rocks formed of the sediments having the grain size of 0.03 mm to 0.004 mm are called silts whereas clays are formed when the sediments of the grain size of 0.004 mm to 0.00012 mm are cemented and consolidated. Silt and clay are soft and weak rocks but they are definitely impervious. Clay rocks are formed exclusively of kaolin minerals. Since clay rocks are not soluble and hence these are least affected by chemical weathering but these are easily eroded away. Pure clay rocks are of white colour but they change in colour when they are mixed with the impurities of other materials. Shales are formed due to consolidation of silt and clay. Shales are formed of thin laminae which are easily separated. Shales are impermeable rocks and therefore they hold mineral oil above them.

Chemically Formed Sedimentary Rocks : Running water contains chemical materials in suspension. When such chemically active water comes in contact with the country rocks in its way, soluble materials are removed from the rocks. Such materials are called chemically derived or formed sediments. These chemical materials after being settled down and compacted and cemented form chemical sedimentary rocks such as gypsum and salt rocks.

Organically Formed Sedimentary Rocks : The sediments derived from the disintegration or decomposition of plants and animals are called organic sediments. These sediments after being deposited and consolidated form organic sedimentary rocks. On the basis of lime and carbon content these rocks are divided into 3 categories e.g. (i) caacareous rocks, (ii) carbonaceous rocks and (iii) siliceous rocks.

(1) Calcareous rocks are formed due to deposition and consolidation of sediments derived from the skeletons and remains of those animals and plants which contain larger portion of lime. Limestone is the most significant characteristic example of calcareous rocks. Limestones are formed in the following manner —

(i) Calcium oxide (CaO) reacts with water (H_2O) to form calcium hydroxide ($Ca(OH)_2$)

$CaO + H_2O$ ® $Ca(OH)_2$

(ii) Calcium hydroxide reacts with carbon dioxide (COD to form calcium carbonate ($CaCO_3$) $Ca(OH)_2 + CO_2$ ® $Ca\ CO_3$ (limestone) + H_2O

The calcareous rocks are collectively called as carbonate rocks or simply carbonates. Limestones ($CaCO_3$) or calcium carbonate, magnesium carbonate ($Mg\ CO_3$) and dolomite ($CaMg\ (CO_3)_2$) are important carbonate rocks. Limestones are found in both the forms- thinly bedded and thickly bedded. They are formed of both fine sediments as well as coarse sediments. The most dominant minerals are calcite (of hexagonal shape) and argonite (of orthorhombic shape). Since limestones are formed of chemically soluble materials and hence these are most susceptible to chemical weathering as follows

(i) Carbon dioxide (COD after being dissolved in water form carbonic acid (HZ CO_3)

$$CO_2 + H_2\ O \rightleftharpoons H_2\ CO_3$$

(ii) Carbonic acidreacts with limestone ($CaCO_3$) to form calcium bi-carbonate ($Ca(HCO_3)_2$)

$$H_2CO_3 + CACO_3 \rightarrow Ca\ (HCO_3)_2$$

Though limestoens are very weak rocks in humid regions because these easily dissociate when they come in contact with water but these become resistant rocks in hot and dry climate because of the fact that limestones have uniform and homogeneous structure and hence these are not affected by differential expansion and contraction due to temperature changes. The rocks having the carbonates of both calcium and magnesium are known as dolomites which are less soluble than limestones. These carbonate rocks, after weathering and chemical erosion, give birth to karst topography. Chalk is another form of carbonate rocks but it is softer and more porous than limestone. Chalks are formed due to precipitation of carbonate materials which are derived from micro-organisms like foraminifera.

(2) Carbonaceous rocks are dominated by carbonic materials which represent vegetation remains. These rocks are formed due to transformation of vegetations because of their burial during earth movements and consequent weight and pressure of overlying deposits. The initial form of carbonaceous rocks is peat which is of a dark gray colour. Vegetation remains can be seen with the help of microscope. The other subsequent forms of carbonaceous sedimentary rocks are lignite, bituminous and anthracite coals with greater proportion of carbon and darker colour. Coals are also found in stratified form wherein coal layers are known as coal seams. Carbonaceous rocks are more important economically than geomorphologically.

(3) Siliceous rocks are formed due to dominance of silica content. Siliceous rocks are formed due to aggregation and compaction of wastes de rived from sponge and radiolarian organisms and diatom plants. Geyserites are also deposits of silica around geysers. Geyserites have different colours e.g., white, gray or pink due to impurities of deposition of various types of sediments.

Classification on the Basis of Transporting Agents : Sedimentary rocks are also classified on the basis of transporting agents or geological agents (e.g. running water or rivers, wind, glaciers, oceanic currents and sea waves). These agents of transportation obtain different types of sediments and deposit them in suitable places where sediments are consolidated and cemented to form sedimentary rocks of various sorts. Based on major transporting agents, sedimentary rocks are divided into argillaceous, aeolian and glacial rocks.

(1) Argillaceoue rocks are also called as aqueous rocks because these are formed in water areas. Aqueous word has been derived from Latin word 'aqua' which means 'water'. Aqueous rocks are called as argillaceous rocks because of the dominance of clay in the rocks. In fact, the word argillaceous has been derived from Latin

word 'argyll' or 'argill' meaning thereby clay. Argillacous rocks are characterized by their general softness. These are essentially impervious rocks. Argillaceous rocks are further divided into 3 sub-types on the basis of the places of their formation. (i) Marine argillaceous sedimentary rocks are formed due to deposition and consolidation of sediments in the oceans and seas mainly in their littoral zones. The process of sedimentation in marine environment is well ordered and sequential in character. In other words, the size of particles decreases progressively from the coastal lands towards the seas or the oceans e.g., the order of the particles from the coast lands towards thc sea is of boulders, cobbles, pebbles, granules, sands, silts, clay and lime. It is evident that as we go away from the coast lands towards the sea, the size of sediments becomes so fine that they are kept in suspension with oceanic water. Sandstones, limestones, dolomites and chalk are the most important examples of marine argillaceous sedimentary rocks. (ii) Lacustrine argillaceous sedimentary rocks are formed due to deposition and consolidation of sediments in lake environment. Generally, the sediments are deposited at the floor of the lakes. The lacustrine rocks may be seen in 3 conditions viz. (a) if the lake becomes dry, (b) if the floor of the lake is raised due to earth movements and (c) if the whole lake is filled up with sediments. It may be pointed out that there is no ordering in the size of sediments as is the case with the seas and the oceans. (iii) Riverine argillaceous sedimentary rocks are those which are formed due to deposition of sediments in the riverine environment. The sediments may be deposited in the beds of the rivers and in the flood plains. Such deposition includes alluvia which are dominated by clay. Alluvia are deposited on either side of the alluvial rivers during floods. It may be pointed out that alluvial deposits are renewed almost every year. Alluvial deposits develop polygonal cracks due to their exposure to insolation.

(2) Aeolian sedimentary rocks are formed due to deposition of sands brought down by the wind. Preexisting rocks are greatly disintegrated due to mechanical weathering in the hot and dry regions. This process results in the formation of immense quantity of sands of different sizes. Winds pick up these sands and deposit them at various places. The particles are further comminuted into finer particles due to attrition while they are being transported from one place to another. Continuous deposition of sands results in the formation of different layers but these layers are not well consolidated as is the case with the argillaceous rocks. Sometimes, there is complete absence of layers in the airborn or aeolian sedimentary rocks. Loess is the most important member of this group. Loess is, in fact, the heaps of unconsolidated fine materials. There is general absence of laminae and layers in the loessic formation. These are soft and porous rocks. Water can easily infiltrate in the loessic deposits. Thus, loess is easily eroded away. Thus, most outstanding

characteristic feature of loess is that the entire loessic mass may stand like a vertical cliff or wall. The best example is observable on the left and right banks of the palaeochannel and valley of the Narmada river at Dhunwadhar falls (Bheraghat) near Jabalpur (M.P.) where the loessic banks rise 20 to 25 m from the valley floor and form complete vertical free-face cliff section. The sediments are so loosely arranged that they can be removed even by using fingers. The most extensive loessic deposits are found in north China where the thickness of sediments is of several hundred metres. The deposits are of yellow colour and are rich in lime and hence these look like fine loam soils. The Yellow river (formerly Hwang Ho) and its tributaries easily erode the loessic deposits and hence the river becomes overloaded and causes frequent severe floods. It may be pointed out that the Yellow river of China carries the largest amount of sediments (1640 million tonnes per year) in the world. The river is called 'Yellow' because of the yellow colour of the sediments which are derived through the erosion of yellow coloured Chinese loess.

(3) Glacial rocks-The materials deposited by glaciers are called glacial drifts which are deposited in four conditions and therefore there are four types of morainic deposits viz. (i) laterlal moraines, when glacial materials are deposited on either side of a glacier, (ii) medial moraines, when glacial sedimentary materials are deposited along the joining glaciers ('The lateral moraines of joining ice streams merge and form a single medial moraine in the middle of larger flows-' F. Press and R. Siever, 1978), (iii) ground moraines, when the glacial materials are deposited in the bed of the glacier and (iv) terminal moraines (these are formed when the glacier is ablated and materials are deposited therein).

Metamorphic Rocks

Meaning and Characteristics : 'Metamorphic rocks include rocks that have been changed either in form or composition without disintegration' (P.G. Worcester, 1948). Metamorphic rocks, as the word 'metamorphism' implies, are formed due to changes in the forms of other rocks. Originally, the word metamorphism has been derived from the word 'metamorphose' which means change in form. In fact, metamorphic rock means complete alteration in the appearance of preexisting rocks due to change in mineral composition and texture through temperature and pressure. Metamorphic rocks are generally formed due to change in form of sedimentary and igneous rocks. Some times, even previously formed metamorphic rocks are again metamorphosed.

It may be mentioned that the process of metamorphism simply means change in form but in geology this is used for specific meaning and condition. For example, the form and composition of a rock may change during the process of metamorphism but there is no disintegration and decomposition of the rock. When already formed metamorphic rocks are again metamorphosed, the process is known as remetamorphism. This becomes possible only when the temperature becomes exceedingly high and the weight and pressure of overlying rocks becomes enormous due to orogenetic movements. When the rocks are metamorphosed to the greatest intensity, the process is known as intense metamorphism. Dharwarian sedimentary rocks of peninsular India have suffered intense metamorphism.

The change in the form of the rocks during the process of metamorphism takes place in two ways viz. (i) physical metamorphism pertaining to changes in textural composition of the rocks and (ii) chemical metamorphism, leading to changes in the chemical composition of the rocks. Some times, both the processes of metamorphism become operative together. It may be pointed out again that during the process of metamorphism there may be complete alteration in the form of the rocks, the form and nature of minerals may change, old minerals may be rearranged and changed in new minerals, new minerals may be added, preexisting minerals may be transformed into other forms due to melting caused by very high temperature, preexisting crystalline rocks may be recrystallized but there would be no disintegration and decomposition of the rocks in any circumstance.

Some times, the form of the rocks is so changed due to intense metamorphism that it becomes difficult to find out the original form of the rocks. Some rocks, after metamorphism, become harder than their original forms, such as marbles from limestones and quartzites from sandstones. Marbles and quartzites are relatively more resistant to erosion than their parent rocks, limestones and sandstones. Fossils of sedimentary rocks are also destroyed during the process of metamorphism.

'Unlike igneous rocks, the texture of metamorphic rocks is the result of recrystallization or conversion of one mineral to another in the solid state' (Press and Siever, 1978). Foliation, defined as streaking or parallel arrangement of the constituent crystals (of the metamorphic rocks) which generally cut the rocks at an angle to the bedding planes of the original sediments of the parent rocks' is the most common characteristic feature of metamorphic rocks. The coarse-grained metamorphic rocks are imperfectly foliated (e.g. gneisses from granites) while fine-grained metamorphic rocks are perfectly foliated (e.g. schists from shales). The property of metamorphic rocks to part or split along the bedding planes is known as fissility. The structure of the presence of numerous closely spaced parallel planes of splitting is known as cleavage. In fact, cleavage is a special type of foliation which denotes the tendency of a rock to cleave or break or split into moderately thin sheets or laminae. Schistocity reefers to the growth of larger crystals and segregation of some minerals into lighter and darker bands.

Agents of Metamorphism :—

(i) Heat is the most important factor for the development of metamorphic rocks from pre-existing parent rocks. It may be pointed out that mineral composition is entirely changed due to intense heat but the rocks are seldom melted. The required heat for metamorphism is available during vulcanicity when hot and molten magmas ascend through the crustal rocks.

(ii) Compression resulting from convergent horizontal movements caused by endogenetic forces causes folding in rock beds. Thus, the resultant pressure from compressive forces and consequent folding changes the form and composition of parent rocks. This factor becomes operative during mountain building.

(iii) Solution-Chemically active hot gases and water while passing through the rocks change their chemical composition. Magmatic water and water confined in the beds of sedimentary rocks also help in introducing chemical changes in the rocks.

Types of Metamorphisms

The agents and factors of metamorphism some times operate separately and some times work together. The processes of metamorphism may be classified on the bases of

(i) the nature of the agents of metamorphism and

(ii) place and area involved in metamorphism.

(1) On the basis of the nature of agents

(i) Thermal metamorphism (due to heat)

(ii) Dynamic metamorphism (due to pressure)

(iii) Hydro-metamorphism (due to hydrostatic pressure)

(iv) Hydro-thermal-metamorphism (due to water and heat)

(2) On the basis of place or area

(i) Contact metamorphism (localized in area)

(ii) Regional metamorphism (involving larger area)

(3) Composite classification

(i) Contact or thermal metamorphism

(ii) Dynamic and regional metamorphism

(iii) Hydro-metamorphism

(iv) Hydro-thermal metamorphism

Contact Metamorphism

Contact metamorphism takes place when the mineral composition of the surrounding rocks known as aureoles is changed due to intense heat of the intruding magmas. This process of metamorphism is called contact metamorphism because of the fact that metamorphism occurs when the rocks come in contact with the intruding magmas. This process is also called as thermal metamorphism because the rocks are changed in their forms due to high temperature of the introducing magmas. Such metamorphism occurs during volcanic activity when the physical properties of the surrounding rocks are changed due to intense heat of the rising magmas of dykes. Some times, the rocks coming in contact with the intruding magmas are also changed in their chemical composition due to some water and water vapour associated with the intruding magmas. Limestones are changed to marbles due to contact metamorphism.

As stated above, the rocks surrounding the igneous intrusions are altered due to intense heat of magmas. The margins of the altered rock around igneous intrusions are called aureoles the width of which (i.e. the dimension of metamorphosed rocks) depends upon mainly two factors e.g. (i) temperature of intruding magma, and (ii) the depth of magma intrusions in the crust.

Regional Metamorphism : When the rocks are altered in their forms in extensive area the process is called regional metamorphism. Such metamorphism is also known as dynamic metamorphism because pressure plays dominant role in the alteration of the form of the rocks though temperature is also an important factor. The sedimentary rocks are folded due to compressive forces during the period of mountain building. This process results in intense pressure and heat which ultimately alter the original form of the concerned rocks. Dynamic metamorphism leads to crystallization in the rocks and if the rocks are already crystallized they are recrystallized. Regional metamorphism is a characteristic feature of mountainous area. Regional metamorphism is further divided into two sub-types viz. (i) Dynamic regional metamorphism, when the rocks are metamorphosed due to compressive forces and resultant high pressure caused by convergent horizontal movements, and (ii) Static regional metamorphism, when the rocks are metamorphosed at greater depth due to intense pressure and weight of overlying rocks (superincumbent load).

Hydro-Metamorphism : The alteration in the composition of the rocks due to hydrological factor takes place in a number of ways e.g. (i) When the chemically active water (solvent) passes through the country rocks, there occur several chemical changes in the rocks due to varied chemical reactions. (ii) The storage of immense volume of water in big reservoirs exerts high pressure on the underlying rocks and thus the rocks are altered in their forms due to pressure of overlying huge volume of water. Such type of metamorphism is known as hydrostatic metamorphism.

Hydro-Thermo-Metamorphism : The minor alteration in the physical and chemical composition of the rocks caused by the weight and pressure of water mass and chemically active hot gases and water vapour is called hydro-thermometamorphism which is, in fact, geographically less important.

The Classification : The classification of metamorphic rocks is easier and less complicated because generally these are classified on the basis of those original or parent rocks from which they have been formed. It is obvious that the parent rocks in relation to metamorphic rocks are sedimentary and igneous rocks. Some times, the process of metamorphism becomes so intense and the parent rocks are so greatly metamorphosed that it becomes very difficult to trace the true nature of the rocks before their metamorphism. Besides such conditions, metamorphic rocks are divided into 2 broad categories.

(1) Meta-sedimentary or para-metamorphic rocks are those metamorphic rocks which are formed due to alteration of the forms of sedimentary rocks, e.g. marbles from limestones, quartzites from sandstones, slates from shales and clays etc.

(2) Meta-igneous or ortho-metamorphic rocks represent those metamorphic rocks which are formed due to changes in the form of igneous rocks, e.g. gneisses from granites, serpentine from gabbro, basic granulites from amphibolites, eclogite from basaltic rocks etc.

Metamorphic rocks are also classified on the basis of foliation into (i) foliated metamorphic rocks e.g. slates, gneisses and schists and (ii) non foliated metamorphic rocks, e.g., quartzites, marbles, serpentines etc.

Important Metamorphic Rocks : Marbles are generally formed due to changes in limestones because of temperature changes. Limestones are transformed into marbles due to contact thermal metamorphism during volcanic activity. Limestones are also metamorphosed due to dynamic regional metamorphism wherein calcium carbonates and other finer particles are changed into calcite. In fact, the metamorphism of limestones to marble involves a number of changes in the mineralogical characteristics of limestones. For example, the reaction between calcium carbonate of limestone during the process of metamorphism produces a new mineral known as wollastonite or calcium silicate. The colour of marbles depends upon the nature of parent limestones. If the original limestones are devoid of any impurities, the resultant marbles become pure white in colour. The colour changes due to impurities of other materials in the parent limestones. The marbles of Carrara region of Italy are pure white while the marbles exposed along both the banks of the magnificent and stupendous gorge of the Narmada river at Bheraghat near Jabalpur (M.P.) show different grades of colour though white and pink colours dominate.

Dolomites and chalks are also metamorphosed to marbles due to excessive heat but these have only local importance. Marbles are more resistant to erosion than their parent limestones. Besides, they are economically valuable rocks because they are used as building materials for the construction of very important buildings as monuments. For example, Tajmahal of Agra and Dilwara temple of Mount Abu (Rajasthan) have been built of marbles.

Schists are fine-grained metamorphic rocks and are characterized by well developed foliation. The word schist has been derived from French word 'schiste' and German word 'schistose' which means to split. When shale sedimentary rocks are subjected to intense compressive force and consequent folding and pressure, the clay and other minerals of the original shale rocks are changed to mica minerals due to high pressure and temperature and thus shales are changed to schists. During the process of regional metamorphism the schists get foliated. Schists are named on the basis of dominant minerals. e.g., mica-schists, hornblende schists, quartz schists etc. Mica schist is the commonest type of schist rocks because it is formed from argillaceous shale sedimentary rock which is a very common rock and is abundantly found on the earth's surface. Micaschist is composed of muscovite, biottle, plagioclase and some times garnet. Hornblend schists are formed from basaltic rocks and contain hornblende, plagioclase and some quartz minerals. Green schists are composed of green minerals such as hornblende and chlorites, provided that the rocks are well foliated. If the schists rich in green minerals are poorly foliated, they are called greenstones. The term metabasite is used to name those schists which are formed from basalts or dolerites.

Slates are formed due to dynamic regional metamorphism of shales and other argillaceous rocks. Slates are characterized by the 'presence of numerous closely-spaced parallel planes of splitting or cleavage but the splitting planes of slates are not parallel to the bedding planes rather they form angle with the bedding planes. Some times, the angle between the splitting planes and bedding planes becomes obtuse angle. Such structure of slates is known as slaty cleavage which is formed due to compressive pressure exerted on the rocks.

The cleavage is always at right angle to the direction of compression. Slates, if subjected to further intense metamorphism due to immense compression, are changed to phyllites or fine-grained mica-schist, 'Slates, in fact, may be regarded as a special type of fine-grained schist' (S.W. Wooldridge and R.S. Morgan). Slates are not as much resistant to erosion as are schists and gneisses. They are of varied colours.

Gneisses are coarse-grained metamorphic rocks which are formed due to metamorphism of conglomerates (sedimentary rocks) and coarse gained granites (igneous). Feldspar is the most dominant mineral of gneisses. Like schists, gneisses are also foliated rocks but the foliation is open and is some times absent. There are several types of banded gneisses, which some times pass into augen gneiss. The process of granitization or granitification means the transformation of mica-schist to gneiss. Gneissic rocks produce, after weathering and erosion, rounded topography.

Quartzites are generally formed from sandstones which are dominated by the abundance of quartz mineral. During the process of metamorphism the voids within the sandstones are compacted due to excessive compression and heat and are also filled with silica, with the result quartzites become very hard and resistant to erosion. When quartzites lie over weaker sedimentary rocks like shales or limestones as caprocks, they form stupendous wall-like escarpments. Kaimur escarpment along the left bank of the Son river (in M.P. and Bihar), Bhander escarpments (Satna and Parma districts of M.P.), Rewa escarpments facing the Ganga plains etc., have been formed due to resistant caprocks of quartzitic sandstones resting over shale lithology. 'The term quartzite is also extended to sandy rocks which have been subjected to cementation by silica deposited from solution. Such rocks are generally softer than the true metamorphic quartzites and often behave more like normal sandstones, breaking down into sandysoils' (S.W. Wooldridge and R.S. Morgan, 1959).

9

Significance of Data

Logical Ground

Concept : What principle or principles should the geographer have in mind in selecting phenomena for consideration in the geography of an area? Of all the vast heterogeneity of material and immaterial phenomena that may be observed in any area, which are significant for geography? If the study is made for some special purpose, practical or otherwise, obviously that purpose will provide the measure of significance for different kinds of phenomena. Our interest here, however, is with studies whose purpose is simply to increase geographic knowledge, studies, that is, in pure geography, in the only proper sense of that expression-namely, in contrast with applied geography. We examine this question from the point of view of furthering geographer:, knowledge not "for its own sake" simply, but on the assumption that the furtherance of geographic knowledge will produce values that we need not demonstrate. On this basis we may not permit our selection of data to be determined by any ulterior purpose, whether that be to fit that' data to particular techniques of study or to enable geography to lay claim to the title of :'science" in a particular sense that would be otherwise unattainable. Geography must aim to provide geographic knowledge, even though the nature of some of that knowledge should cause it to lose any hope of being called a "science" in the sense that certain other branches of knowledge are called "science," just as the geographer must adapt his methods of study to the material of geography rather than select the material suitable to his methods: To state the conclusion directly, the basis for selection of data must be derived logically from our fundamental concept of geography.

Among geographers who' disagree in their fundamental concept of the field we can hardly expect agreement as to the basis for selecting and rejecting different phenomena to be studied. Fortunately, however, as we observed earlier, there is a marked agreement among the large number of students who accept what we have called the chorographic concept of geography. The largest number of them, including most of the geographers of Germany as well as many in this country, use a form of statement similar to that which Hettner derived from Richthofen: "Geography studies the areal differentiation of the earth's surface-the differences found in the different continents, lands, districts, and localities".

Everyone who has travelled twenty miles from his home knows that differences exist from place to place ; these differences are therefore "naively given" facts. Further, any thinking person is aware that these differences are not phenomena of independent circles of facts, but are interrelated to each other, so that the human curiosity is immediately quickened with a desire to know what these relationships are.

A person whose knowledge of the world was limited to a local area and who might assume that all parts of the world were much like his own, would, no doubt, be most impressed with differences in those elements, both material and immaterial, which vary greatly within a small area. He might observe great variation in slopes, in drainage, in certain aspects of soil, and in the productivity of farms and the relative prosperity of farmers. If he lived in the Lower Rio Grande Valley he would certainly observe the differences in the cultural character of the population on either side of the river. For many a peasant of Eastern Europe, or of many parts of China, the greatest differences in the world are those between the countryside in which he lives and the local town or city; indeed-these differences between the rural and the urban environment are among the greatest differences that the earth's surface shows. They involve not only the differences in density of population and in number and character of buildings, but also in the character of the population, the contrast between "urban" and "rural" types—the "city slicker" and the "country yokel"—a contrast to which Passarge gives much attention.

Anyone who has traveled somewhat more extensively, either in fact or vicariously through reading, knows on the other hand that certain elements which may be approximately uniform over a considerable area vary greatly in different parts of the world. These might include climate, natural vegetation, major landforms and soil types, major types of agriculture, and many cultural characteristics of the population.

All of these considerations are so well known that the geographer may without hesitation start from them. Further, he would recognize in a general way that these differences are interrelated so that, where several basic features were fairly uniform over a considerable area or region, many, though by no means all, of the other features are also fairly uniform. He would therefore recognize, perhaps quite unconsciously, that each of many such areas has a distinctive character in the total ensemble of its phenomena, a character which extends over an as yet undefined but considerable stretch of territory, and which is different in some important way from that of any neighbouring area.

For illustration, let us consider a German peasant youth from the neighborhood of Breslau who should happen to travel slowly to the southeast, up the plain of the Oder. For some distance he would observe characteristics much the same as those he knew at home. Fields of rye, wheat and oats, potatoes and some sugar "beets on the flat plain of loess soil, would all seem familiar to him; the towns, each with its rectangular market centre, its double oval of streets around the oldest part of the town, marking the inside and outside of the former city wall, and the German conversation which he hears, all this tells him that he is still in his own region, his *"Heinwtgebiet,"* even though the details be unfamiliar. Farther up the plain, however, (keeping to the east of the Oder) he will note very significant changes. The farms are obviously less productive, he sees little wheat or sugar-beets, and many large stretches of pine-forest; as an intelligent peasant he immediately notes the sandy character of the soil, and needs no geographer to explain the relationships involved. But he also notes—quite possibly first of all that the people here who look like himself and his folks at home, who live and work much as they do, are to him very different, for they are Polish-speaking, and he will certainly consider this the most remarkable observation he has yet made.

Let us go with him one step farther. Wandering across the unfenced fields he may unwittingly cross a line which, however clearly marked on the political map and in the landscape "observable" as a series of small white posts—is somewhat difficult to observe in fact in the fields. (Since he himself is an observable object this may prove dangerous for him, but we can assume that he is not observed.) It will not be long before he will realize that he has now entered a very different kind of country. For, not only do the peasants speak Polish, but so do most of the people in shops, and in nearly all of the schools and other public buildings, and if he should settle down to live here he will find that the produce of his land is not to be sold in Breslau but in Katowice, that, not only economically but also socially, he and his family are cut off from things in Germany and tied to things in Poland, that their lives will be dominated not from Berlin but from Warsaw, that his sons will see military service not in the Rhineland, but perhaps on the borders of Russia. In sum, it will not take him long to realize that the most important difference between the area to which he has come and that which he has left, is that this is Poland, that was Germany.

It is instructive to consider a similar, though imaginary, situation in the field of history. Various writers have used the fictional device of putting a person of our times temporarily back into an earlier period, though still in the same place. What are the phenomena which such a person would be able to contribute to the study of history? The historian would ask him for those aspects of the past period which could be regarded as characteristic of that period, *i.e.*, not incidental differences which had no relation to other characteristics of the time, and for those events which might have significant relations to events of previous or subsequent periods. Obviously it will often be difficult to say what things are historically *significant*, but the historians have long since recognized that there is no categorical rule for eliminating this difficulty; it must be met, as best it can, in each particular instance.

Basis of Data Selection

Hettner has stated the geographer's criteria for the selection of data in terms of two conditions that correspond logically to those governing an historical study. His exposition, presented in translation below, was first published in 1905, and has since been accepted by a large number, if not indeed the great majority, of German geographers. We have previously noted that his views on this question have been followed by Gradmann, Hassinger, Banse, Soleh, and Maull, among many others.

"One condition. . . is the difference from place to place together with the spatial association of things situated beside each other, the presence of geographical complexes or systems-for example, the drainage system, the system of atmospherical circulation, the trade areas and others. No phenomenon of the earth's surface is to be thought of for itself; it is understandable only through the conception of its location in relation to other places on the earth. The second condition is the causal connection between the different realms of nature and their different phenomena united at one place. Phenomena which lack such a connection with the other phenomena of the same place, or whose connection we do not recognize, do not belong in geographical study. Qualified and needed for such a study are the facts of the earth's surfaces which are locally different and whose local differences are significant for other kinds of phenomena, or, as it has been put, are geographically efficacious." (This is dearly a development from Richthofen's criterium: "insofar as they have a recognizable relation to the earth-surface".

"The goal of the chorological conception," Hettner continues, "is the recognition of the character of the lands and localities from an understanding of the mutual existence and mutual effectiveness of the different realms of nature [reality] and their different forms of expression, and the conception of the entire earth's surfaces in its natural [real] formation in continents, lands, regions, and localities.

"Only in the application 'of both these points of view lies the character of geography; anyone who has not taken them into his flesh and blood has not comprehended the spirit of geography; just as an historian who does not inquire after the temporal sequence of things and the inner connections of different groups of development, has not grasped the spirit of history. With this conception, to be' sure, the choice of material presumes a previous ; consideration of the causal connection of the phenomena; with the development of knowledge, whole groups of facts can be won for geography or may be abandoned and the circumference of geography will be differently conceived according to subjectively different evaluations' of causal connections. But we find just such fluctuations in the historical and the systematic sciences and no objection to this principle of choice of material is to be understood as a result of them. The choice of material depends not on a single fact, but always on an entire group of facts, which one has learned to regard as causes or effects of other geographical groups of facts. Geography does not take up the particular facts only when it has recognized their geographical conditionality, but establishes their geographical circumstances descriptively, before it goes on to the causal investigation; and it can easily happen that it must introduce facts whose causal connections are not yet clear.

Present Day Geography

"The diversity of the material according to this conception is indeed large and becomes ever larger, for the progress in knowledge discovers in an ever increasing number of groups of facts a conditioning in the nature of the locality, and therefore geographical character. Present-day geography includes events as well as patterns and material conditions, facts of social life as well as of nature; but it includes all these objects only under the chorological point of view, and can therefore pass by many characteristics and peculiarities which, for the material and historical sciences, are perhaps the very most important. It can omit, not only all conditions which are everywhere on the earth the same, or whose local differences recognize no kind of rule of distribution, but also all those things whose local differences have no relation, at least so far as our knowledge goes, with those of other groups of phenomena."

We may note, in passing, a somewhat similar expression by the founder of modern geography in France, Vidal de la Blache: "Geography is the science of places and not that of men; it is interested in the events of history insofar as these bring to work and to light, in the countries where they take place, qualities and potentialities that without them would remain latent".

This distinction, or limitation of the things to be studied, unlike that of the limitation to material things, is founded on the logical position of geography as a science, as well as on its historical development. For "the necessity of a chorological science of the earth surface results from the two conditions, (1) that facts of one and the same circle of phenomena which are spatially close to each other do not lie isolated beside each- other, but are mutually related, and (2) that the facts of different circles of phenomena united at one place on the earth stand in causal relationship and together determine the character of the area." It is therefore proper for the geographer to extend his study "over all phenomena of the earth's surface so far as these two points of view are applicable to them".

The geographer, we may say, considers the arrangement and relationships of the phenomena within a region; to think geographically is to think of phenomena not as individual objects in themselves, but as elements determining the differential character of areas. Phenomena which exist in an area but in no relation to other phenomena in that area, such as the general phenomena of earth magnetism (where unaffected by local minerals), may be said to have no areal relationships, and are not therefore of concern to geography.

Criteria Application

We may draw a number of conclusions of practical value in applying the criteria stated, if we first emphasize three particular concepts involved in the criteria. These are (1) the *interrelation* of different kinds of phenomena that are directly or indirectly tied to the earth; (2) the *differential* character of these phenomena and the complexes they form, in different areas of the earth; and (3) the *areal expression* of the phenomena or complexes. On the basis of our fundamental definition of geography as the study of the areal differentiation of the world, it follows logically-and has commonly been followed in practice-that only such phenomena as are significant according to all three concepts are to be described and interpreted in geographic study.

The requirement of interrelation, or *Zusammenhang,* of phenomena provides us, as Gradmann has explained, with a specific method of selecting geographic objects of study, whereby we can set aside "the dilettante and arbitrary method, formerly much in favour, of putting mere curiosities together. The individual fact enters with a degree of importance that increases with the extent to which it is interlaced, or many sides and internally, with neighbouring circles of phenomena, both forwards and backwards as cause or as effect".

In other words, the answer to Schluter's argument that the phenomena of non-material culture are not related in detail to the details of the natural elements, as are the material facts of the cultural landscape, is to accept the argument as a statement that is in relative degree true and to apply it to just that degree. But factory buildings, cities in general, or the distribution of population, are facts that are not related in detail to all the details of the natural elements. Insofar as the cities of our South show distinctive characteristics different from those of the North and the differences are related to other regional factors, they are significant to a regional comparison of South and North, but they are not so important for such a comparison as the distinctive agriculture of the South, which is much more intimately related to its particular climate, soils, and population structure. Social phenomena that show relatively little relation to other regional characteristics need not be formally excluded; their relatively low degree of interconnection automatically makes them of little geographic significance, a conclusion that Richthofen clearly indicated.

Consequently we may agree in part with Penck: insofar as the state is considered as a social organization of people, it is a social phenomenon whose characteristics-whether described in terms of the form of government, the personality of its rulers, or its youthful, mature, or old—age nature-is a cultural phenomenon dependent largely on other cultural phenomena which may have but little relation to facts tied to any particular earth area, and therefore will call for little attention from the geographer. But the state is not only a social organization, it is at the same time the organization by man of a particular section of the earth's surface, and in this respect is inevitably closely interrelated to basic earth facts.

At this point some readers may protest that environmentalism is once again "raising its ugly head" and those in whom the traditional overemphasis on relationships in American geography has developed a phobia against the very idea of considering phenomena in interrelation will part company with us without noting perhaps that Schluter and Penck depend upon the concept in their expositions. Our historical survey showed that geographers have from the beginning been concerned with the study of phenomena of different categories 'in interrelation, and no "landscape purist" can interpret the facts found in his landscapes without discussing "relationships." To insist that for phenomena to be geographically significant they must be causally interrelated with other regional phenomena, is not to define geography as the study of relationships; if we say that a house cannot be built of bricks without mortar, we' are not saying that a house consists of mortar. (Presumably it is this misunderstanding that caused Michotte to refer to Hettner as "defending, with certain variations, the same ideas" as Brunhes—namely, the study of the interdependence of phenomena as the definition of geography.

Differential Character

The second requirement listed, the differential character of the phenomena with reference to the various parts of the earth, is based not merely on our fundamental definition as derived from the historical development of the field, but equally well, if one will, on common sense. Just as the historian of the Roman era does not need to the us that at that time people found it necessary to spend a considerable part of their lives sleeping, that they put on more clothing in the winter than in the summer, or that mothers commonly showed an affectionate interest in the welfare of their offspring, neither is it necessary for the geographer studying any particular district of the world to tell us things about it that are true for any land area of, the world. If he is describing a farming area, we may assume that the population live in some sort of permanent structures built above ground in which the livestock are not housed-only in the exceptional cases where conditions are different is it necessary to describe them.

If these statements appear ridiculously obvious they are nevertheless required as contrasts to certain quite logical conclusions that have been derived from the concept of geography as the study of the visible landscape. That we are not indulging in dialectics is shown by some of the detailed work of Grano, the student who has perhaps most logically pursued the landscape concept. He has made a detailed study of a small rural district in Finland, Valosarri, to which the geographer who has never seen that country might naturally turn in order to learn the character of such a district there in contrast, say, to one in Iowa. He will find that Grano has not only mapped the individual buildings with care, but on several separate maps has shown the surfaces of the ground covered by these buildings as rainless areas, and as areas protected from wind and cold. Likewise he will learn that in this district of Finland there is less wind in the forests than in the fields, that birds can be heard more often in woods and meadows than near the houses, and that there are different kinds of Odors in each of these places. This description of Grano's study is not intended to belittle it, but rather to show the difference in kind between a detailed landscape study as logically worked out, and the point of view in regional study that most students would regard as geographic.

This latter point of view, dominated by the concept of areal differentiation, we may illustrate by the contrast in geographic interest offered by a farm house and the farm fields. Though in an absolute sense the house might be regarded as of equal if not greater importance than the fields, in terms of areal differentiation of the different parts of the world, it offers much less of significance. Regardless of where the farm is located we know a great deal about the house before we see it, and to see it will tell us relatively little about the character of the area. Merely from its name we know that it is a man-made structure elevated above the ground, that includes physical facilities used by the family for sleeping, cooking, eating, relaxation, and as a retreat from unpleasant weather out of doors, and that some members of the family perform much of their productive work within it. In major degree the forms of houses are everywhere adapted to these functions, so that the differences between them are either minor, or are of minor significance in relation to other regional factors. They are minor, that is, in contrast with the notable and significant regional differences in the use of the farm fields-in crops produced, methods of cultivation and harvesting, yields attained, and ultimate manner of consumption.

It may be objected that this point of view is bound to give an unbalanced presentation of reality in any area, but this is true only if the reader of such a study lacks the universal concepts whose description is omitted because assumed. Unquestionably there is some danger here, just as for the ignorant student, history may give an impression of life in the past as made up entirely of wars and political and economic changes. To be sure, the fellahin of Egypt have sown, cultivated and reaped the land flooded by the Nile through every year of its many milleniums, but the historian is not called upon to describe the annual cycle of Egyptian life for every period of its history, any more than he needs to record the fact that in any particular generation, a million or so babies were born, although that historical fact was absolutely essential to all the following history of Egypt. In other words, history assumes countless unnamed universals and similarly in geography we may assume that a house shuts off sun, rain, wind, and cold, more or less effectively; that farmers work in the fields by daylight and spend the nights in their houses, etc.

On the other hand, the emphasis on differential factors is not to be exaggerated falsely into a concern for the rare and unusual feature, as Krebs has noted. An exceptional feature that attracts our attention, whether because it is seldom found anywhere, or because it is of general importance in other regions—a Swiss chalet in the Appalachians, for example—is significant to the geographer only insofar as it is significant in the area he is studying. In writing the geography of the Po Plain, one must resist the temptation to dwell on the very small districts of minor rice production, simply because they are rare in Europe, or one will lose the significance of the far more important and significant combination of corn-wheat-and-dairy agriculture with the horticulture of vine, fruits, and nuts. Similarly, Kemp has recently emphasized that geographers concerned with the Balkans should pay less attention to the scattered tobacco fields, or the rose gardens of Kazanlik, and far more attention everywhere to corn (maize).

This last consideration is closely related to the third concept involved in our criteria: phenomena are significant in geography to the extent that they have areal expression. Although all might agree on this bare statement—we have previously noted its expression by Schluter and Sauer-there is a notable difference in its interpretation. Schluter evidently measured this attribute in terms of the physical extent of the object concerned. Similarly Penck, in ruling studies' of distribution out of geography, explains that geography is concerned not with plants but with forests, not with men themselves, but with their effects on the earth surface. But Krebs reminds us that, in Ritter's classic phrase, it is not the objects that fill areas that are themselves of concern to the geographer, but the areas. No particular category of object, not even the trees that together form a forest, nor the fields that form the major part of a rural scene, actually fin the area that we unhesitatingly map as characterized by that particular phenomenon. Furthermore, some objects that occupy but a minor fraction of the surface, may be of such great importance in their effects on other features, that their significance in areal differentiation is our of all proportion to the space they occupy physically. We have previously noted, for example, that if the types of houses in rural Louisiana furnish a reliable guide to the culture that has been developed throughout each of its different sections, Kniffen is entirely justified in giving such great attention to objects that Schluter would exclude because they occupy so little space. Similarly, though the actual area on which the Negro population of our South stands, sits, or lies, is an insignificant fraction of the total area of the South, the presence of so many people of that particular social group is a factor of the greatest significance in the agricultural and urban character of the South in contrast with almost all other regions of the world.

Schluter has, in fact, accepted this viewpoint, at least in part, in one of his discussions. In order to permit the geographer to study the scarcely visible facts of population density, he postulates as necessary for geographic data that the phenomena must each have "a specific location and a specific areal extension on the earth surface". But it is obvious that we are not interested in the area covered by individual human beings, but rather the area of the earth in which such and such kinds of people are found in such and such numbers-it is the character of the area itself, that is given it by the presence of the human beings. If we can consider that as a real character of the area we can equally well consider the low standard of living of the population of that area, in contrast to higher standards in other areas, as a characteristic of the area.

On much the same basis we can approach the difficult problem of what size of areas may be considered. Penck has argued that one can speak of a geography of a city, but not of a market place, but it is not clear what difference in principle is involved. If the geographer however keeps in mind the question of relative significance, he may regard size as but one of the important attributes to be considered. An understanding of the rural geography of Mexico necessitates a knowledge of the peculiar structure of Mexican rural communities. If Dicken was justified in concluding from reconnaissance that in many important respects the village of Galeana is representative of the character of many similar communities in a particular, fairly large, area of Mexico, then his detailed study of that small and unimportant village has more significance for the geography of Mexico than would the study of such an exceptional city as the oil port of Tampico.

Special Techniques

It may be instructive to apply Sauer's argument, concerning the relation of a special geographic technique to the material to be studied in geography, to that which is included by the criteria here discussed. If the geographer has any specific technique, what is it, and does it fit him to consider all the multitudinous variety of phenomena here included?

If we examine the product of geographical work, as contained in the volumes of any of the leading geographical journals, including therefore work of many different kinds, what particular technique is disclosed as common to all, and yet relatively little used in other fields of knowledge? Surely it is the technique of cartographical presentation, the one technique which geographers have developed in a great variety of rich detail. Nowhere is this better illustrated than in Bowman's unique exposition of the points of view, techniques, and kinds of data employed in current geographical work.

Geologists, to be sure, are also dependent on maps, but for most parts of geology the maps are but minor means toward the ultimate purpose of understanding the history of the earth. Though the making of the standard topographic maps is entrusted in this country in contrast to most others—to the Geological Survey, the work is placed under the direction of a "geographer."

The geographer's claim to a special technique in maps hardly seems to require demonstration. Many other disciplines use maps, just as they must so, at times, use the historical method, but no one questions that, the latter is the distinctive technique of history. The field of history, to be sure, may not claim a monopoly in the historical method and neither may geography claim cartography as an integral part of its field. Nevertheless, workers in other fields commonly concede without question that the geographer is the expert on maps, whether in making or in using them. This is the one technique on which they most often come to him for assistance and often in such cases they come to realize that the geographer can do far more with maps than they anticipated. "The most important contributions of geography to the world's knowledge," James holds, "have come from an application of the technique of mapping distributions and of comparing and generalizing the patterns of distribution".

It is no accident that geographers above all others should be concerned with maps. How otherwise can the geographer assemble the facts of spatial relationships for study, how otherwise present them in their spatial relationship, except on maps? (Insofar as the photograph can do this for relatively restricted areas, it is an important supplement to the map.) Consequently we can often judge the geographic quality of a man's work by the effectiveness with which he has presented it on maps. One wishes that those editors of geographical publications who all too often reduce a valuable detailed map to the size of a half-page, or less, would post the following statement of Hettner's, written in 1905, over their editorial desks: "In consequence of the development of cartographical methods of presentation, verbal description has lost its original importance and serves now only to complete and explain the maps".

It requires, finally, no discussion to prove that the cartographical technique of the geographer can be applied to nonmaterial phenomena as well as to material phenomena. It has been so used hundreds of times. Indeed, the distinction in the study of a particular territorial problem, between the work of a geographer on the one hand and that of a political scientist on the other, is no more clearly indicated than by the relative degree and effectiveness in the use of maps. This distinction is not merely in the maps themselves, but rather in the use of the maps in the study—the true geographer cannot help but study the problem in terms of maps.

An excellent example of the value of this technique is offered by Bowman's *New World*. While this deals with problems which many political scientists have studied, its extraordinarily effective use of maps has made it a well-nigh unique contribution. Clearer examples may be found in the detailed studies of individual border areas, of which the German and French geographical literature could provide dozens of examples. Thus, the writer's discussion of the boundary problem of Upper Silesia depends primarily on the use of 16 maps.

It should not be supposed that this technique is limited, in political geography, to the study of boundary problems. The fact that it is of great importance in other parts of political geography, as well as in various other topics in what maybe called "social geography," has been demonstrated many times in well known works by Ellsworth Huntington, Griffith Taylor and Mark Jefferson. More particularly, illustrations are offered by the writer's study of the problems of sectionalism in pre-war Austria-Hungary, by Milojevic's study of similar problems in Yugoslavia, by Wright's study of regional tendencies in national elections in this country, and by the writer's study of the racial geography of the United States.

So important, indeed, is the use of maps in geographic work, that, without wishing to propose any new law, it seems fair to suggest to the geographer a ready rule of thumb to test the geographic quality of any study he is making: if his problem cannot be studied fundamentally by maps—usually by a comparison of several maps-then it is questionable whether or not it is within the field of geography.

10

Current Thoughts

Phenomena of Geography

The question as to what phenomena in any area are proper grist for the geographer's mill has always been a difficult one. Those who have understood and accepted the concept of geography as-like history an integrating rather than systematic field, recognize that there is no escape from the multiplicity of diverse phenomena which combine to form the character of any region. While any of these phenomena may form, in themselves, proper material for the systematic studies of any of the natural or social sciences, any of them may also be appropriately studied by the historian insofar as they show differences significant to his problem of integrating differences with reference to time. Similarly they are appropriate to the geographer, regardless of their intrinsic character, if they show differences-in themselves or in their relations to other phenomena-significant to his problem of integrating phenomena as found in space.

On the other hand the smaller group who have taken their lead from Schluter, Brunhes, Passarge, and Sauer would limit the geographer's examination of a region to material features, both natural and cultural, excluding any nonmaterial cultural features. Geography then includes, as it always has, man and his works, but only his material works—commonly those that are visible. Although the limitation may appear arbitrary, the motivation underlying it commands our respect. As stated by one of its first exponents, Schluter, what is needed in geography is "limitation in the subject-matter ,and objectivity in the observation".

In contrast with most of the methodological problems current in geography, which we have already found were repeatedly discussed in earlier periods, the present issue appears to be relatively new. Both Humboldt and Ritter, as well as their predecessors, and their followers for a generation or more, considered without hesitation many kinds of phenomena that this limitation would exclude from the field of geographic study. During the past thirty years, however, the question has been the subject of a great number of discussions in the German literature. More recently it has been introduced into this country, and, though largely confined to oral discussions and textbooks, has had, as we shall observe, significant effects on the character of geographical research. Consequently it becomes necessary to subject it to detailed examination.

It is common practice to assert this limitation without going to the trouble of providing any justification for it. Many have been led to believe that it represented the point of view of the entire modern movement in Germany including those representatives like Hettner, who in fact have most vigorously opposed it. Sauer, writing on geography as a social science, in the Encyclopaedia of Social Sciences, contrasts two groups of geographers one group is concerned with the study of relationships, which he calls "Human Geography," while "the other group directs its attention to those elements of material culture (later referred to as 'visible') which give character to area.

German Geography

In reality the situation in German geography is very nearly the reverse of what American geographers have been led to suppose-not with respect to the chorographical concept itself, of course, but with respect to this particular limitation of the phenomena. Though Schluter suggested the limitation as early as 1899 and presented the case for it more fully in his celebrated paper of 1906 he recognized in 1920 that it had "found little attention and still less approval" and eight years later he was no more encouraged. Likewise Passarge, who has followed him in certain respects, has frequently given expression to his wrath at critics who, though praising his productive work, fail to accept-worse yet, even ignore-his methodological views. The only other German writers on methodology who appear to support this—limitation are Tiessen and Penck, to whom should be added Grano, Brunhes, and Michotte in other European countries. Other writers either ignore the suggestion entirely or have expressed their objections to the thesis directly. Among the latter may be listed: Hettner, Philipson, Supa Oberhummer, Sapper, Gradmann, Hassert, Heiderich, Friedrichsen, Volz, Lautensach, Hassinger, Banse, Maull, Solch, Waibel, and Burger, to whom we may add the two philosophers who have most carefully considered our field, Graf and Kraft, Pod the Italian geographer, Almagia. Finally if we were to subtract from he list of supporters of this limitation those who have repeatedly ignored it in practice, none would be left except (presumably) Grano and Michotte.

In view of these facts we might be justified in discussing this thesis as of but little importance in current geographic thought. On the other hand a small number of American geographers have repeated the statement of this limitation so persistently that it appears to have become entrenched in current thought. Objections that have been raised to it in connection with political geography, or references to Hettner's attack upon it, have brought no response other than mere repetition of the thesis as decreed by higher authority. Consequently it seems necessary to examine this thesis thoroughly from the point of view of the field of geography as a whole.

The restriction itself is somewhat difficult to examine because of the lack of agreement among its various proponents as to just what is to be excluded. The original conception appears to have been an outgrowth of the meaning of *Landschaft* as "landscape," but, as we saw, this term has a somewhat different meaning for almost every individual geographer. In theoretical statements most of the "landscape purists"-to use the convenient term that Dickinson suggests for those who limit the objects of geography in this way—limit geography to objects that are *sinnlich wahrnehmbar,* or perceptible to the senses, whether those of sight, sound, smell, or feeling, Grano therefore protests that Hettner is not meeting the issue when he compares the concept of the "character" *(Wesen)* of an area simply with the form or picture *(Bild)* of the area.

But if Hettner's argument does not apply to the generalized statement of the landscape purists, it does apply to the point of view that their detailed discussions present. For in these, they speak almost exclusively of the visible aspect of objects, and when they consider them together in an area, it is clearly the *Landschaffsbild,* the visible landscape, that they have in mind. To be sure, any material objects not included on this basis, such as climatic conditions, are added on the grounds of their perception by other senses, but the impossibility of combining these things under a single unit concept is reflected by the fact that none of these writers have discovered any way of expressing this total other than by the term they have create for it, or by such loose expressions as "in general, visible," or, finally, by simply saying "area" and later telling us that "area': does not include everything in the area. Even Grano himself, who has attempted to work out this concept more logically than any other student, is in practice contorted by the visibility of objects. Thus he finds the colour of clothing more important than its form, and men, in general, form a part of the *Landschaft* where they appear in masses that constitute a visible form. These conclusions are logical only if we are to determine what is significant in geography on the basis of how much impression they make on our senses-in which case, as Grano properly concludes, everything beyond our immediate vicinity is considered only in terms of visibility; but this is simply the *Landscnaftsbild.* If the essential point is simply that the objects be somehow directly perceptible to anyone of the senses, or more simply, must be material objects, how can we say that form and texture of clothing is less important than colour? Are men in the mass any more material than when they are scattered?

Grano would apparently answer this question by means of an additional limitation, which was originally stated by Schluter and has been accepted by Sauer among others, but not by all of the landscape purists. According to Schluter, the geographer's attention should be directed toward "those expressive features that have areal extent, as distinct from individual objects". On this basis he would include settlement forms but not house types, whereas the followers of Brunhes and Passarge have been stimulated to give great attention to just that feature. Schluter recognizes that the same principle reduces men to "minute grains in the landscape," and all of these students have had difficulty in defining the proper place for man in geography. For Schluter the problem is solved by the fact that our special interest in man acts as a magnifying glass, and thereby human geography is made of equal importance with physical geography. On the other hand, the magnification apparently must not be sufficient to permit us to observe the colour of men's skin, nor may our ears observe the language he speaks.

Though Grano's conclusion appears more consistent, we may question whether it is actually as logical as it seems. If men in the mass form a phenomenon that is physically perceptible and covers a certain areal extent, both of these attributes are nothing more than the arithmetic sum of the perceptibility and areal extent of each of the individuals composing it. Scatter them as you will, they still cover, in total, exactly the same amount of area, and if we approach each and every one separately they are all material objects of which our total perceptions may be no less than our perception of them in mass. It is only on the basis of the part they form in the visible landscape sensation-our psychic response-that there is any difference.

For this reason, perhaps, Passarge finds no difficulty in including man in the description of a *Landschaft*. "Farmers plowing-perhaps in colourful costume—shepherds singing, merry vintagers or reapers," all these may be included in the form of the *Landschaft*. On the other hand, he bewilders his readers completely by asserting that men as such do not play any part in forming the *Landschaft (nicht Landschaftsbildner)* and must not be used to delimit the *Landschaft* areas. Somewhat different statements in earlier publications are even less clear, so that it is not surprising if Gradmann and Waibel should have interpreted his statements differently from the way in which they are interpreted by himself and his students.

Nevertheless, difficulties of this sort should not by themselves cause us to dispense with a concept which might otherwise be valuable. But certainly before we become involved in what would evidently be a very difficult problem of definition, we will wish to test the concept in general by three tests- of more fundamental character. Is it logically founded on general principles of science or of geography as a chorographic science? Does it represent an historically consistent development in the field of geography? Does it provide a basis on which general agreement may be possible in order to restrict a field now too wide to be effectively studied by a single group of students? Let us examine the general concept in the light of each of these questions.

Logical Limitation

None of the American geographers who recommend, or even insist upon, this limitation has presented a thorough consideration in which we might examine its logical foundation. Not only in presenting their own work, but in some cases also in criticising the 'work of others, they have contented themselves with categorical statements for which no foundation is offered other than the authority of various geographers, here and abroad. Examination of the writings of these authorities fails to reveal any logical defence of the limitation. Consequently in order to examine the reasoning underlying their thesis for we presume that it must have some rational basis—we are limited to scattered arguments such as those brought out in discussions in the several symposia held by this Association in recent years. Though the statements were in many cases made extemporaneously, their authors had the opportunity to check them before publication. In any case, it will not be the purpose to catch them up on any unfortunate phrases, but to endeavour to understand the thought' they wished' to present.

In one of these symposia, both Trewartha and Finch, speaking separately, emphasized a distinction between "observable features" of area and other features that presumably are-not—observable. The geographer is concerned directly with the former, these are "primary geographic facts"; be is only indirectly concerned with the latter, "secondary geographic facts," to which Finch also referred as "deduced facts required for explanation of the observable features". These statements may welt appear to express a fundamental distinction in the logic of scientific reasoning, but do they express it correctly ?

Unquestionably we must distinguish clearly between facts as facts—i.e., as observed phenomena, "the primary data of all thought. . . not to be derived from anything else"—and what we presume to be facts, on the basis of reasoning. Thus, in the example cited by Finch, if the presence of hard and soft rocks in the Jura had been observed directly by examination of the rocks themselves, it would be a primary fact; if however it was deduced from the form of the ranges, it is a fact of secondary quality, even though it is presumably *"observable."*

It might be objected that only material facts can be directly observed and that therefore the difference in kinds of objects leads to the logical distinction in kinds of facts. But our "observation'-' of material facts is by no means limited to direct means; on the contrary the more exact and certain methods of determination are often indirect—notably in the most exact sciences, such as astronomy and physics. That the distinction between indirect observation and deduction from reasoning is not always a sharp one need not concern us here. It is necessary only to recognize that such facts as the following are sufficiently established by indirect methods of observation to be recognized as primary facts in any science concerned with them: that the United States has a republican form of government rather than a monarchical form, that many of the inhabitants of our South before 1861 were slaves, or that the people of Arabia are largely Mohammedan in religion.

Logical Distinction

Consequently we conclude that the sound logical distinction between primary and secondary facts is irrelevant to the difference between material and immaterial facts in geography. Our knowledge of natural vegetation in many areas of the world must be classified as of secondary quality-—deduced from primary facts of soils, climate, vestigial remnants, etc. whereas our knowledge of the distribution of peoples of different language and customs consists or-primary facts. In other words, as Hettner has observed, all facts—in the proper restricted sense of primary data-whether material or immaterial, must be perceived in one way or another by our senses to be recognized as facts. The term "observable" or the alternative, "perceptible to the senses," provides therefore no specific criteria for geography.

It is significant that white the proponents of this limitation are ultimately forced in discussion to define that which is-to be studied simply as material objects, they persistently retain the concept of "observable." Thus Finch and Trewartha say "observable material features". Apparently the redundancy is employed in order to emphasize that only those things are to be studied that are observable directly and these are material things. That a science of geography should limit its direct consideration to these objects is defended, in the various arguments of the landscape purists, on either or both of two different assumptions, one concerned with the nature' of science in general, the other concerned with the nature of geography as an "observational science."

The first of the assumptions is most clearly presented by Grano. Although he recognizes that the material and immaterial phenomena of an area form a unity *(Einheit)* he insists that geography must consider directly only the material facts and leave the immaterial to sociology. By limiting ourselves in this way "we can comprehend and describe our objects naturally in the way that a normal *(gesund)* person sees them." This he claims to be an essential characteristic of science in general. In other words, geography must be made to conform to an assumption of science that excludes the social studies. In general, we may note that this argument, whether stated or not, underlies most of the arguments of those who propose this limitation. Those who insist that geography must define itself in terms of geography and leave the question of whether it is .. "science" in any particular sense for secondary consideration, might ,dismiss these arguments as ırrelevant; but it is significant to note the paradox into which these arguments lead.

This paradox appeared particularly clearly in a discussion, during the symposium previously referred to, between Finch and Colby. "How an area *is,*" Finch objected, "involves subjective ideas. You *see* only things which are *objective*". Actually just the reverse is the case. Any given section of the shell of the earth's surface, containing material and immaterial facts, is a piece of objective reality. The immaterial phenomena are no less objective than the material objects. It is simply a problem of method to find objective means for observing them. When drought turned the wheat fields of the Great Plains into a "dust bowl the widespread bankruptcy and the mental anguish of the farmers were facts no less objective than the number of inches of rainfall deficit for that summer or the decrease in humus content of the soil over many years. None of these immaterial facts could be observed directly with any degree of accuracy; each could be observed with reasonable certainty by indirect methods. On the other hand, even the *sober* scientist may on occasion see things that are not objective-as in the case of a mirage. In any case, even if We may assume that most of the things we see are objective, we do not see them objectively; indeed one of the most fundamental axioms of any science is that "things are not what they seem."

In passing, we may note that the assumption concerning the nature of science in general does not apply even to the natural sciences, not even to the most exact physical sciences. The latter are not concerned merely with the study of objects but also, and directly, with the study of forces which cannot be seen or observed directly by any sensual perceptions. Physicists did not delay their study of X-rays until they had convinced themselves that those were composed of particles of matter. If, according to tradition, it was the impact of the apple on his head that led Newton to the concept of the force of gravity, it was not that force itself that he observed by his sense of feeling, but simply the moving apple. Physics studies gravity; that is, only by indirect means of observation; it is not a material thing.

One may ask why similar examples are not equally obvious in physical geography. Unquestionably we would have a clear example had the earth's shape been far less spheroidal. If the force of gravity varied over different parts of the earth surface to the extent of even a third or a half, physical geography could hardly have avoided recognizing this as one of the most important of all geographic factors. It would have been reflected in major differences in *most* of the physical objects the geographer would study, but could never be observed directly; it would' not be a material object. Any interested in following this thought might consider whether a wind is a material object, directly observable-whether it is not rather a phenomenon of motion of air particles, so that we do not actually feel the phenomenon of motion, but only the impact of the air particles.

Since these latter considerations, however sound in theory, may seem abstruse, we will attach no great importance to them. The essential conclusion is that under the broad view of science as the pursuit of knowledge of reality by objective means, science, in general, is not required to limit itself to any particular category of phenomena. The business of the scientist is to pursue knowledge wherever objective means of study permit, regardless of whether they be direct or indirect; if immaterial phenomena can be determined by indirect, but objective, means of measurement, they are phenomena for some scientists to study. There remains, however, the second question-does the particular nature of geography require it to exclude immaterial phenomena?

Sauer's View : Sauer has insisted that geography must "claim a certain field of *observation,*" within which, however, "we are to observe, describe, and explain according to the best methods at hand". An area of the earth surface is a certain field of observation, but the study of an area would logically include the study of all significant phenomena in it, of whatever kind, observed by the best means at hand. Though the visible landscape, in the very limited sense in which that constitutes a reality-the external surface-might be considered as a field of observation, neither Sauer nor any other students wish to limit geography to that extent-for very good reasons, as we shall see later. But the total of material objects selected out of the earth surface cannot, as we found, be properly regarded as *a* field of observation. Robbed of the connecting element of immaterial phenomena, it breaks down into as many different fields of observation as there are different categories of material objects represented. To give this selection of heterogeneous objects an arbitrary single name, "landscape," does not make them into a single field of observation but merely obscures from us the fact that they are not that. This is further attested to, not only' by the absence of any word-whether in English or German-that represents in common thought this total collection of material objects, but also by the inability of the students concerned to agree even approximately on what the concept they are trying to express includes.

Consequently we find that the supporters of this limitation fall back on the argument that the particular science of geography is a science in which phenomena are studied by a particular method or technique, and therefore it selects from the phenomena of area those that are encompassed by its technique of observation. Since Grano appears to be the only geographer who has ever actually used his ears and nose in the field, we may assume (*for* the sake of simplicity-it is irrelevant to the argument) that this technique is restricted to visual observation. But why, asks Hettner, should geographers be limited in their manner of perceiving things? What other science restricts itself in this way?. If the astronomer should learn that the physicist had invented techniques for observing stellar changes without use of either the human eye or the telescope, must the astronomer decline to use them? Does a geologist trained in field work hesitate to use methods of microscopic chemistry to study rocks which he has brought in from the field, defining geology, so to speak, as the science of the hammer? (Hettner). Sauer claims that historians restrict themselves to written records. Though popular thought might accept such a statement, historians do not accept the limitation. Indeed, Trewartha has offered us a paradoxical study of early French settlements on the Upper Mississippi for which the geographer has studied chiefly the written records, whereas the historians investigated (in addition) the remains of ancient fireplaces. In general, historians do not hesitate to utilize whatever evidence of the past they can find, including oral traditions, maps, pictures, as well as ruins, tools and implements, indeed "all the material relics of human activity that dot the earth".

Actual Techniques

When we consider the actual techniques employed by geographers, including the landscape purists, this theoretical argument appears to be completely dissipated. It is not necessary, evidently, that the objects of study in geography should actually be observed by the senses, but merely that they should be, or have been in the past, "observable" (in this way). To take a specific case, the argument would appear to be that the written account of a French officer of the eighteenth century describing where on the Mississippi he located a particular fort, produces a "primary geographic fact" whereas his statement that his soldiers were Frenchmen and the surrounding natives were Indians would provide us only with "secondary geographic facts." To science as a whole the important distinction between these two sets of facts would be that the latter appears to be highly reliable whereas we may not be at all sure about the former.

Similarly, no one has suggested that the geographer's knowledge of climatic conditions should be based on sensual perceptions whether observed by himself or by others. On the contrary it is axiomatic that such observations are most unreliable. We depend exclusively on indirect methods of observation and obtain thereby data far more reliable than the results of our direct observations of any class of material phenomena in geography. Even more reliable however are our data concerning the areal extent of political sovereignty in most parts of the world.

In short, geographers have long recognized in practice that, for many of the features that all agree must be studied, we cannot depend upon any direct form of observation. On the contrary, as R. E. Dodge observed in the discussion of this question, our direct observations may be much less reliable and significant than those obtained indirectly. No one has yet suggested in what way these indirect methods of observing material objects are significantly different from the indirect methods of observing immaterial phenomena. Indeed Finch has recently recognized that the means by which we 'gain awareness of "the activities and forces of human dynamics" are "only slightly different" from direct observation.' Further he states that these phenomena "are recognized by all regional geographers," but whether as primary or secondary geographic facts he does not specify.

There remains one final logical argument which might appear to confine the attention of geography to material objects. Grano argues that material objects, in contrast to immaterial objects, have spatial extent, and that the before they alone are significant in geography as the study of area. Similarly Michotte states categorically that "if geography is really a 'chorological' science, its object can necessarily be only material," because-as he adds in a footnote objecting to Hettner's view—the concept of space is essentially tied to the idea of 'matter'.

Michotte's discussion of this question comes at the end of a long essay concerned primarily with more fundamental questions. Perhaps for this reason he did not examine this question, as he did the previous questions, in terms of specific examples of geographic work-the method of reasoning which he recommends in contrast to "a discussion of principles, which is the origin of errors committed in these matters by laymen and even a certain number of geographers". If we grant, as of course we must, that the concept of space must be tied to a material basis, does it follow that the differences in contents or characteristics of different spaces are limited exclusively to material phenomena? Whatever is produced or used by man, whether a house, a tool, a language, a custom, a political allegiance, or an idea, is specifically located on the earth surface at the point where he produces or uses it. That these things vary in their ease of transport is irrelevant-indeed, a tool, particularly a type of tool, or even a type of house, in fact, more often transported than is a language. The immaterial phenomena known to science are tied to specific physical objects, men, and can therefore be located specifically with reference to space-at the points on the earth surface occupied by the men with whom they are associated. This association, to be sure, is in many cases but transitory and the geographer, interested in the more enduring characteristics of regions, will ignore such cases for the same reason that he would ignore the crop production of an abnormal year. But it is an error to associate the attribute "transitory" with immaterial phenomena in contrast to material phenomena. The languages and customs of China are far older than the houses of its inhabitants; if the same crop, rice, is found in the fields decade after decade, that is primarily not because of any particular characteristic of the land, but rather because the custom of regarding rice as the staff of life endures under the impact of foreign ideas.

"The proper object of geography," Michotte wrote, in the part of his essay dealing directly with human geography, "should consist in 'delimiting' and in 'describing' the various 'terrestrial spaces' characterized (characterisces) by a particular form of settlement, by a particular form of house, etc. . . .". But if an area may be characterized by the form of houses that are most common within it, may it not also be characterized by the language of its inhabitants, or by any particular custom that is characteristic of that area in contrast to others? If, as Vidal says, the concept of country is inseparable from its inhabitants, then it involves not merely the physical facts of their distribution and physical characteristics, but any other characteristics that distinguish them as an areal group from the groups of other areas. The only objection to this argument that I find in Michotte's entire essay is the limitation, assumed without any explanation, of the "distinctive characteristics" of an area to those that meet the eye.

Unquestionably, extensity in space is an essential requirement of whatever is to be studied under a definition of geography based on the choreographic concept. But we must examine more closely the manner in which geographic phenomena have spatial extensity. As Humboldt and any number of others since have observed, geography is not concerned with individual plants and animals, but rather with "the plant and animal cover" of the earth's surface. Obviously the word "cover" cannot be taken literally: a plant society, much less an animal society, does not actually "cover" any area. We do not ignore the vegetation of the desert because it occupies less space than the intervening bare places. We do not hesitate to map types of house forms in rural areas, even though they occupy but a very slight fraction of the total area concerned: if every house in a particular rural area is of the same type and that is distinct from the houses of neighbouring areas, geographers may recognize a geographic fact whose significance is independent of the slight space filled by the total of individual houses.

In other words the concept of spatial extensity does not require that the individual phenomena actually cover a particular area of the earth but simply that they are generally characteristic of that extent of areal space. The Magyar language is an immaterial phenomena that is no less a characteristic of a particular area in the Middle Danube Basin, and is no less definitely tied to this particular space, than is the peculiar type of farm wells, or the giant agricultural villages.

Many of the proponents of this limitation appear to follow this line of reasoning, but unconsciously and therefore only partway. Thus Schluter does not limit man to those instances where he can be observed in the mass, but considers the density of human population, apparently because that phenomenon is visible, in an abstract sense. If this is not plain hocus-pocus, it at least indicates that the principle of objective perception is not to be followed strictly.

Grano has, in fact, recognized by inference at 'least that nonmaterial phenomena may have areal extent in the sense here defined, but rules them out of consideration because it would be too difficult to use them in determining the limits of regions-a problem which he regards as of primary importance. The areal extent of many of the phenomena that he excludes such as language, religion, or literacy—can often be determined with greater degree of certainty and accuracy than those that he regards as "objective." Of all maps of individual earth facts, the political map of organized states is the most reliable and exact. Language distribution may be difficult to map correctly, but it is far less difficult than, say, the distribution of landforms, not to speak of natural vegetation.

If geography could confine itself to non-human aspects of the earth, it would perhaps be free of the difficulty of studying immaterial phenomena. Once geographers are agreed as in general they have always been that they will study human or cultural geography, they are committed to the study of things cultural as well as natural culture is basically immaterial and manifests itself both in immaterial and in material results, both of which are subject to scientific observation. If culture can be geographically significant in its material manifestations, it would be most extraordinary that it should not be geographically significant in its more fundamental, immaterial, aspects. Unless that can be demonstrated it is essentially illogical for geography to confine itself to the material, to consider the subject culture only in terms of its material manifestations.

Historical Development

Far from being a consistent step in the development of the field, the general adoption of this restriction in geography would represent a radical departure in geographic work. Branches of the field as old as geography itself, and which geographers have cultivated with especial interest, would be excluded, notably the geography of peoples and political geography. East is expressing a traditional concept in geography when he states: "The state structure of the world. . . no less than its physical structure, provides phenomena for geographical analysis, classification and interpretation. States, like physiographic regions, have their origin, histories, individualities and relationships". Does anyone really think it is of no concern to the geographer studying Bohemia to know which parts are German in character and which are Czech, even if houses and farms look alike? One does not have any particularly nationalistic interest to feel there is something missing in Passarge's *Landschaftskunde* of a part of the South Tyrol in which he examines house architecture minutely but gives no indication as to whether the people are German or Italian.

Undoubtedly geography has gained by the elimination of much miscellaneous material that has properly been taken over by economics, astronomy, and geophysics. But before we throw out wholesale time-honoured sections of the field it would be well to consider certain of the consequences.

In the first place, who is ready to take over these fields which we are to disown, who will care for their proper cultivation, or are they merely to be dropped in the void between narrowly defined sciences? In the case of political geography, for example, I have elsewhere shown that there is no reason to suppose that political scientists are prepared, either by their interest or their training, to take over that part of the field that geographers have long cultivated. On the other hand, when students like Grano inform us that economic geography as well as political geography can develop just as well outside of geography proper, one may be permitted to ask what entitles them to speak for those fields. If political geographers were demanding their independence, the question would assume a different aspect: but so far as I know, no student of political geography has suggested that his subject should be given independence. Likewise the situation would be different if political geographers were seeking admission into the field. On the contrary, political geography has been a part of geography as long as geography has been studied. As Penck at least has recognized, the new concept produces a "new geography" different from that of the classical geography. The issue, therefore, is whether the concept of geography may be changed to exclude branches of the field in spite of the objections of those most familiar with those branches.

From the point of view of those who may be interested in political geography it is no mere academic question which we are here raising. Most of its principal students have recognized that a continued sound development of political geography is dependent on the continuance of its two-thousand-year-old position as an intimate part of geography. Most of the fundamentally valuable work in the subject has been done by students who were primarily geographers, such as Ratzel, Vallaux, Sieger, Solch, and Maull, as well as Penck, Supan, Vidal de la Blache, and Bowman [specific references in 216; to these should be added Vidal's study of the states and nations of Europe]. Furthermore, a whole host of geographers have made, and should continue to make, excellent studies in political geography as part of the general work in geography, whether these be published separately, or simply as essential parts of their regional studies. A full list of those who have made studies in regional political geography would include most of the geographers of Germany and France. That American geographers should have less reason to make such studies is natural since their special field of interest in North America presents few such obvious territorial problems. Fortunately those who specialize in Hispanic America have not hesitated to treat the geographic problems involved in such disputes. All these studies provide valuable material for the student who wishes to specialize in political geography because they are based on more intimate knowledge of certain regions than he could hope to attain.

If now we consider the possibility of abandoning these special aspects of cultural geography-including the geography of peoples as well as political geography-from the point of view of the general student of geography, is it certain that our field would lose only outlying portions and not, perhaps, elements which playa direct role in its central core? Both Hassinger and Whittlesey have shown how a consciousness of political geography can enrich the work of the general regional geographer. For many regions of Europe, the student attempting general regional study will be seriously handicapped if unfamiliar with the principles of political geography and the geography of peoples.

Finally it is only fair to point out the contribution which political geography has made toward a clearer understanding of certain problems of geographic theory. This may seem to be a strange claim to make for that branch of the field which is most commonly considered as the "wayward child of the geographical family"—to repeat an expression of Sauer's which appears to have made a lasting impression. Undoubtedly the nature of the material in this field, as well as the extraneous influence of national interests, exposes its workers to the danger of loose and tendentious reasoning, but it is evident that other branches of the subject could offer keen competition in this direction. On the other hand, the obvious difficulties to clear thinking may stimulate the student to examine his assumptions and arguments more critically than in other fields. Further the fact that the conclusions may have practical importance—in the political world-and that the conclusions of students of different countries may come into conflict, may lead to a sharpening of the line of reasoning. Finally, it may be that the oft-repeated challenge from other geographers to justify the field of political geography has required its workers to consider at greater length than other students, the theoretical problems of the entire field.

Whatever the reasons may be, some of the sharpest and soundest considerations of geographic theory have developed from, the study of political geography. Particular mention may be made of Sieger's analytical study of the nature of boundaries, and Solch's study dealing both with that subject and with the essential character of regions—a work of the first order to which repeated reference is made in this paper.

Geographers who contemplate casting overboard whole divisions of their subject might well recall the period in which all of human geography was ignored by many geographers and consider the unbalanced situation to which geography was thus brought. Where can we find a systematic treatment of the cultural elements of geography comparable, say, to that of the natural elements which forms the great body of Finch and Trewartha's volume. The contrast results not merely, as these authors suggest, from the greater complexity of cultural features and the failure of the specialized social sciences to provide us with ready-made classifications comparable with those of specialized earth sciences. In fact these authors have fortunately not accepted the classifications of landforms from the geologists nor of climates from "non-geographic" climatologists, but have used those of geographers (including themselves). Similarly we have no reason to expect that the agriculturists will provide us with a classification of farms suitable to geography, or that any economist will do the same for industrial landscape features. Likewise if these geographers should later realize that political features are significant in geography they would find no satisfactory classification of these had not some gecgraphers continued to study them geographically. A concrete example is found in connection with the problem of political boundaries, which has certainly not been ignored by political scientists and historians. Nevertheless nothing approaching a satisfactory classification was developed until the problem was seriously considered by geographers, including Sieger, Penck, Maull and Soleh.

There is a certain air of unreality in all of the discussions calling for the exclusion of these particular branches of geography. The history of science shows repeated development of new independent branches out of older stems, but it seems doubtful that this has ever come from expulsion by those less interested in the development of those branches, rather than from inner compulsion within them. Furthermore, the "landscape purists" themselves—in Europe-at feast-do not accept personally the limitation that they' urge for geography. Thus Schluter, in the same year in which he presented his program for limiting human geography to physical "landscape features," did not hesitate to discuss the very difficult concepts of "nation" and "nationality," in a geographical publication; some years later he discussed "the influence of geographical conditions on the distribution of population and the development of culture," including the state and has since examined "the arealy distributed societies" and the relations *(Einklang)* of state and land. Penck's acceptance of the pure landscape thesis seemed to represent a renunciation of his wartime interest in political geography. (His significant rectoral address on the theory of polical boundaries was followed by actual work on the German-Polish frontier.) Yet he contributed again, in the same year, to the political geography of Germany. Likewise Tiessen was moved by the peace treaties not only to examine their results on the political map, but also to define the problems of political geography. Passarge has repeatedly considered both theoretical and practical problems concerning the geography of peoples and political geography. Brunhes published two books in political geography. Grano and Michotte, so far as I can find, are the only exceptions.

In this country there has been relatively little incentive to develop either the geography of peoples or political geography. Nevertheless one finds that many of those whose general statements appear to limit the field to the study of material features, do not hesitate to include discussions of non- material aspects of regions. Thus, one might suppose from the introductory statements of Preston James—though these are stated in no dogmatic form-that the geographer is limited to "things, organic and inorganic," but in the detailed treatment of areas, he maps the Negro residence district of Vicksburg, discusses the political activities of the nomads, the mixture of cultural ideas in the Mediterranean, "the cradle of Western civilization", and bases his discussion of Manchuria on its character as a "cradle of conflict". In general, sound geographical interest in whatever phenomena are characteristic of a region will overcome any arbitrary defined limits.

Nevertheless the resultant situation is very confusing either to the specialist in nonmaterial branches of geography or to the general student examining such aspects in his particular regional problem. Puzzled as to just what he should not do in this shadowy border area, can he look for guidance to these authorities? Since they disagree so radically, it will be necessary to examine briefly the suggestions of each.

Schluter's Changed Approach : Schluter apparently changed his mind on this question several times. At one time he wrote definitely of the "inner" or "true geography" and of its outer branches. More recently he appears to have welcomed *Geopolitik* as the solution of the problem, as far at least as political geography was concerned, though this still left such fields as the geography of peoples unprovided for. That *Geopolitik* can provide a suitable substitute for political geography will hardly be accepted by geographers or political scientists outside of Germany, or indeed by few political geographers in Germany.

Brunhes's Concept : Brunhes is best known in this country for his presentation of the "essential facts" of geography, from which one can readily quote statements which specifically limit the field. It is commonly overlooked that once he gets "beyond the essential facts"—and it is only then that he considers regional geography-the range of his observation is well-nigh unlimited, including epidemics, physical aptitudes, moral habits and social rules, property rights, collectivization, social organization, stock companies, and social anarchy in large cities. All these may be studied by the geographer as long as he can see any relationships between them and "the facts of the terrestrial world". The well-nigh unlimited range which this allows him is most clearly shown in his two books in the field of political geography; Brunhes himself describes the former work as "a study vigorously and consistently geographic".

The concepts of Passarge appear to lead to similar conclusions. His view of geography is by no means easy to understand, because he appears to change it constantly and to publish immediately every change of thought. Anyone attempting to follow his 'discussions will agree with Hettner that "the route of thought from the brain to the printing press might well be somewhat longer". If such students as Sapper have failed to understand Passarge correctly, as he insists, it is reckless for a foreign student to attempt to analyze his theoretical concepts. But the influence of his ideas-whether because of the attractive possibilities which they appear to offer, or because they are accompanied by effective illustrations in detailed studies-has been so marked in this country, that it is necessary for us to examine exactly what they involve.

Fortunately, for our purposes, experience has taught Passarge, "that it is difficult for the beginner to follow the train of thought in *Landschafts-kunde*, and consequently he has gone to some pains to explain himself more fully and to illustrate his explanation with a specific detailed example. As nearly as the writer can judge, there appear to be three grades of geography. The first, which is "pure landscape study" *(reine Landschaftskunde)*, is limited strictly to those objects perceptible to the senses. Whether or not it includes animals and men is not entirely clear, but certainly they are included in various aspects of the study. A geographic description *(stadtlandschaftliche)* of Madrid includes the tortoise shell combs of the senoritas and the coquetish angle of the mantilla.

Secondly, there appears to be a form of *Landschaftskunde* which we must presume is not pure, but which considers the extent to which the cultural phenomena (including perhaps man himself) of a *Landschaft* depend on its character. Thus the devout Catholicism of the Tyrolese mountaineers is considered as a result of the dangers of mountain life, just as, in an earlier study Passarge examined the characteristics of the Jewish people as "dependent on the nature of the *Land*, the *Landschaft*, and, in a wider sense, of the *Umwelt*'. Similarly, in his brief summary description of the "major landscape belts" of the world, he notes that America "lacks the farmer class of the Old World, which because of tradition and custom clings fast to the hearths of their fathers, the dependable anchor of a people," and finds in New York "the most imposing and at the same time the most horrible of all 'office buildings, the skyscrapers. . . where men are pressed together as though in an ant-heap, joyless, peaceless, breathless, caught in the chase for the dollar".

Finally Passarge writes of *Landerkunde* (alternately, *Landeskunde)*, a term hitherto used in the German literature to refer to regional geography, but to which he chooses to give a quite different meaning. For Passarge, "the *Land* is an artificially bounded area which generally takes no account of the landscape-areas" but is simply a political, historical, religious, or folk area-most importantly, the state area. His *"Landeskunde"* is concerned "primarily with the dependence of cultural conditions (both material and spiritual) and population conditions (including character and talents) on the artificially bounded land-area (commonly a state)". *"Landerkunde,* carried as the crown of *Landschaftskunde*, brings a presentation not only of the present area but also of its development, the history of men, their state, social, economic, material and spiritual culture-goods. *Landschaftskunde* is thus the trunk of the tree of *Erdkunde*, which unites the roots and the crown-physical *Erdkunde* and *Landerkunde*".

On this basis, Passarge elsewhere defines political geography as the study of "mutual relations between the area (the sum of *Landschaften)* and the political organizations". He has illustrated this point of view in several articles and particularly in his book on the political geography of the Near East, which is, he says, "not *Geopolitik* but an attempt to give an example of what specifically should be understood by political geography".

According to Passarge, therefore, all parts of geography, excepting only "pure *Landschaftskunde*," are defined as the study of relationships, the very concept which "modern geographers" so vigorously oppose. If it is scientifically dangerous to define the core of geography in terms of relationships which are to be discovered, surely this objection applies with equal, if not greater, force to those branches of the field that deal with nonmaterial aspects of areas. Passarge's view of geography appears as an unsatisfactory hybrid which few will accept. It is not his concepts but only his terms which are new and the introduction of new terms, or the arbitrary use of old terms in new ways, inevitably makes for confusion.

Turning to those American geographers who have spoken in favour of the limitation to material things, we find that Sauer was admittedly uncertain as to what to do about political geography . We are still at a loss to know what he would do about this and other nonmaterial aspects of the field-such as the geography of races and peoples, which he does not mention-since he has not yet illustrated his methodological study by any complete regional work.

A discussion of the problem at the 1937 meetings of this Association made clear the dilemma to which Sauer's statements logically lead: we cannot exclude political geography from geography but it does not fit under our definition of geography so that it is difficult to see what to do with it. Since the dilemma was caused by the adoption of a purely arbitrary definition, in no way founded on either the logic or the history of the subject, one might suppose that the way out was obvious. Partsch has written of those who "find an attractive importance in creating out of the depth of their own judgement a limitation of the problems, and a division of the material, of geographical science, so that from a Judgement seat erected on their own authority they may look down on the workers who have at any time dug their spades in the field of geography, issuing certificates to those who have, in fortunate prescience, worked according to the point of view of their successors, denying them to those who have deviated in their conceptions of their problems".

Unified Field

In spite of these logical and historical objections, many of those who have accepted the restricted view of geography will hesitate to abandon it, since it appears to offer a basis for restricting the multiplicity of things that the student of geography must consider. Such a need has been expressed by many geographers. Douglas Johnson, in discussing the prospects of geography as a science, spoke of "the necessity of effecting some restriction of the vast field now claimed by geography". Granted that, from many points of view, a restriction would appear desirable, we must face the possibility that any such hope is in vain. If geographers are to study areas of the world, they must recognize that each area, like the world itself, is so full of a number of things, that these students who wish to see it in more simple terms may necessarily be as unhappy as kings. Most historians recognize that this is one of the inescapable characteristics of their subject, preventing it from becoming the sort of science which some of their number also might prefer. Yet on the whole, as Crowe notes, "historians bear their side of the burden with singular *sang-froid*". Nevertheless, since geographers have been troubled for a long time by this complexity of their problem, any suggestion that appears to give hope of a solution will not be dismissed on purely logical grounds. Will the suggested limitation of the field provide a basis on which we can reduce the necessary "inner" field of geography and at the same time retain that inner core in complete unified form?

Such a basis for regional geography will hardly be found in Brunhes' system, in which regional geography is classified as "beyond the essential facts," together with considerations in political and social geography based on the principle of relationships. Helpful as his outline of the essential facts may be, it is evident that he himself does not regard it as providing a complete basis.

Likewise we will hardly look to Passarge for a more restricted and unified system for geography. He has, to be sure, discussed his concepts more frequently than any other student of this group, and he and his students have published a great number of valuable regional studies based on his ideas [he lists many of these in 268. Even if we can ignore his studies in political and social geography, we found that his "pure *Landsehaftskunde*," however narrowly he may define it theoretically, 'was shown by his model example to contain far more than could possibly be included in any restricted concept of geography.

Sauer claims that the limitation, primarily to visible objects, will reduce the field to one in which the special experience and training of the geographer in geomorphological study "provides the necessary technique of observation and a basis for evaluation" an idea which is no doubt in the minds of those who urge a return to more thorough training of geographers in physiography. Granted that such a preparation is desirable and necessary, this should not lead to "neglect of other, even more important parts of geography," as Hettner observed in a criticism of the "Davis school" of physiography. On what this is it assumed that the special technique of the geographer is that developed in geomorphology? This is to place a very old science on the shoulders of a younger one, and one whose position" with reference to another science, geology, is by no means clear. Granted that progress in geography was greatly furthered by the development of the study of landforms, it has likewise been furthered by the progress in climatology which utilizes an entirely different technique.

Parenthetically we may be permitted to suggest that the concept promulgated by the "landscape purists" may possibly represent a belated outgrowth of the period of the late nineteenth century when geographers—following Peschel and Richthofen—were primarily concerned with landforms. The result was a greater emphasis on forms than on functions and an assumption that geography was exclusively concerned with studying the appearance of things observed with the eyes. A geomorphologist, like Penck, could accept the addition of crops and houses, since these constitute part of the landform in a full physical sense-but the men who cultivate the crops and build the houses must be excluded. In this country a further support for the limitation of cultural phenomena studied in geography to things of material culture has apparently come from the contact of the California school with the cultural anthropologists. The limitation imposed upon the student of prehistoric, illiterate peoples, whose culture can only be studied in the remains of their material products, was apparently transferred to geographers, who are told that they may study the culture of a living area only in terms of its inanimate products-but may use those, even though they be pieces of pottery that can have but the slightest significance in the landscape. This chance combination of rather widely separated ideas has led to a peculiar paradox. It would hardly be claimed that training in geomorphology makes the student of geography better able to distinguish between a dozen different kinds of Indian potsherds than to distinguish between core areas and peripheral areas of a state.

Consideration over Techniques : If we consider the techniques used by present geographers, including the landscape purists, it is clear that they are of a wide variety of kinds. Every regional geographer must depend to a large extent on facts which he has not collected by observation; those who did observe them may have used techniques which have little to do with geomorphology. Mackinder meets the exaggerated emphasis on the study of the lithosphere, as he puts it, with the claim that geographers are really more concerned with the hydrosphere. Even if this view represents exaggeration in a different direction, certainly the regional geographer must study climatic data, to evaluate which, one would hardly use the techniques of geomorphology.

Much the same is true of a large part of the "observable material data" used in any studies other than those of very small districts. Granted the importance of studies of small districts, such as Finch's study of Montfort, based on direct detailed observations, we do not add the facts observed in many different small areas in order to describe and interpret larger regions; rather we depend, and will continue to depend, for data concerning crops, animal culture, and population, on census materials which were certainly not collected by any direct observations. To be sure, they ultimately rest on somebody's observations the farmer presumably has "observed" how many cows he has and how many children, but then all facts must be observed in some way, in order to be recognized as facts.

In Pfeifer's discussion of methods in economic geography, specifically recommended by Sauer as the proper application of the concept of landscape to that field, we find that we must study the course of economic events, the production and amount of transported commodities and that "the way of investigation must be fulfilled both by landscape observation and by the evaluation of statistics and the consideration of research in economics.

Actually Sauer's outline, as well as his subsequent work, clearly requires an increase rather than a restriction in the techniques which the geographer must utilize. He must use "the additional method. . . the specifically historical method" since the study must be oriented always on development. The geographer therefore will use all available data "in the reconstruction of former settlements, land utilization and communication, whether these records be written, archaeology, or philologic".

One can only wonder how these methods would be used in the development of a complete regional study since the theoretical proposals have not been accompanied, as Hettner recommended to any wishing to reform the methodology of geography, by the presentation of "a work that makes his views real and shows their superiority to the conception hitherto held". In the regional studies that Sauer has subsequently published, as well as in those of various of his followers, we found earlier that the non-geographical methods, historical and anthropological, tend to predominate over any specifically geographical method. The most striking illustration of the effect of applying the historical mode of presentation to geography is to be found in Trewartha's study of the early French trading posts of the Driftless Hill Land. Whereas the author-whose field studies of Japan are outstanding-has repeatedly insisted that geographers should confine themselves to "observable features," his first chapter in a regional study is not only strictly historical in outline but is based almost entirely on the written records, since even vestigial traces of the early French occupance are scarcely to be found in the present landscape.

That geographic work cannot be considered in terms of one special technique, but inevitably requires the use of many different techniques, was not questioned by Schluter. Even if we were to exclude the study of man, it "would still present a vari-coloured picture in which were mixed such different colours as those of meteorology, hydrology, geology, botany, and zoology." In addition "the works of man . . . enter—as ingredients into the *Landschaft* and become part of its nature," but these "cannot be understood without historical, economic, or ethnological research. . . . That geography thus stands between the natural and the social sciences, however, is not the essential point, but rather that it seeks to produce universal connections between different sciences. . . . In fact, one can now say, without great exaggeration, that, for the true geographer, the transition from the physical to the human part of his science involves no greater jump than from climate to land forms, or from that to the vegetative cover. In each of these cases it means to be transposed into a new realm of thought; in each case, however, broad bridges lead across from one to the other".

On the other hand the geographer (or at least the "pure" geographer) is not to cross any bridges that may lead from the consideration of man's settlements to that of his nonmaterial creations-language, customs, or states, Schluter justifies his "narrower conception" as "determined by the need for concentration", Unless this be done, a complete and unified study of a region is impossible. Incidental considerations of nonmaterial features which may reflect the nature of the land destroy such unity, and a complete study of them would make the geographer's task impossibly large, indeed "would greatly disturb the cohesive unity of regional presentation," However the "narrower conception" is not so narrow as one might suppose. It includes the numbers and density of population, its composition according to sex and age groups, its growth and migrations, and the development of the entire economic structure, from production through intermediary agencies to consumption including the actual movement of goods through trade.

Beyond these, Schluter recognizes that a complete interpretation of visible landscape features may require a consideration of nonmaterial factors, and they may be introduced on this basis, but only on this basis But what, as Hettner asks, is the value of excluding such features at the front door if they are to be brought in whenever significant through the back door? This is to disrupt the natural unity of a study and then to attempt to patch it up as one goes on. Similarly Penck apparently would not permit the geographer to study directly the distribution of people of German culture although he emphasizes that the *Kulturlandschaft* bears the particular mark of the people *(Volk)* who have developed it. In other words, as Burger notes, the geographer may map the "cultural landscape" as the areal expression of cultural ability of its inhabitants, but may not map the direct areal expression of people of that culture.

Grano has objected vigorously to Hettner's analogy of the open back door: only those nonmaterial factors whose differentials are interrelated to those in material phenomena are to be admitted, and only in order to explain the latter. But he does not answer the argument that on this basis he would ultimately consider all the immaterial phenomena that others, like Hettner, would have considered in the first place—with the difference that the belated consideration represents an attempt to restore unity to a study first torn apart. Grano himself starts with the assumption that all the phenomena of our environment, material and immaterial, form a unit *(Einzeit)* , and he recognizes that "the spiritual and social milieu. . . is in direct and causal connection, and in reciprocal relation to the physically perceptible milieu," but insists that only the facts of the latter are to be studied directly in geography, those of the former belong to sociology. This conclusion would appear to be not only the reverse of logical, but essentially irrelevant. Its frequent repetition by American geographers in one form or another is not an argument, but merely a reiteration of the point at issue. If the study of immaterial facts belongs to the various social fields, the study of material facts belongs to physics, geology, zoology, soil science, etc.-and also, in part, to such social sciences as economics. If this line of argument is followed, geography ends up with no objects whatever excepting those hitherto overlooked by all other sciences.

Logical Conclusion

Although this would appear to be the logical conclusion to be drawn from any argument that immaterial phenomena in an area belong to sociology, and therefore not to geography, fortunately none of these students in practice resign their claim to any areal phenomena. The only important consequence of their thesis is that in their procedure they resolutely limit themselves to certain kinds of things at the start-not to the primary facts of science in general, but to a collection of primary and deduced facts of a particular category but later in their study bring in facts of another category whether primary or deduced. Crowe therefore seems quite justified in his comment on the "vagueness of the landscape idea. As soon as a clear definition is given, it becomes apparent that the bulk of 'landscape' philosophy is a process of removing from one's hat the very things one has carefully put into it".

The point at issue may be illustrated by a consideration of some region in South-eastern United States. Suppose the regional geographer limits himself at first to straight description of material things. He would describe the cotton fields, plantation houses and the huts of field-hands, carefully avoiding any mention of skin colour and, particularly, of cultural heritage from slave times. These would be considered, if at all, only in order to explain the material things seen. But any reader familiar with the area will wonder why the writer thus postpones, or even omits, the consideration of what is one of the most significant *characteristics*—second in importance perhaps to no other characteristic whether natural or cultural—of our South as compared with other parts of the country, namely the fact that a large part of the population are Negroes and descendants of slaves. In other words any feature of any area, no matter how known, which is significant to the whole complex of material features, is itself a characteristic of the area, to exclude which, even in description, will result in an incomplete picture of the region.

One conclusion appears certain: if immaterial phenomena are going to be needed ultimately in a regional study, they had better be considered as they are encountered in the field, along with the other phenomena, and not first rigidly excluded and then, as the work in the study shows them to be needed, brought back out of the vague memories of the human observer, as distinct from the notes of the careful scientist.

There would, however, appear to be one way in which things that are strictly material form a unified object, the study of which might conceivably be regarded as the core of geography. As previously suggested, one could confine oneself to the study of the actual visible landscape, in the sense "of the external surface form of the earth. Although I do not find that any student has suggested that such a study forms the core of geography-unless possibly Penck—and there is certainly no intention of proposing it here, the concept is nevertheless significant. For it appears to form the concrete base on which the vague concepts of "landscape" have permitted various students to construct different conceptions of a core of geography of geography, as a whole.

The landscape as we defined it earlier, is an objective reality and forms a continuous object which, for the world as a whole, is a single concrete object. It therefore constitutes a unit basis for a field of study. To be sure, the moment the study passes beyond bare description the student must leave the landscape itself, must go beneath it, even to state what its form represents-to translate the outer foliage of a forest into the forest, the outer surface of buildings into different kinds of buildings, etc. If he does not do that, his study is exclusively concerned with the surface forms-meaning the plane forms in the surface. Such a study would be, not merely literally, but in any other sense, superficial, as Kraft has observed of "landscape" studies in general. Our interest in houses, factories, and forests cannot be confined to their surface form; only in the limited field of aesthetic geography could such a restriction be justified. Our very use of such words as house,' barn, factory, office building, etc., indicates that we are primarily concerned with the internal functions within these structures; the external form is a secondary aspect which we use simply as a handy means to detect the internal functions—and should use only insofar as it is a reliable means for that purpose.

Interpretation of Patterns

Further, the interpretation of the pattern and slope of the landscape forms requires us to look under the features that produce that surface. We must observe not merely the surface of the soil under the vegetation but also the subsoil beneath that, and even the country rock still farther down—whether it be the basis for a residual soil or merely the supporter of a transported soil. The location and character of the bedrock may be essential to our understanding of such surface manifestations as slumping, springs, and artesian wells. The interpretation of a mine structure on the surface may require us to go thousands of feet beneath the surface. It is even more generally important that we leave the landscape in the opposite direction to interpret the effects of climatic conditions.

Such departures from the object of study into external objects that concern it are necessary in any science. In most sciences however they appear as incidental, and in total of perhaps less importance than that which is included in the objects of the science itself. In the case we are suggesting, it is obvious that the reverse is true. The study of the visual landscape would consist very largely in the study of things not included in the landscape itself. Further, though these things 'are studied in other sciences, just as each of the objects that contributes its surface to the landscape may be studied in some other science, no other science stands ready to supply the geographer with the necessary information concerning the areal differentiation of these non-landscape phenomena in their relations either to each other or to the landscape. The geographer would, therefore, be no more exclusively at work when he described the landscape than when he studied the other features necessary for its interpretation.

On the other hand, may we regard the landscape as so much more important than the non-landscape features that affect it, as to justify our considering the landscape as the core of our field? To ask this question is to answer it. Only from the point of view of aesthetics or of visual sensations can we regard the external form of a forest as more important than its contents, the surface buildings of a- coal mine as more important in the area than the underground workings, or the contour, of the land as more important' than the precipitation that falls upon it.

We conclude therefore, that while a study which 'limited its purpose to the full interpretation of the actual landscape—the external surface of the earth beneath the atmosphere, which we found to be the only concept of the term that formed a concrete empirical object—while such a study would ultimately involve the consideration of most of the geography of an area, the landscape itself is not the core of the area, but merely an outward manifestation of most of the factors at work in the area. This outward manifestation is in no proper sense a centre, core, or heart of the area, it is not even necessarily the most important manifestation of the area, but is only that manifestation which we can most readily observe by the easiest, but not the most reliable, of methods—i.e., by simply looking at it. We may therefore recognize the value of this manifestation in reconnaissance study, where we need methods of observing large areas, even though we may depart from it as soon as we attempt more careful and detailed observations—with thermometer, rain-gauge, census data, or what not.

Finally, insofar as the actual landscape does not reveal the presence of certain factors that we know from other means of observation are present in the area, that fact in itself is not evidence that those factors are insignificant in the area; *a priori* we can only conclude that in that case the landscape is not a complete guide to the contents of the area. Whether these non-represented features are significant or not, can be determined only by examining their relation to other features. For example, the sensible temperature of any area, at different times of the year, is a characteristic of that area and it is a characteristic of considerable importance to the human beings in that area. To question whether it is geographic or not is to beg the question. It is a significant areal characteristic that has but little if any representation in the landscape. In practice it would seem safe to say that any student who confined himself to the landscape forms and whatever factors produced them would not have occasion to examine the difference between actual 'and sensible temperatures. Likewise the North American student who has not happened to study across the Rio Grande or in other continents, will seldom discover any important manifestation of political geography in the landscape. In reality, however, this very lack of differentiation within the United States is in no sense to be regarded as the normal situation, in comparison with which the landscape differences between areas of English and French culture, or areas of Western European and Eastern European culture, or areas of American and Mexican culture, represent abnormalities introduced by a strange external factor. On the contrary this relative uniformity of cultural forms in the landscape of the United States might much more properly be regarded as one of the strangest of landscape phenomena, resulting only from the fact that the entire area has been developed by civilized man, as one single political unit. But this political unit is a fact of area for which the student of the landscape itself would hardly discover any manifestation other than this apparently negative one.

We conclude therefore that the chorographical concept of geography not only does not require the exclusion of any particular category of phenomena, but that on the contrary the full study of areas of the earth surface inevitably requires the geographer to study many immaterial as well as material phenomena, and that to the student of area there is no logical basis for giving preference or higher rank to either of these groups of phenomena, or in general, to any particular category of phenomena. In the study of areal differentiation, all features that are significant to areal differentiation are on the same plane.

Practical Consequences

Our examination indicates that the limitation of geographic study to things observable by the senses is founded neither in the logic nor the history of the subject and does not provide a basis for restricting even a central core of the field which would be unified and complete in itself. A study limited to the actual visible landscape, the external surface of the earth, would constitute a unified subject of study significant only from the point of view of man as an observer of his surroundings. Since he observes them also in certain other ways, such a study would form but a part, though in this case certainly the major part, of the full study of the relation of an area to man as the source of his sensations. Hettner classifies such a subject as a special field of geography, under the title of "aesthetic geography." That it is closely related to psychology is obvious but there appears to be no practical need to decide whether it belongs more properly in the one field or the other.

It is true that a few geographers have been interested in this study since the very beginnings of modern geography. Humboldt endeavoured to consider it objectively in terms of the external phenomena rather than the psychological sensations that they produce, but later critics have questioned whether he succeeded. In any case he regarded this aesthetic geography as but a part of geography. The scientific atmosphere of the late nineteenth century gave little encouragement to geographers interested in this field, but a few, such as Ratzel, Oppel and particularly Wimmer, endeavoured to develop it. It was in the aesthetic sense, according to Friedrichsen, that Wimmer introduced the term *Landsehaftskunde* into geographic terminology. "The descriptive geographer," he wrote, "is nothing other than a landscape painter and map drawer in words".

Beauty of Nature

Among modern workers, the English geographer, Sir Francis Young husband, has been particularly interested in the study of the beauty of nature in geography. Two of his articles have been published in German, as *"Dos Herz der Natur"*, and have apparently aroused more interest in Germany than did the original addresses in England. Banse and Volz in particular have followed a similar path. Banse definitely maintains the concept of *Landschaft* as the form or picture of the landscape, representing the outward manifestation of the total milieu.

Although such students as Schluter and Passarge might not accept the classification, various of their critics-including Friedrichsen and Vogel—find that the point of view they present is essentially an aesthetic one. In the case of Passarge this judgement would appear to be confirmed by his notable dependence on photographs (not air photographs) rather than maps. His study of the "city landscape" of Madrid has but one very simple map of the city, showing little more than its areal growth, but uses four photographs. The study which he presents as a model of his work is beautifully illustrated with 31 photographs and four panorama drawings, but includes only two maps, a section of the ordinary topographic map and a small sketch map of a minor part of his area.

An outgrowth of this point of view is the discussion by Banse and certain other German geographers of the importance of aesthetics in geography. This has led some to raise the question discussed in the early, part of this paper, whether geography was properly a science or an art. Hettner distinguished between aesthetic geography on the one hand, and geography as art on the other. The former as a part of geography should be objective (and for that reason alone not likely to be accepted by artists even if we should wish to turn it over to them) whereas the latter is subjective, essentially artistic, so that one can hardly follow Banse in wishing to make it a part of geography.

While geography is not to be restricted to any particular form of observation, this is not to say that the concept of landscape, as we have defined it, is not of value for geographers. There may be great value in focussing the attention of the regional student on the material features he sees in an area. Especially on entering a region he is faced with such a host of facts that he may well be puzzled as to where he is to start; as a rule of thumb he might well start with what he sees before him. If he limits himself to that, however, his study is likely to be incomplete. On the other hand this is certainly not the only proper procedure. Excellent studies have been made, for example, by starting with a detailed examination of population distribution, the data for which were not obtained by direct observation. It may also be useful to assume as a rule of thumb that visible landscape facts are generally among those that are most likely to be significant to the character of an area, although this may not necessarily prove to be the case, and that nonmaterial facts are more likely to be found to have no geographic significance, though again the reverse may be the case in any particular instance.

Whatever value this rough rule may have in field investigation, it does not provide an adequate criterion for the selection and rejection of phenomena that are to be considered. Even though we should exclude all nonmaterial phenomena and limit the consideration to sensible, perceptible phenomena-overlooking for the moment the difficulties that the proponents have had in defining what that includes we would still have far too many facts to consider. While the general criterion limits these to concrete facts, it provides *no* objective basis for selection from among them. Following the concept logically, Grano, as we found, determined the selection on the basis of the impression that the different features make upon our senses. On this basis the weather-beaten old grain elevator that yesterday was hardly noticed in the landscape may appear tomorrow as the most prominent object, thanks to its new metal covering. Though any landscape painter would accept this statement as a truism, presumably few would regard it as expressing a proper measure of geographic significance.

It may be suggested that in place of this aesthetic standard the geographer uses the standard of areal extent of landscape objects. The geographer's business is to describe and interpret the things that make up the landscape, and therefore he considers primarily those that constitute areally its most important parts. Though this would seem logical, it is significant that few students follow it in practice. Schluter suggested this standard in his theoretical discussions, but to illustrate the systematic study of cultural landscape forms, he chose bridges, a feature that would have to be rated on this basis as one of the least important of cultural phenomena. Further, as we have seen, he applied a special magnifying glass to human beings in order to prevent them from vanishing as cultural features of almost no areal extent.

In each of these cases Schluter appears to have drawn a compromise between the logic of geography as the study of landscapes and the logic of geography as the study of areal differentiation in which, fortunately, the latter had the preponderant influence. Most geographers make much the same decision in considering cities. Though these cover but a very minor part of the earth surface, no theoretical considerations can convince us that they are not of great importance in geography. Possibly the fact that most geographers live in cities may likewise have saved these features from any strict application of the measure of relative areal extent. On the other hand, other types of highly specialized areas are often overlooked simply because they are small in area and, though intensive in development and thoroughly distinctive in character, this distinction may not be readily apparent in the landscape.

As a specific example we may note the tendency to neglect the special characteristics of mining areas. Where the mining is carried on underground it shows but minor effects in the landscape, prominent though the structures and sidings at the mine-pits may be. But anyone familiar with an area in which mining occupies the attention of a considerable portion of the population recognizes that in countless ways the character of the area is different from, say, otherwise similar but purely agricultural areas. While this difference might be measured in terms of the quantities of minerals shipped out, and consequently in the actual movement of trains, many of the special characteristics are nonmaterial. But they are nonetheless real and geographically significant; the social and ethnological character of the population, the labour organizations operating in what appear to be rural areas, and even the contrasted social attitudes obvious in the streets of the towns—all these are specific characteristics associated with a particular district as a result of its mining activities. The very fact that these areas possess special characters warrants greater, rather than less, consideration in proportion to any measure of their intrinsic importance, but even in terms of population they are vastly more important than the nomadic grazing areas of the world, which seldom lack of detailed consideration.

Material Objects

In selecting out of all the multitude of various material objects in the landscape, no one appears to have attempted to apply the standard of areal extent as a basis of distinguishing the more important from the less important. All material objects occupy some area and therefore any of them apparently may be studied, though it appears to be agreed that they should be *immobilia*—things that stay in the same place. Though Schluter, in theory, ruled out the study of house types, others who accept his theory in general have fortunately disregarded this logical conclusion and have given us significant studies of rural house types, although, in terms of areal extent, these would have to be regarded as of very slight importance in the rural landscape. On the other hand, houses are obviously far more important in the urban landscape than in the rural, and the theory of the landscape purists would appear to justify studies of individual cities which include maps showing, by blocks, the percentage of colonial, mansard, or neo-Spanish styles, or even a "geographic" study of the different types of filling stations within a city.

All of these suggestions-which it is hoped no one will regard as other than absurd-would be justified by the statement with which Kniffen, in theoretical discussion, defended his study of house-types: "the primary concern of the cultural geographer is with toe nature, genesis, and distribution of the observable phenomena of the landscape directly or indirectly ascribable to man". In his actual study, to be sure, Kniffen has indicated that he is concerned with house types for reasons that are in no way based on the landscape or on material phenomena, but rather on the character of immaterial culture, Nevertheless we may accept his statement quoted above as the logical and necessary conclusion from the concept of geography as exclusively the study of landscape.

The landscape as an observable reality consists of individual objects (or their surfaces) which together form a whole significant only in aesthetic terms. This we found was true even if the concept be stretched to include all material objects in an area. Viewed as parts of an observable landscape their direct relation to each other has no significance except as parts of a "picture" ; to study their indirect relations carries us immediately out of the visible landscape and sooner or later also to nonmaterial factors. Although the rules of the landscape purists permit this, it is obvious that they discourage it. In any case, if the landscape features are justified solely because they are observable features in the landscape, the geographer, being neither artist nor psychologist and therefore not interested in the landscape as either a picture or a total sensation, is logically led to study each of these objects, or categories of objects, in its own right, in terms of 'its "nature, genesis, and distribution."

That the theoretical conclusion stated is not without significance in practice is indicated by Scofield's study of house types in Tennessee. These are classified as to types, and examined as to origin and development, with no indication of any geographic significance other than the mere fact that houses are "landscape features".

If there is one point in methodology on which almost all modern geographers are in agreement, it is that the study of the nature, genesis, and distribution of any single type of object -is not a part of geography but rather belongs to the systematic science that studies those objects. It is ironical that Sauer, who erroneously accuses Hettner of wishing to include the studies of distribution in geography should have developed a concept of geography as the study of "the landscape" which leads logically to that very conclusion. It may be claimed that no systematic science has happened to take up the study of house types, but we can be sure that any success geographers have in that study will stimulate the appropriate students to concern themselves with it and we shall have simply another case of a field opened up by geography to be taken over by another science. Insofar as our students of house types are concerned largely with morphology, genesis, and distribution only-in other words primarily with the objects themselves-they justify Leighly's logical conclusion that the professional geographer has nothing to add here to the work of the scientific historian of art and culture forms.

The all-important distinction that we are here emphasizing is exactly that represented by the inversion of Ritter's classic, if somewhat awkward, phrase describing geography as the study of the areas of the earth surface filled with earthly phenomena. The study of the material objects that physically fill areas is not the study of areas, it is not chorology; since it does not study all that is in areas, it logically disintegrates into the study of material things as found in areas. The study of "the phenomenology of landscape" proves to be nothing more than the study of individual phenomena found in landscape; the structure, origin, growth, and function of landscapes proves to be the structure, origin, growth and function of each material object in the landscape.

We may note in much of the work of the landscape purists an additional result in the tendency to emphasize structure or form alone. This is natural enough since our sensual perceptions can observe only forms; origin, growth and function must either be deduced from present forms-which is by no means a reliable method—or observed by indirect methods of observation involving often invisible and even immaterial factors. Furthermore, if one adheres with any consistency to the concept of "landscape," the objects in it are members of the whole only in terms of form. The landscape, as we have noted, does not grow as a whole; separate things within it grow separately.

Concentration on Physiognomy : The concentration on physiognomy might also be regarded as in part the result of the importance of geomorphology in the training of modern geographers. To the geomorphologist the forms of the earth surface presented significant problems because they had to be explained as the end product of a physiographic history. Much less attention was given to the functional relation of different landforms to other earth features; this contrast may be noted particularly in Penck's discussion of observation in geography, of 1906, as well as in later work of Schluter. If, however, as many students now feel, the geographer is not called upon to explain the genesis of landforms, a description of forms that does not give a corresponding attention to functions appears barren. In the work of Schluter, Passarge and Grano, various critics have observed that the excellent analyses of the physiognomy of the landscape are not balanced by due consideration of its physiology-the interrelation of the phenomena. Similarly, Finch's discussion of the presentation of the geography of regions devotes many paragraphs to form and but one line to functions.

Indeed, a strict interpretation of the "pure landscape" view in geography might well rule out functions entirely, except as they express themselves in forms. But since no geographer wishes to exclude the functions of the things he describes, the tendency in many cases is to assume *a priori* that the form expresses the function-a factory looks different from an apartment house, a building with the form of a barn is not a residence. But on New England farms one may find that summer residents have converted a barn into a house without changing notably its external appearance, indeed with but minor changes in its internal form. The writer once spent several hours riding through the streets of the famous silk centre of Lyons looking for silk-mills, only to discover, finally, that the silk was manufactured on small machines enclosed in buildings which could not be distinguished from the ordinary tenement buildings which they had once been and perhaps still were. Would he have been justified, on this basis, in dismissing the industry as unimportant in the geography of Lyons? As Colby writes, in answer to a recent questionnaire, "some very important human enterprises are housed in Victorian buildings. Why judge the significance of an institution by its mansard roof?"

Urban Landscape

It is particularly in urban landscapes, as Waibel observes, that form may be an inadequate indicator of function. No doubt the well trained urban geographer should be able to observe the difference in appearance of a city that is predominantly commercial from one that is primarily manufactural, but it would be extremely difficult actually to measure the difference in degree—since all cities are in part both. A warehouse may be changed into a factory, or vice versa, without undergoing any obvious change in external form, but the change in function may be of great significance in the life of the city. In sum, as Dickinson notes, "function and form are not necessarily in harmony".

Fortunately the importance of functions is so obvious that they appear to triumph over limitations of definition. In the treatment of cities, thanks perhaps to the influence of city planners on urban geography, even the geographers who adhere most rigidly to the limitation to material things analyze areal structure primarily in terms of function, only secondarily in terms of form of buildings.

The landscape concept of geography leads naturally to the emphasis on form rather than function not only in the consideration of individual objects but also in that of combinations of objects in the landscape. The current enthusiasm for "patterns" has been characterized orally by some critics as simply one of the latest fads in geography. The writer is not one of those who would deny significance to pattern; on the contrary he has found that the thorough study, for example, of the street pattern of cities leads to significant conclusions as to the basic factors in their development; this was the case both in the study of a city in this country, Minneapolis, and of European cities, in Upper Silesia. (In the latter case the results were achieved, it is only fair to add, because of the objection raised by Preston James that it was impossible for a city to have "no pattern.") Platt's series of studies of patterns of occupance in scattered districts in Hispanic America has shown the value of studying patterns particularly as an approach to differences in functions.

On the other hand, the mere presentation of patterns without further consideration of them is description in its simplest and most uncritical form. Unless the patterns presented are subjected to further study we have no assurance that they are of any important significance; the mere fact that they cover large areas on a map, and excite the interest of anyone who takes pleasure in designs, does not establish their relative importance in comparison with the host of phenomena which we are forced to ignore as unimportant. For example, the pattern of the railroads of western Ohio gives some indication of the location of the principal trade centres, but its most striking characteristic apparently is that railroads run in almost every direction, more or less indiscriminately a reflection to be sure of the character of the surface. But if we could procure the data of freight movements over all these routes and map them by lines of proportionate widths, the resulting pattern would have far greater significance in leading to an interpretation of the relations between rural communities' and urban centres within the area, and particularly of the area as a whole in relation to the Pittsburgh-Cleveland coal and steel area and the Atlantic Seaboard on the one hand, and to Lake Erie and the Northern Interior on the other.

In other words the significance of patterns depends entirely on the extent to which they depict significant relations in the location of different places in reference to each other. It is therefore most significant that many critics have found that in much of the work of the landscape purists this particular aspect is often neglected. These writers do not necessarily neglect to mention areal connections to be sure, but the constant emphasis that geographers must concentrate their attention on concrete material things naturally leads to a tendency to minimize the purely geometric factor of location though most other geographers would agree with Burger in regarding it as the particular element that is more distinctively geographic than any other.

Illustrations of this effect may be found in the work of many different writers who have adopted the landscape concept. Thus Schluter was able to state that in physical geography at least, the concept of location was of very minor importance. Sauer's various "functional diagrams" provide no place for relative location, nor is the importance of this factor suggested in his subsequent studies, either in the text or in the small number of maps. Likewise in Grano's work, Waibel correctly notes that "relations of location *(Lagebeziehungen)* are completely eliminated".

Much the same is true of the work of Passarge. In *Die Landschaftsgurtel der Erde,* the "landscape belts" of the world, including their "cultural landscapes," are not only described with great effectiveness but also explained in some detail with practically no consideration of the location of these cultural landscapes in relation to each other or to the sea. Even the development of cities appears to be primarily a question simply of site—a harbor, a ford or convenient river crossing, a mineral deposit, etc. If we may judge from the model example which he presents, the same is true of the *ulandskundliche"* method in the study of a small area. We have already noted the very slight use of maps in this example. From careful examination of the text one would suppose that the cultural landscapes around Meran, in the South Tyrol, can be studied in detail without the slightest consideration of the relative location of that district either as a tributary valley of the Adige, leading to Italy, or in its more significant relations to northern Tyrol and the German area of Europe as a whole, by way of the great Brenner route or the Reschen-Scheideck.

A striking instance of the shift in point of view is offered by the use to which Mark Jefferson's widely known maps of proximity to railroads have been put both by James, and by Finch and Trewartha. In his original study, Jefferson entitled his maps simply "Europe within 10 miles of a railroad," etc. (such areas being shown in white against the black background of areas farther from rails). His text makes clear that it is the proximity of certain areas, in contrast to the remoteness of others, with which he is concerned. James introduces these maps however as maps of *pattern*, without suggesting any significance in pattern other than that of density; but he does suggest to some degree the importance of proximity. Finch and Trewartha, more strictly logical, have renamed the maps "Density and patterns of railroads," because it is with these characteristics of railroads "that the geographer is chiefly concerned".

Actually Jefferson's maps show density very poorly since they allow for no distinctions in areas where railroads are wen developed, as he himself dearly realized; likewise in these same areas-including most of Western Europe and eastern United States-they show nothing whatever of patterns. Pattern, at least, is shown far better on any ordinary map of rail-lines. Jefferson's maps, in fact, are not maps of railroads at all, but rather, as his own titles clearly indicated, they are maps of *areas having a certain location with reference* to *railroads.* Are geographers to ignore this characteristic simply because it is invisible, immaterial? Economists have long considered the density of railroads and some, like Ripley, have speculated on patterns; only a geographer would have developed the idea of mapping proximity. The value of the idea is indicated by its widespread adoption by other students in relation to roads as well as to railroads, and it has already entered at least one European atlas.

In general we may say that, just as the study of the individual objects in an area in terms of form, without function, prevents us from putting them together significantly as areal phenomena, 30 the study of the structure or physiognomy of the landscape, without consideration of the location of the differently located features in reference to each other, the factor that underlies the *physiology* of the patterns, produces a morphology that is barren. Crowe has commented on the tendency in modern geography to retreat upon *things,* to become a morphological analysis of all *sorts* of objects in areas that nobody has previously thought worthy of study.

Visible Objects

The reader can hardly fail to have observed that underlying all the arguments of those who wish to consider only material objects objects directly observable by the senses is the spirit of physical science. Material things and particularly visible objects are the sort of phenomena that students trained in' physical sciences know haw to deal with. Geographers with that background-which includes *most* geographers today-would naturally prefer to have to study only such definitely tangible *things*. Nonmaterial facts are grudgingly admitted of necessity where they affect the material things, but the purpose is clear: the less they are to be considered the better. In particular, if one could limit the consideration of an area to the form, or appearance, of its landscape, the decision as to what phenomena are significant appears easier. In reality, however, we found that the concept of geography as the study of the landscape, or simply of material things in areas, provides no usable standard for the selection of phenomena to be studied, and students appear to select whatever observable objects excite their interest.

On the other hand, as Burger observes, to decide what is significant to the "character" of an area to answer Colby's question "not how it looks, but how it is" appears "too uncertain to the' natural scientist accustomed to a sharp corporeal division of the materials of science. The critical scholar will overcome this uncertainty and in the particular case can judge what is significant and what is not. *To* be sure, the uncritical and superficial, who does not sufficiently engage himself in the study of his area, has not yet actually 'experienced' it, will succumb to the attempt to describe everything possible as regionally significant and geographically important. But of him, methodology can take no account".

It therefore appears paradoxical that many of the students who have expressed their theoretical adherence to the landscape concept should have endeavoured to study features of the landscape whose significance in the total character of an area is important chiefly in terms of immaterial culture—namely settlement forms and house types. Whether they claim that they study these simply because the)' are visible features in the landscape or recognize that they are actually attempting to study the geography of culture, *their* readers win not be deceived, it is culture they are pursuing.

To some geographers these topics may seem of but minor importance, but others appear to regard them as of almost first importance. In a paper read at a recent "Round Table on Cultural Geography," Fred B. Kniffen included the study of individual buildings of different kinds as among the *responsibilities* of the cultural geographer, and Leighly, apparently, would have the whole of cultural geography concentrate upon the study of the localization of such "cultural 'immobilia". One might be justified in ignoring theoretical claims that are apparently based on no consideration of the history of geography which indeed to many will appear to lead us completely out of geography into something like historical architecture for which few, if any, of us are prepared. On the other hand the painstaking and illuminating field studies that have been made in these topics particularly by European geographers, but also by Hall, Kniffen, and others in this country require serious consideration. In particular, though such studies are clearly legitimized by the doctrine of "observable material things," it is necessary to consider whether, under that iron rule, they can *become* significant in geography.

Study of the Forms

The systematic examination of the forms of houses and settlements appears in this country to be still in that elementary stage which is marked by enthusiastic observation and classification of types with little consideration of the question of the significance of these cultural features to geography. Many, of the corresponding studies of European students have been motivated from the start by the desire to study the distribution of culture itself, in its differential national forms. Our students, however, have hitherto paid but little attention to the recent work on *house* types produced by European students.

Since we are here concerned with a branch of geography that is relatively new to American geography-though we noted an early example in Ritter's studies of Asia, and Meitzen's classical study was published as early as 1895 -it is somewhat difficult to judge just what its significance may be to geography as a whole. If, as some think, the study of settlement and house forms may give us a valuable key to the study of cultural geography in general, it is appropriate to consider whether its development will be encouraged or discouraged by a particular concept of geography. For this reason, and not without hesitation, the following considerations are offered.

For some students, the study of house types is justified simply on the grounds that houses can be seen in the landscape. It is obvious, however, that the geographer cannot study everything he can see in the landscape; there must be some basis for selection. It is not surprising, therefore, that many students regard house types and settlement forms as simply the latest fads in geography—and not without reason. Leighly has suggested that a Rhenish castle overlooking the Columbia River would be worthy of geographic study and we have actually had detailed descriptions of individual villages on the grounds that they were unique in the regions where they were found. Indeed it has been suggested that there are as many village forms in a region as there are villages and each one warrants special investigation. If one considers the total number of villages in the world and calculates the mountainous pile of manuscripts that would thus be called for, Leighly's nightmare of the surface of the earth plastered with topographic descriptions becomes by comparison a pleasant dream.

It would however be unjust to make too much of the natural results of enthusiasm, or to emphasize those studies of house-types that have been exclusively concerned with classification and genesis of forms. In much of the work of these students the trained reader recognizes at once geographic quality which need not-though it could-be defined. Furthermore, many readers may feel that in the wealth of detailed material presented are included new concepts, suggestions, or conclusions that should add to his general understanding of geography and to his own ability to work in it. Unfortunately this reader, at least, has found it' difficult to derive such ideas or conclusions from these studies, in part, perhaps, because of the lack of any adequate statement of conclusions in many cases, but also, perhaps, because the author himself may not have thought out what conclusions, if any, could be derived from his study—i.e., just what the significance of the study might be.

Of the students of these topics in this country, Kniffen appears to have most seriously considered their geographic purpose. Seeking a "logical approach to culturo-geographic regions" he has classified and mapped by isoplethic methods the types of houses in Louisiana, not simply to portray the material physical character of these observable material objects, but as an "attempt to get at an areal expression of *ideas* regarding houses—a groping toward a tangible hold on the geographic expression of culture".

It should be noted in passing that the purpose stated by Kniffen represents by no means the only possible significance which such studies may have to the field of geography. The different forms of both buildings and settlements may be significantly associated with differences in climate, relief, vegetation, or bed-rock, or with differences in any number of material cultural features, and may therefore add to the distinctive characteristics of regions. On the whole, however, the geographic significance of such features in themselves—as one group of physical features among many others—is that which they might have "in providing a tangible hold on the geographic expression of culture."

Amplifying his written statements in personal discussion with the writer, Kniffen makes clear that they mean just what they imply. Culture is made up of ideas; an idea may have geographic expression—i.e., areal differentiation—but that expression is in itself intangible. We need to find a tangible representation of the idea, a representation; furthermore, whose geographic expression correctly portrays the geographic expression of the idea. Thus, we might have a general impression that in each of several different sorts of areas in Louisiana a particular idea regarding houses predominated. This is a geographic expression of cultural differences, but it is an intangible one, we can't get a hold on an idea. If the actual houses, which we can observe, classify and map proportionately, are a true expression of ideas regarding houses, we have a technique for a geographic study of an idea.

The ultimate objective, then, is a geographic study of nonmaterial aspects of culture-indeed one can properly say, of the very culture itself, as well as the more outward and material manifestation of culture.

Justification of Kniffen's Study **:** This detailed analysis of the fundamental point of view represented by Kniffen's study is justified, not only by the well-merited attention which geographers have given it, but also by the fact that many who have discussed this study have not been fully aware of the ultimate purpose upon which it is based. Indeed, the author himself, in considering the theoretical nature of cultural geography in a later round table discussion, appears to deny completely the essential approach of his previous paper. In place of the approach from the point of view of regional differentiation of *ideas,* as in his field study, he bases his theoretical consideration of cultural geography on the familiar dictum, already quoted, that its "primary concern is with the nature, genesis, and distribution of the observable [meaning 'in general, visible'] phenomena of the landscape directly or indirectly ascribable to man." Consequently, though "the characteristic expression of a religious faith-churches, wayside shrines, and cemeteries-is of primary geographical importance," he would "consider it of little geographical concern that 90 per cent of the people of a given area belonged to a certain religious faith, if that fact had no expression' in the landscape".

In other words, not all ideas that have areal representation are to be included in geography, but only those that are expressed physically in the landscape; the test of geographic significance is the presence of physical forms. We may agree that not all cultural ideas are geographically significant, not even all of those that we on map areally, without necessarily concluding that the sole and sufficient test is their physical landscape representation. We have seen that the character of some cultural landscape features reflects little more than the absence of cultural ideas and, on the other hand, elements of the first importance may have little landscape representation. Take for example the areas immediately north and south of the boundary between Germany and Switzerland." Though the language is the same through the whole range of political ideas the Swiss area is notably different from that in Germany. If these differences are especially marked today, they have been well defined for centuries. Whatever representation these differences may have in the landscape, however, are sufficiently obscure as not to be perceived by a group of geographers who made a reconnaissance trip across the line. On the other hand, any discussion of political questions with the inhabitants on either side reveals immediately a wide difference, not merely in political interests, but in the fundamental political mentality of the populations.

The test of landscape representation is but one test, and by no means the most reliable test, of the differential character of culture of different areas. Furthermore, the emphasis on the physical representation in the landscape leads easily to an emphasis on the form of the representation. Indeed, one critic has argued that, though Kniffen's study of house types is primarily concerned with forms rather than with functions, he should have treated them exclusively in terms of form. The forms of wayside shrines, as seen along the trails of the Austrian Alps, for example, tell us something of the artistic ideas of the people, their ideas of what can be made of wood and stone, but surely it is far more significant to observe what they tell us of the religious ideas of the people.

It may be appropriate at this point to suggest that the assumptions of individual geographers concerning the nature of their field may be arrived at by one or more of three different methods. Some may attempt to analyze from a broad philosophical point of view the logical character of geography. This method, as we have seen, is capable of leading students completely astray, unless they test their logical conclusions against those of their numerous predecessors, and also by a thorough study of what geographers in general have actually tried to do. A much larger number of students, starting only with a very general feeling for geography-what may be called a geographic sense-develop their motivating concepts as a result of their actual research. Given the initial assumption—which is not guaranteed by the mere fact that somehow they chose this particular field-such students may arrive at a point of view which does not differ greatly from that of the most successful members of the first group. This fact is often obscured because in any direct discussion of the nature of the subject they may express apparently quite contrary views, but if one judges by their mature works, one finds that the actual differences in point of view are minor. The third group, finally, are those who have taken their concepts of geography on authority from others, usually from some member or members of the first group.

Like all simple psychological classifications, this one is unreal. Every student belongs in part to all three groups. It is to be hoped therefore that those American geographers who have made a significant start in the study of the geography of culture will not permit obedience to *ex cathedra* pronouncements to prevent the sound development of their own thought in, and as a result of, their work. On the contrary, all methodological pronouncements must, as Hettner repeatedly insists, be tested in the light of actual solutions of geographic problems.. Michotte's warning merits repetition once more: a discussion of principles," divorced from the study of specific examples of actual problems, "is the origin of the errors committed in these matters".

In the particular case under discussion, the geographer appears to be attempting to ride two horses pulling in somewhat different directions. If geography is the study of the visible landscape, the forms of rural houses constitute one element (though a minor one) in his object, but the *ideas* that people have about houses are of only secondary importance, needed only to interpret why particular forms of houses are found. To "attempt to get at an areal expression of *ideas* regarding houses" would not appear to be a proper major objective of the geographer. If, on the other hand, the geographer assumes, without conscious consideration of methodological principles, that he wishes to establish culturo-geographic regions and that for this purpose he must secure a "tangible hold on the geographic expression of culture," the attribute of visibility in the landscape is but one test, neither a necessary nor even a sufficient test, of the significance of the phenomena either to culture or to its geographic expression (its areal differentiation).

There remains the question whether in studying the areal differentiation of cultural ideas, the geographer is restricted to techniques that depend on examination of material things. Though we may find no logic in this distinction it might still represent good sense. In the discussion following Kniffen's Round Table paper to which we last referred, Stanley Dodge suggested that in cultural regional geography we are concerned with areas of uniform thinking and that we might use that as an index. But obviously we have no direct way of measuring what people think, we must arrive at that indirectly. Examination of their material products offers such a method. But is it the only possible method, or is it even certain that it is in all cases the best method? These are the critical questions, for if we can study that which we should study, by methods better than those now used, no authority in all of science, either in our own field or in other fields, shall forbid us to use the better methods.

People's Ideas about Houses : We need not restrict ourselves to people's ideas about houses, since it is obvious that such ideas are by no means of major importance in the total complex of their ideas, nor are they necessarily the best key to that complex. For example, Kniffen found that the houses in the prairie area of Louisiana indicated that the population there had Midwestern ideas regarding houses a fact easily explained by their Corn Belt origin. Does this indicate that these people now have Midwestern ideas in general? Possibly so, but equally possibly they may not in the slightest. They may have become entirely "Southern" in their points of view on everything else in life, but have simply continued to build houses in the inherited style (or to live in inherited houses) without even knowing that their houses were' Midwestern until a geographer came to tell them. Isn't it possible that an equally reliable if not more reliable, guide to their thoughts might be gained from a study of election returns?

If it were found that year after year these areas voted Republican, one might well feel that Was a better key to their particular culture. I trust that no one will suggest that geographers who can train themselves to observe and classify such complicated phenomena as house forms are not equipped to study election returns. Similarly, as Ekblaw suggested, the prevalence of certain groups of folk-songs and customs in different parts *or'* his state might provide Kniffen with another key to cultural regions. If the classification and distribution of these latter phenomena have already been studied by students of other fields, so much the better. The geographer, who would not hesitate to appropriate the findings of the geologist and soil scientist in order to correlate them with his own, should rejoice whenever social scientists have provided him with similar materials ready for his use in cultural geography.

In brief, the geographic study of forms of houses and settlements can be justified as more than a fad, as a contribution to cultural regional geography, primarily if it provides a method of observing indirectly that which cannot be observed directly, namely the areal differentiation of essential cultural characteristics-the ideas, attitudes, and feelings of man. These characteristics manifest themselves in a number of different ways that are subject to observation, classification and proportionate mapping. Our thoughts are expressed 'not only in the things which we make but also in the manner and content of what we speak, sing, or dance, in what we write, vote, or tell to the census collector. Though immaterial, these phenomena are no less observable than material objects, in many cases they represent more direct manifestations of our ideas than our buildings, not to mention the settlement forms whose remote origin we may have completely forgotten. The excellent regional atlases which German geographers have recently been producing demonstrate the value of mapping such nonmaterial culture phenomena as well as the material house types and settlement forms. The student of cultural geography who permits an arbitrary rule to forbid him from studying these observable expressions of culture is depriving himself of the possibility of achieving a well-rounded and, so far as possible, complete achievement of his fundamental purpose-the interpretation of cultural geographic regions.

It may be well to summarize the conclusions which have been reached in the course of the rather lengthy discussions of this section. The doctrine that the field of geographic study must be limited to areal phenomena which are "in general, visible" admits of no precise statement other than a limitation to material features. Although enthusiasts have been preaching the doctrine for over thirty years it has been accepted by a relatively small minority of geographers; still fewer maintain it in practice. This is not surprising since it is founded neither in the logic nor in the historical development of geography, nor does it offer the basis for a more restricted though unified field. On the contrary, it would disrupt even a "central core" of "pure geography" by placing certain aspects of area outside of the geographer's field of study, except as he later finds it necessary to investigate them in order to interpret what he has studied. While it cannot exclude those fields of geography which have ever been concerned with primarily immaterial features, it would thrust them into ill-defined outer reaches of geography with no better guide to their objectives than that of the study of relationships. By reducing the element of relative location to a purely secondary position it tends to neglect the very essence of geographic thought-integration of phenomena in spatial associations. In consequence there develops an uncritical enthusiasm for patterns as such, regardless of their significance, and an overemphasis on form In contrast to function which tends to slight those characteristics of areas whose importance is not represented proportionately by material objects.

Although the point of view of this doctrine may encourage the study of settlement forms, if strictly interpreted it would prevent these studies from achieving the significance which they may well have as part of a more rounded study of the geography of culture. The extreme results of this point of view are represented either by the tendency to look for objects not studied by other fields, which would reduce geography to the consideration of the least significant things, or, by the tendency to enter into competition with artistic workers in the subjective comprehension of landscapes.

Bibliography

Agnew, J.A. : *The City in Cultural Context,* Allen & Unwin, New York, 1984.

Aitchison, J.W. : *The Farming Systems of Wales : A Study of Spatial Variability,* University College of Wales, Department of Geography, 1970.

Alexander, B. : *Geography : Realms, Regions, and Concepts,* John Wiley & Sons, New York, 2000.

—— : *Human Geography : Culture, Society and Space,* John Wiley & Sons, New York, 1999.

—— : *The Earth : An Introduction to its Physical and Human Geography,* John Wiley & Sons, New York, 1995.

Alexander, J.W. and Gibson, L. James : *Economic Geography,* Prentice Hall, New Jersey 1979.

Arvill, R. : *Man and Environment Crisis and the Strategy of Choice,* Penguin, New York, 1967.

Balchin, W.G.V. : *Geography : An Outline for the Intending Students,* Routledge & Kegan Paul, London, 1970.

Bansil, P.C. : *Agricultural Problems of India,* Vikas Publishing House, New Delhi, 1977.

Bansil, P.C. : *Agricultural Statistics in India,* Arnold Heineman, New Delhi, 1974.

Bater, J. H. : *Russia and the Post-Soviet Scene : A Geographical Perspective,* New John Wiley & Sons, New York, 1996.

Bennett, D. : *World Population Problems,* Park Press, Campaign, 1984.

Berger, W.H. : *Abrupt Climatic Change,* D. Radiel Dordrecht, Boston, 1990.

Bhatia, B.M. : *Poverty, Agriculture and Economic Growth,* Vikas Publishing House, New Delhi, 1977.

Bose, A. : *India's Urbanization,* Kanishka Pub., New Delhi, 1980.

Brian, J.L. : *Geography of Market Centers and Retail Distribution*, Prentice-Hall, New Jersey, 1967.

—— : *The Human Consequences of Urbanization*, St. Martin's Press, New York, 1973.

Brun, S.D. : *Cities of the World : World Regional Urban Development*, Harper & Row, New York, 1983.

Brush, John E. : *The Morphology of Indian Cities*, Popular Prakashan, Bombay, 1962.

Burton, Robert, W.K. and Gilbert, F.W. : *The Environment as Hazard*, Oxford University Press, New York, 1978.

Calder, C. C : *An Outline of the Vegetation of India: An Outline of the Field Sciences of India*, Indian Science Congress Association, Calcutta, 1937.

Carlson, L. : *Geography and World Politics*, Prentice Hall, Englewood, 1958.

Chandna, R. C : *A Geography of Population*, Kalyani Publishers, Ludhiana, 1992.

Chauhan, V. S. and Gautam, A. : *Comprehensive Geography of India*, Rastogi Publications, Meerut, 1996-97.

Clark, W.A.V. : *Migration and Development*, The Hague, Mouton, 1975.

Clarke, J.I. : *Geography and Population : Approaches and Applications*, Pergamon Press, New York, 1984.

Claval, Paul : *Introduction to Regional Geography*, Blackwell, Maiden, 1998.

Cole, M.M. : *The Savannas : Biogeography and Geobotany*, Academy Press, London, 1986.

Cressey, G.B. : *Asia's Land and Peoples*, McGraw-Hill, New York, 1963.

Crother, G. : *India, A Travel Survival Kit*, Berkley, U.S.A.,1990.

Datt Ruddar and Sundharam, K. P. M. : *Indian Economy*, S. Chand & Company Ltd., New Delhi, 1991.

Deshpande, C. D. : *India : A Regional Interpretation*, Indian Council of Social Science Research, New Delhi, 1992.

—— : *India: A Regional Interpretation*, Northern Book Centre, New Delhi, 1972.

Detwyler, Thomas R. : *Man's Impact on the Environment*, McGraw-Hill, New York, 1988.

Dev, S. M. and Ranade, A. : *Persisting Poverty and Social Insecurity : A Selective Assessment*, Oxford University Press, New Delhi, 1999.

Dickinson, R.E : *The Regional Concept*, Routledge Kegan Paul, London, 1976.

Dikshit, R.D. : *Geography and the Teaching of the Environment*, Department of Geography, Poona, University, 1984.

Dubey, R. N. : *Economic Geography of India*, Kitab Mahal, Allahabad, 1998.

Edgar, A. : *Plants, Man and Life*, University of California Press, California, 1967.

Ehrilich, P.R. : *Eco-Science,* W.H. Freeman & Co., San Franciscess., 1977.

Eisner, W.B. : *The Middle East : A Physical, Social and Regional Geography,* Methuen, New York, 1978.

Enyedi, Gyorgy : *Natural Geography,* Applied Geography in Hungary, Budapest, 1964.

Farmer, B.H. : *Green Revolution,* MacMillan, London, 1977.

Febure, L. : *Geographical Introduction to History,* MacMillan, London, 1925.

Fitz Gerald Brian P. : *Science in Geography,* Oxford University Press, London, 1974.

Futehally, I. : *Our Environment,* N.B.T., New Delhi, 1988.

Gansser, A. : *Geology of the Himalayas,* John Wiley & Sons, London, 1964.

Gilbert, Alan, : *The Meha City in Latin America,* United Nations Library Press, New York, 1996.

Gill, K. S. : *Evolution of the Indian Economy,* N.C.E.R.T., New Delhi, 1993.

Goudie, A. : *The Human Impact on the Natural Environment,* Basil Blackwell, Oxford, 1986.

Gregory, D. : *Ideology, Science and Human Geography,* Hutchinson, London, 1978.

Grigg, D.B. : *Geographical Impact of Migration,* White and Woods, Longman, London, 1980.

Haggett, P.: *Network Analysis in Geography,* Edward Arnold, London, 1969.

Haq, M.U. : *Human Development in Changing World,* Oxford, New York, 1992,

Hathaway, D.E. : *Government and Agriculture : Public Policy in a Democratic Society,* MacMillan, London, 1963.

Herskovits, M.F. : *Cultural Anthropology,* Alfred A. Knopf, New York, 1955.

Husain, M. : *Agricultural Geography,* Inter India Publications, New Delhi, 1979.

—— : *Evolution of Geographical Thought,* Rawat, Jaipur, 1995.

—— : *Mapping of Environment and Sustainable Development,* Jamia, New Delhi, 1996.

—— : *Nagaland : Habitat, Society and Shifting Cultivation,* Rima Publishing House, N7ew Delhi, 1990.

—— : *The Valley of Kashmir,* Rajesh Pub., Delhi, 1998.

Isard, W. : *Location and Space Economy,* MIT Press, Massachusetts, 1965.

Jacobson, J.L. : *Planning the Global Family,* World Watch Institute, Washington, 1987.

Jhingran, A. G. : *Geology of the Himalayas,* Oxford University Press, New Delhi, 1981.

Johnson, B. L. C : *India : Resources and Development,* Arnold -Heinemann, London, 1980.

—— : *South Asia,* Heinemann Educational Books Ltd., London, 1969.

Johnson, B.L. : *India : Resources and Development,* Barnes & Noble, New Jersey, 1979.

Johnston, R.J. : *Philosophy and Human Geography,* Edward Arnold, London, 1983.

—— : *A Dictionary of Human Geography,* Black-well, Oxford, 1990.

—— : *City and Society : An Outline for Urban Geography,* Hutchinson, London, 1984.

Joshi, H. L. : *Industrial Geography of India,* Rawat Publications, Jaipur, 1990.

Kosinski, L.A. : *People on the Move,* Methuen, London, 1975.

Kostabade J.T. : *Essentials of World Regional Geography,* Sunders College Publishing, New York, 1992.

Kumar, A. : *Chotanagpur Highlands : A Study in Synchroniety,* Thinker's Library, Allahabad, 1985.

Kumar, P. : *Madhya Pradesh : Ek Bhaugolik Adhyayan,* (Hindi), Madhya Pradesh Hindi Granth Academy, Bhopal, 1996.

Lamb, A. : *The China-India Border : The Origin of the Disputed Boundaries,* Oxford University Press, New York, 1964.

Lewin, K. : *Field Theory in Social Science,* University of Chicago Press, Chicago, 1951.

Lewis, G.J. : *Human Migration,* MacMillan, London, 1982.

Mackinder, Halford, J. : *Democratic Ideals and Reality,* Holt, New York, 1919.

Majumdar, D. N. : *Races and Cultures of India,* Asia Publishing House, Bombay, 1958.

Mannion, A.M. : *Global Environmental Change,* Longman, New York, 1991.

Marshall, B. : *The Real World,* Houghton Miffilin Company, Boston, 1991.

Mash, G. P. : *Man and Nature Physical Geography as Modified by Human Action,* Charles Scriber, New York, 1964.

Maxwell, K.E. : *Environment of Life,* Dickenson Publishers, Co., California, 1973.

Michael, C. : *Rural Settlement and Land Use : An Essay in Location,* Hutchinson, London, 1966.

Misra, V. C : *India-Regional Studies,* 21st IGU Publication, New Delhi, 1968.

Miller, G.T. : *Living in the Environment,* Wadsworth Press, Belmont, Calif., 1994.

Misra, J. P. : *Bharat Ka Bhoogol* Geography of India, Vidya Sagar, Allahabad, 1997.

Misra, R.P. : *Development Issues of our Time,* Concept, New Delhi, 1984.

—— : *National Development,* Vikas Publication, New Delhi, 1965.

—— : *Regional Planning : Concept, Technique and Policies,* Concept, New Delhi, 1992.

Muller, S.H. : *The World's Living Languages,* Oxford, New York, 1964.

Myrdal, Gunnar : *Rich Lands and Poor,* Harper & Brothers, New York, 1957.

Nielsen, N.C. : *Religions of the World,* Oxford, New York, 1983.

Ojha, S. S. : *Geography of India,* Bhaugolik Adhyayan Sansthan, Allahabad, 2001.

Pannikar, K. M. : *A Survey of Indian History,* Popular Prakashan, Bombay, 1964.

Paul, A. : *World Regional Geography : A Question of Place,* Harper & Row, New York, 1977.

Pedelaborde, P. : *The Monsoon,* Methuen and Co., London, 1963.

Perpillou, A.M. : *Human Geography,* Longman, New York, 1977.

Persson, G.A. : *Resources and World Development,* John Wiley, Chichester, 1987.

Prescott, J.R.V. : *Political Frontiers and Boundaries,* Allen & Unwin, Winchester, 1987.

Qureshi, M. N. : *Himalayan Geology,* Oxford, New Delhi, 1971.

Rao, B.P. : *Bharat : Ek Bhaugolik Samiksha,* Vasundhara Prakashan,Gorakhpur, 1995.

Rapp, A. : *Human Activity and Environmental Processes,* John Wiley, Chichester, 1987.

Ray, S. : *Geomorphology of India,* Firma, K. L. M. Pvt. Ltd., Calcutta, 1978.

Rubenstein, J.M. : *The Cultural Landscape,* Prentice Hall, London, 1983.

Sen, Bandhudas : *The Green Revolution in India : A Perspective,* Wiley Eastern, New Delhi, 1974.

Sen, L. K. : *Planning Rural Growth Centres for Integrated Area Development : A Study in Miryalguda Taluka,* N. I. C. D., Hyderabad, 1971.

Shafi, M. : *Ganga-Yamuna Doab, in India : Regional Studies,* National Committee for Geography, Calcutta, 1923.

—— : *Agricultural Productivity and Regional Imbalances—A Study of Uttar Pradesh,* Concept Pub. Co. : New Delhi, 1983.

Sharma R. C : *Readings in General Geography and Geography of India,* Jawahar Publishers, New Delhi, 1992.

Sharma, H. S. : *Environmental Design and Development,* Scientific Pub., Jodhpur, 1988.

Sharma, R.C. : *Settlement Geography of Western Rajasthan,* Rawat Pub., New Delhi, 1972.

Shaw, J.B. : *The Post Soviet Republics : A Systematic Geography,* John Wiley & Sons, New York, 1992.

Shiva, V. : *The Violence of Green Cultivation,* Mapusa, Goa, 1991.

Short, J.R. : *Introduction to Urban Geography,* Routledge & Kegan Paul, London, 1984.

Singh, H. H. : *Geography and Environment,* Concept, New Delhi, 1986.

Singh, J. : *An Agricultural Geography of Haryana,* Vishal Publication, Kurukshtra, 1976.

Singh, L. R. : *The Tarai Region of U.P. : A Study in Human Geography,* Ram Narain Beni Prasad, Allahabad, 1965.

Small, R. J. : *The Study of Landforms,* Cambridge University Press, London, 1970.

Smith, D.M. : *Human Geography : A Welfare Approach,* Edward Arnold, London, 1977.

Srinivasan. T.N. : *The Green Revolution,* Indian Statistical Institute, Calcutta, 1971.

Stamp, L.D. : *Applied Geography,* Penguin, Harmondsworth, 1960.

Sundaram, K. V. : *Geography of Under-development : The Spatial Dynamics of Under-development,* Concept, New Delhi, 1983.

Tarrant, J.R. : *Agricultural Geography* Newton Abbot : David and Charles, New York, 1974.

Taylor, James A. : *Weather and Agriculture,* Pergamon, Oxford, 1967.

Taylor, P.J. : *Geography : Its Nature,* MacMillan, London, 1986.

Turner, Roy : *India's Urban Future,* University of California Press, Los Angeles, 1962.

Vayda, A. P. G. : *Man in the Pacific Islands : Essays on Geographical Change in the Pacific,* Oxford University Press, New York, 1972.

Verma, R. V., *Bharat Ka Bhaugolik Vivechan* Hindi, Kitabghar, Kanpur, 1956.

Weyer, E.M. : *The Eskimos : Their Environment and Folkways,* Yale, 1932.

Willens, A. : *Dry Lands, Man and Plants,* Architectural Press, London,

Wirth, Louis : *On Cities and Social Life,* University Press, Chicago, 1964.

Yantis *: The Encyclopaedia of World Cultures-Africa,* MacMillan, New York, 1993.

Zelinsky, W. : *Geography and a Crowding World,* Oxford University Press, New York, 1970.

Zimmermann, E.W. : *World Resources and Industries,* Harper and Row, London, 1951.

Index

F

G

H

I

K

L

M

N

O

P

R

S

□□□